Artificial Intelligence and Human Existence

Artificial intelligence (AI) has become part and parcel of many aspects of our lives, impacting how we approach tasks and understand the world around us. This book recognizes the impact that AI has on education, computer science, and society. Taking an interdisciplinary approach, it looks at the positive and negative implications and delivers a comprehensive understanding of how the integration of AI is shaping the future of human existence.

This book covers 12 topics that reveal the links between AI and humans. The topics range from the transformational role of technology in education to the ethical considerations surrounding AI's application, and many more. Specifically, it investigates how AI algorithms are pushing the boundaries of inventions in computer science and driving innovations in human–computer interaction and software development, through themes such as 'Computer Science and Technological Frontiers'. It also reflects on the societal impact of AI and offers tools to navigate the changing landscape of technology and humanity. With discussions on the latest advancements in AI and their implications for future innovations, readers of this book will learn of the ethical challenges posed by AI through in-depth analysis, including coverage of equity, privacy, and algorithmic bias.

Artificial Intelligence and Human Existence: Insights on Education, Technology, and Society is for researchers, educators, and policymakers in the fields of human factors, human–computer interaction and computer science, and anyone interested in the links between AI and human society.

Human Factors in Design, Engineering, and Computing

Series Editors: Tareq Ahram and Waldemar Karwowski

This series focuses on research and development efforts intended to promote the comprehensive integration of people and technological systems. To help foster integration of human factors research with emerging technologies, this series will focus on novel methodologies, design tools and solutions that advance our understanding of the nature of human factors and ergonomics with intelligent technologies and services. It will address all aspects of human factors and human-centered design, with a particular emphasis on applications of emerging technologies, computing, Artificial Intelligence and Systems. The series offers a multidisciplinary platform for researchers and practitioners alike, discussing emerging issues in the field of human factors engineering systems, with a special focus on (but not limited to) computing and AI-based technologies. It will be useful for researchers, senior graduate students, graduate students, and professionals in different domains including human factors, engineering design, systems science and engineering, and ergonomics.

Human Factors in Design: Intelligent Tools and Technological Innovations
Edited by Tareq Ahram and Waldemar Karwowski

Human Factors in Engineering: Manufacturing Systems, Automation, and Interactions
Edited by Beata Mrugalska, Tareq Ahram and Waldemar Karwowski

Human Factors in Computing: Social Dimension and Artificial Intelligence-Based Technologies
Edited by Tareq Ahram and Waldemar Karwowski

Human Factors and Emerging Technologies: Integrating People and Technological Systems
Edited by Tareq Ahram and Waldemar Karwowski

Human Factors in Intelligence, Technology and Analytics
Edited by Tareq Ahram and Waldemar Karwowski

Human-Technology: Blending Artificial Intelligence, Computing, and Intelligent Design
Edited by Tareq Ahram and Waldemar Karwowski

Converging Minds: The Creative Potential of Collaborative AI
Aleksandra Przegalinska and Tamilla Triantoro

3D Digital Design in Ergonomics and Human Factors
Tihomir Dovramadjiev

Artificial Intelligence and Human Existence
Edited by Akanksha Srivastava, A.V. Senthil Kumar, Disha Maheshwari and Samrat Ray

For more information about this series, please visit: www.routledge.com/Human-Factors-In-Design-Engnineering-and-Computing/book-series/%20CRCHFIDEC

Artificial Intelligence and Human Existence

Insights on Education, Technology, and Society

Edited by
Akanksha Srivastava, A.V. Senthil Kumar,
Disha Maheshwari, and Samrat Ray

CRC Press
Taylor & Francis Group
Boca Raton London New York

CRC Press is an imprint of the
Taylor & Francis Group, an **informa** business

Designed cover image: Shutterstock

First edition published 2026
by CRC Press
2385 NW Executive Center Drive, Suite 320, Boca Raton FL 33431

and by CRC Press
4 Park Square, Milton Park, Abingdon, Oxon, OX14 4RN

CRC Press is an imprint of Taylor & Francis Group, LLC

ISBN: 978-1-032-90778-9 (hbk)
ISBN: 978-1-032-90909-7 (pbk)
ISBN: 978-1-003-56046-3 (ebk)

DOI: 10.1201/9781003560463

Typeset in Times
by Newgen Publishing UK

Contents

Foreword ix
Preface xi
About the Editors xiii
List of Contributors xv

PART I AI in Education and Academic Transformation

Chapter 1 Impacts of Artificial Intelligence on Education 3
Angela Maria Alves, Gabriel Caumo Vaz, Qian Xun, and Gabriel Gomes de Oliveira

Chapter 2 The Gloomy Aspect of AI: Negative Impact of AI on Society 15
Lokesh Jain

Chapter 3 AI-Powered Academic Knowledge Management: Transforming Academic Research and Teaching 33
Shruti Singh and Ragi T. S.

Chapter 4 AI and Storytelling: Reimagining Learning in Film Institutes of India 52
Priyanka Yadav and Sachin Bharti

Chapter 5 Equity in AI-Driven Education: Addressing Gender Bias and Promoting Inclusive Learning Environment 70
Manisha Singh, Vinod Bhatt, Dev Brat Gupta, and Sonali Srivastava

Chapter 6 Artificial Intelligence in English Literature: Close to Distant Reading 85
Jaydipkumar Devabhai Pandya

Chapter 7 Transforming Classrooms: AI-Driven Innovations in School Education 106
Siddhi Sood and Poonam Angurwala

Chapter 8 Digital Renaissance: AI's Role in Academic Transformation and Career Innovation 124

Aashmi V., Pushpitha Reddy Patur, and Janardhana Kumar Reddy P.

PART II AI in Healthcare and Social Services

Chapter 9 AI-Enhanced Health Communication: A Netnographic Analysis of Indian Pediatrician Influencers on Instagram Promoting Child Health Literacy 147

Sylvia Raha

Chapter 10 AI in Enhancing Social Services for Vulnerable Populations 169

Sheeba Khalid

Chapter 11 AI-Driven Disaster Management: Case Studies and Best Practices 187

Sidharth Verma and Durgesh Tripathi

PART III AI Technology and Data Systems

Chapter 12 The Evolution of Modern Data Platforms: Powering the AI-Driven World 209

Vaishali Mishra, Harshal Darade, A.V. Senthil Kumar, and B. Gayathri

Chapter 13 AI-Driven Data Excellence: Enhancing Governance, Quality, and Business Outcome 233

Vaishali Mishra, Harshal Darade, A. V. Senthil Kumar, and Namita Mishra

Chapter 14 Exploring the Frontiers of Generative AI: Applications, Hurdles, and the Synergy of AI and Human Interaction 254

Rachna Jain and Shikha Sharma

PART IV AI, Society, and Governance

Chapter 15 Ethical Considerations in AI: Balancing Innovation with Responsibility 279

Dhruv Sabharwal and Archan Mitra

Chapter 16 Exploring Gender Bias in Google Search Results: Disparities in Responses to Queries for Men and Women 294

Dinesh Kumar T., Ramalakshmi Lakshmanan, and Vasupradha Srikrishna

Chapter 17 Artificial Intelligence in International Relations: Complexities Involved 308

Ipshita Bhattacharya

Chapter 18 From Threads to Tangles: AI's Disruption of the Social Fabric 323

Yahvi Johri and Akanksha Srivastava

Chapter 19 The Use of Artificial Intelligence in Financial Institutions: A Double-Edged Sword Effect 337

Shruti Ranjan and Ritika Srivastava

Index 347

Foreword

In our three decades of academic and policy engagement, few transformations have rivalled the breadth and velocity of the rise of artificial intelligence. From algorithms curating our news feeds to machines assisting in surgeries, AI has moved rapidly from theory to the very texture of our everyday lives. Yet, in this accelerating adoption, we find ourselves repeatedly circling back to a fundamental question: What does it mean to be human in an age of intelligent machines?

It is this central tension—between technological promise and existential introspection—that lies at the heart of *Artificial Intelligence and Human Existence: Insights on Education, Technology, and Society*, a timely and compelling volume edited by Dr. Akanksha Srivastava, Prof. Senthil Kumar, Dr. Disha Maheshwari, and Prof. Ray. The book brings together 19 meticulously researched chapters by scholars across domains, each offering a unique lens on how AI intersects with the human experience—from classrooms and creative writing labs to government offices and disaster management centres.

The theme of this book is not only timely but urgent. Across the globe, educators, policymakers, technologists, and citizens are grappling with questions of automation, bias, access, and agency. International forums like UNESCO, the OECD, and the World Economic Forum have placed responsible AI and ethical design at the forefront of their policy discussions. Within academia, AI is no longer siloed within computer science departments—it is being interrogated in philosophy, sociology, education, and the arts. This book seamlessly aligns with this shift, embodying the interdisciplinary approach that the moment demands.

One of the most impressive aspects of this volume is its range—both in subject matter and in the diversity of its contributors. We encounter explorations of AI in enhancing inclusive education, enabling better health communication, and managing disasters more effectively. Equally, the book does not shy away from the uncomfortable truths—bias embedded in search algorithms, the disruption of social fabric, and the ethical ambiguity of using AI in financial systems and governance. These contrasting perspectives are not contradictions; they are precisely what make this collection so intellectually robust and grounded.

This editorial vision deserves particular commendation. Her curation balances technological insights with humanistic concerns, scientific enquiry with critical reflection. Under her guidance, the book succeeds not only as a scholarly compilation but as a coherent narrative—one that urges us to think deeply, act responsibly, and remain open to dialogue.

Why does this book matter now? Because AI is not a future problem—it is a present reality. As nations formulate AI strategies, as educators integrate AI tools, and as industries redesign workflows, there is a growing need for critical and compassionate understanding. This book responds to that need by providing a knowledge base that is both academically sound and socially responsive. It is a resource for researchers, a conversation starter for classrooms, and a guidepost for policymakers.

To the reader—whether you are a student, scholar, practitioner, or simply a curious mind—I offer this book as an invitation. An invitation to question, to connect disciplines, to challenge assumptions, and to envision futures where intelligence—both artificial and human—works in harmony with justice, creativity, and care.

Let this book be not just a mirror of where we are, but a map of where we might go.

Prof. Dr. Sibaram Khara
Vice Chancellor
Sharda University, Greater Noida
Uttar Pradesh, India

Preface

In a world that is being redefined at an unprecedented pace, few forces have influenced our lives as deeply and persistently as artificial intelligence. When we first encountered AI tools in my own classroom a few years ago, we were struck not just by their speed but by the questions they raised—about learning, decision-making, and what it means to be human. What was once the realm of futuristic speculation has now become an integral part of our everyday experiences—shaping how we learn, make decisions, connect with others, and even perceive reality itself.

This book, *Artificial Intelligence and Human Existence: Insights on Education, Technology, and Society*, has been conceived at a pivotal moment in this ongoing transformation. It is not merely about how AI works, but about what AI means. It invites readers to reflect on the philosophical, ethical, educational, cultural, and institutional implications of a technology that is no longer at the edge, but very much at the centre of modern life. Whether we fully realize it or not, AI has quietly moved to the centre of our lives. And what that means—well, we're still figuring it out. Some of the ideas in this volume were sparked by conversations we've had with colleagues, students, and even sceptical friends outside academia. The volume serves as a space for critical engagement, offering perspectives that span disciplines, geographies, and experiences.

Divided into four thematic parts, the book begins with an in-depth look at AI's transformative role in education and academic practices—examining its impact on pedagogy, research, inclusivity, and curriculum innovation across diverse disciplines, including literature and film studies. The second part turns to healthcare and social services, offering insights into how AI is enhancing communication, disaster response, and support for vulnerable populations. The third part investigates the technological foundations of AI, including modern data platforms, generative AI, and their implications for governance and business outcomes. The final part critically engages with the societal and ethical dimensions of AI, addressing gender bias, international relations, financial institutions, and the broader disruptions AI brings to the social fabric. Together, these chapters offer a rich and thought-provoking account of AI's promises and pitfalls, encouraging readers to reflect on the responsibilities that come with innovation.

Throughout the book, contributors interrogate both the possibilities and the pitfalls of AI. There is a strong recognition that while AI brings powerful tools for progress, it also demands caution, responsibility, and inclusivity. The essays explore how AI is influencing human relationships, reshaping knowledge systems, reconfiguring power structures, and raising profound questions about identity, equity, and agency.

The strength of this volume lies in its Interdisciplinarity. By bringing together scholars, educators, and researchers from diverse fields, it creates a rich mosaic of insights that go beyond technical analysis. The chapters speak to one another—not always in agreement, but always in dialogue—offering readers a holistic understanding of AI's impact on human life. Whether examining AI in education, governance, literature, or social services, the authors present arguments grounded in both theory and real-world application.

As the editors of this volume, we were deeply inspired by the passion and clarity with which each contributor engaged with the theme. What you'll find here is a genuine attempt to make sense of where AI is taking us—and how we might shape that path together. We are also grateful to the peer reviewers whose thoughtful feedback helped sharpen the arguments and elevate the quality of the scholarship.

We truly believe this book couldn't have come at a better time. As societies, institutions, and individuals navigate the evolving AI landscape, we need tools that help us think critically and act wisely. We hope readers find this book not only informative but also thought-provoking—and maybe even a little unsettling in all the right ways, a resource for educators, a guide for policymakers, and an invitation for all readers to reflect more deeply on the future we are collectively building.

About the Editors

Akanksha Srivastava is Professor and Head of the Department at the School of Education, Lingaya's Vidyapeeth, Haryana, India, and has over 17 years of experience in teacher education, educational technology, and academic leadership. Her areas of expertise span AI in education, digital pedagogy, educational psychology, and inclusive learning in alignment with NEP 2020. As an internationally certified soft skills trainer and MOOC developer for SWAYAM, she also actively supervises doctoral research, contributes to curriculum design, and mentors educators in adopting tech-enabled teaching methodologies.

A.V. Senthil Kumar is working as Principal, Nehru Institute of Information Technology and Management, Coimbatore, India. He has worked as Professor and Director, PG and Research Department of Computer Applications, Hindusthan College of Arts & Science, Coimbatore, Tamilnadu for more than 15 years and as Senior Grade Lecturer in CMS College of Science and Commerce for 14 years. To his credit he has industrial experience for five years and teaching experience of 29 years. He has also received his Doctor of Science (D.Sc in Computer Science). He has to his credit 88 Book Chapters, 234 papers in International and National Journals, 88 papers in International Conferences in International and National Conferences, and edited 17 books and 3 Text books. He is as Associate Editor of IEEE Access. He is an Editor-in-Chief for many journals and Key Member for India, Machine Intelligence Research Lab (MIR Labs). He is an Editorial Board Member and Reviewer for various International Journals. He is also a Committee member for various International Conferences. He has guided 16 Ph.D scholars and guiding 3 Ph. D scholars now.

Disha Maheshwari is Assistant Professor at School of Education, Sharda University, India. She has a PhD from the University of Delhi, India. She was a Commonwealth Scholar at the Centre for Applied Linguistics at the University of Warwick, UK, and her research interests span Gender and Education, Discourse and Identity, and Higher Education Pedagogy. She currently teaches at Sharda University, India, where she offers courses in Gender Studies, Communication, and Qualitative Research Methodology, including Pedagogy at undergraduate, postgraduate, and doctoral levels. Her research engagements include working as Research Assistant on the British Council-funded ELT Survey of India with Dr. Richard Smith (University of Warwick), and as Project Associate on the University of Delhi's R&D-funded project Children and Teachers as Co-researchers: Exploring Young Learner Classrooms in India. She is the co-editor of *Subverting Social Inequalities: Gender as a Social Reality* (Apple Academic Press, CRC Press, Taylor & Francis Group), and her work has appeared in several national and international academic journals.

Samrat Ray is a senior researcher at Peter the Great Saint Petersburg Polytechnic University, Russia, specializing in the economics of innovation, entrepreneurship, and managerial aspects of economics. He is associated with the Industrial and Trade Management Institute of the University, which handles cross-border industrial and economic incubation for product discovery and patenting. He has over 15 years of experience dealing with academic leadership and thought direction in the UAE, Russia and across Europe and possesses a PhD from Polytechnic University, Russia. He has written over 100 Scopus-indexed papers and has been granted seven international and national patents in his name.

Contributors

Poonam Angurwala
Sri Guru Ramdas College of Education
Ludhiana, India

Sachin Bharti
University School of Mass Communication, Guru Gobind Singh Indraprastha University
Delhi, India

Vinod Bhatt
School of Advanced Sciences and Languages, VIT Bhopal University
Bhopal, India

Ipshita Bhattacharya
The Bhopal School of Social Sciences, College Bhopal
Bhopal, India

Gabriel Caumo Vaz
School of Electrical and Computer Engineering, State University of Campinas
Campinas, Brazil

Harshal Darade
Independent Researcher Seattle, United States

Jaydipkumar Devabhai Pandya
Surendranagar University Gujrat, India

B. Gayathri
Bishop Heber College
Tiruchirappalli, India

Gabriel Gomes de Oliveira
Renato Archer Information Technology Center,
Campinas, Brazil

Dev Brat Gupta
School of Advanced Sciences and Languages
VIT Bhopal University Bhopal, India

Lokesh Jain
Department of Computer Science Vivekananda Institute of Professional
STudies- Technical Campus (VIPS-TC)
Delhi, India

Rachna Jain
Maharaja Agrasen Institute of Management Studies, GGSIPU Affiliated
Delhi, India

Yahvi Johri
University of North Carolina Chapel Hill, United States

Sheeba Khalid
Amity University Lucknow Campus
Lucknow, India

Janardhana Kumar Reddy P.
Department of Education Bharathiar University
Coimbatore, India

Dinesh Kumar T.
VIT School of Design (V-SIGN) Vellore Institute of Technology Vellore, India

Ramalakshmi Lakshmanan
VIT School Media Arts and Technology (VSMART)
Vellore Institute of Technology
Vellore, India

Angela Maria Alves
Renato Archer Information Technology Center
Campinas, Brazil

Namita Mishra
I.T.S School of Management Ghaziabad, India

Vaishali Mishra
Expedia Group
Seattle, Washington,
USA

Archan Mitra
School of Media Studies, Presidency University Banglore
Bengaluru, India

Sylvia Raha
Indira Gandhi National Open University Delhi, India

Shruti Ranjan
Christ University Bengaluru, India

Pushpitha Reddy Patur
Independent Researcher Houstan, Texas, United States

Dhruv Sabharwal
Sharda School of Media Film & Entertainment
Sharda University Greater Noida, India

A.V. Senthil Kumar
Principal, Nehru Institute of Information Technology and Management
Coimbatore, India

Shikha Sharma
Maharaja Agrasen Institute of Management Studies, GGSIPU Affiliated
Delhi, India

Manisha Singh
VIT Business School VIT Bhopal University Bhopal, India

Shruti Singh
School of Management Sciences Varanasi, India

Siddhi Sood
DEGSN, National Council of Educational Research and Training
New Delhi, India

Vasupradha Srikrishna
Communication Area, IFMR GSB KREA University
Sri City, Andhra Pradesh, India

Akanksha Srivastava
School of Education Lingaya's Vidyapeeth Faridabad, Haryana, India

Ritika Srivastava
GLA University
Mathura (Noida Campus), Uttar Pradesh, India

Sonali Srivastava
Jaypee Business School Jaypee Institute of Information
Technology Noida, India

Ragi T. S.
School of Management Sciences
Varanasi, India

Aashmi V.
Department of Education Bharathiar University Coimbatore, India

Qian Xun
Unit of Energy Conversion
RISE Research Institutes of Sweden Boras, Sweden

Priyanka Yadav
University School of Mass Communication
Guru Gobind Singh Indraprastha University
Delhi, India

Part I

AI in Education and Academic Transformation

1 Impacts of Artificial Intelligence on Education

Angela Maria Alves, Gabriel Caumo Vaz, Qian Xun, and Gabriel Gomes de Oliveira

1.1 INTRODUCTION

The concern of the relationship between technology and humanity has been discussed by sociologists, civil society, education specialists, and other professionals since the beginning of this century (Meghji, 2024; Trithara, 2024). On the one hand, there were negative projections about the Industrial Revolution that suggested that automation would finally replace manufacturing, increasing poverty and unemployment rates. On the other hand, there were positive projections that expected automation to put an end point to tough working hours and, consequently, to allow people to carry out creative activities, developing arts and philosophy, and improving society. However, none of these prospects have been achieved so far (Edwards, 2009).

Machines have actually replaced many work positions. However, many new positions have been created. Until the rise of the Internet, this balance worked well. The arrival of the global network caused many changes in the way people interact because of both the shortening of distance and the evolution of virtual relationships (Edwards, 2009).

The process described above had a historical watershed on July 11, 2008: the launch of the iPhone 3G developed by Apple Inc. (West & Mace, 2010). From that moment, many gaps have been filled so that we could achieve the current state in the technology segment. As examples, we mention the evolution of communication due to chat and email applications, which became available to the general public and seemed to represent technology at its highest level, and the interaction between humans and machines, which underwent a huge revolution some years before the popularization of smartphones (Grübler, 2003).

Even before digital technologies became a social and political issue, they were already directly affecting the daily lives of people through bank and delivery applications, social media, and other popular tools. The discussion about how these platforms affect social organizations and how they can change the labor market and economy in general has increased in recent years (Waller & Anderson, 2021).

ChatGPT (Generative Pre-Trained Transformer) is another novelty that has stood out since it was launched because it uses artificial intelligence (AI), a disruptive technology, to create texts with precise answers without human intervention. This

DOI: 10.1201/9781003560463-2

technology is intelligent enough to make decisions, no matter what the field of application is (Waller & Anderson, 2021).

Most of the events occurred after 2021, when its main database was created (Lundstad et al., 2023). Although it is still early to know its real effects on daily life, it is of utmost importance to establish guidelines to maximize its use in relevant situations and, at the same time, avoid its use to cheat statutes of companies, universities, and especially elementary schools, which are the base for the development of educationally qualified societies. The quality of education, which includes the quality of universities and research centers, is very important because societies with high-level education can deal better with tough moments such as the Covid-19 pandemic and also find solutions for daily problems of their citizens (Lundstad et al., 2023). From this perspective, this research has two main branches:

- The first part of the article discusses the relationship between humanity and technological developments, particularly those that emerged after the popularization of smartphones;
- The second half addresses the legal aspects of new technologies in the democratic rule of law, society, and academic productions to discuss whether ChatGPT can make education and/or academic work precarious or, otherwise, benefit them.

The objective of the work, which is based on the inductive method, is to build reflections on the problems caused by the rapid proliferation of natural language AI models in educational and academic institutions. The main idea is that educational authorities and society should limit or even prevent these applications for some purposes, thus enabling educational security for the futures of our young people and children.

The research is divided between bibliographic and documentary research. The first originates from the review of works written by national and international authors on how new technologies affect the globalized society of the 21st century. In contrast, the second one faces an epistemological and educational problem related to the outsourcing of cognitive processes to automation, from the perspective of promoting the objective of establishing a basis for tackling the matter.

ChatGPT has emerged as a significant shift in natural language processing and AI. This innovative system, developed by OpenAI, was launched in 2019. Using deep neural network architectures, it can perform text generation tasks with great skill and fluidity. GPT-3, in particular, is one of the most sophisticated versions of this model and has a neural network with 175 billion parameters, making it one of the most sophisticated networks ever created.

Its emergence came about through significant prior training efforts on large volumes of data and texts, enabling it to understand and produce texts in multiple languages, demonstrating remarkable versatility. ChatGPT's ability to understand context, answer questions, translate languages, and produce high-quality original text has revolutionized the way people interact with technology, even enabling the diversity with which this software has been used in a variety of applications, including virtual assistants, automatic translation, customer service chatbots, and, we cannot fail to mention, education. ChatGPT is an excellent example of the increasing convergence between human language and AI. This is creating a new paradigm of digital interaction and communication between man and machine.

In view of the perspective narrated above, this article attempts to closely examine the effects of ChatGPT on higher education methods, addressing the problems associated with the use of this technology in the classroom and its ability to improve communication between teachers and students. In addition, the pedagogical, moral, and social consequences of implementing ChatGPT in higher education institutions will be verified.

This analysis is necessary because higher education is a critical component of academic training, and technological innovations can significantly impact the educational experience. Understanding how professors use and adapt ChatGPT and how it impacts the quality of instruction is essential to making intelligent choices about the future of higher education.

Ultimately, the goal of this article is to increase knowledge about the relationship between AI and higher education, providing useful information for educators, educational institution administrators, and researchers who want to understand and leverage the potential of ChatGPT in the academic context. The following research will address the issues and opportunities that arise with this revolutionary technology, enabling the promotion of important conversations about the continuous improvement of teaching methods in higher education institutions.

Research aims to address the subjective aspects of social phenomena and human behavior using qualitative and bibliographic data. Such bibliographic research consists of a review of all previously published literature, including books, magazines, periodicals, institutional websites, and scientific studies. Subsequently, the objective of bibliographic research is to provide researchers with direct access to all written material on the subject in question, assisting them in the analysis or manipulation of information. Thus, one of the main advantages of bibliographic research is that it allows researchers to study a much larger number of subjects than they could do directly.

1.2 METHODOLOGY

The objective of the qualitative methodology was to understand, interpret, and systematize the approach to topics such as the impact of AI on education. To this end, the main scientific databases, such as Science Direct, Scopus, and Web of Science, were used, among others. In addition, the authors did not restrict the period of analysis, conducting a broad search with articles published from 1942 to the current date. The schematic diagram represented in Figure 1.1 uses a scientific database to examine previously collected articles. It can be divided into two parts: the conceptual level (bottom) and the data level (top). It is important to note that the design of the data schema must be carried out by specialized professionals, as demonstrated in this work.

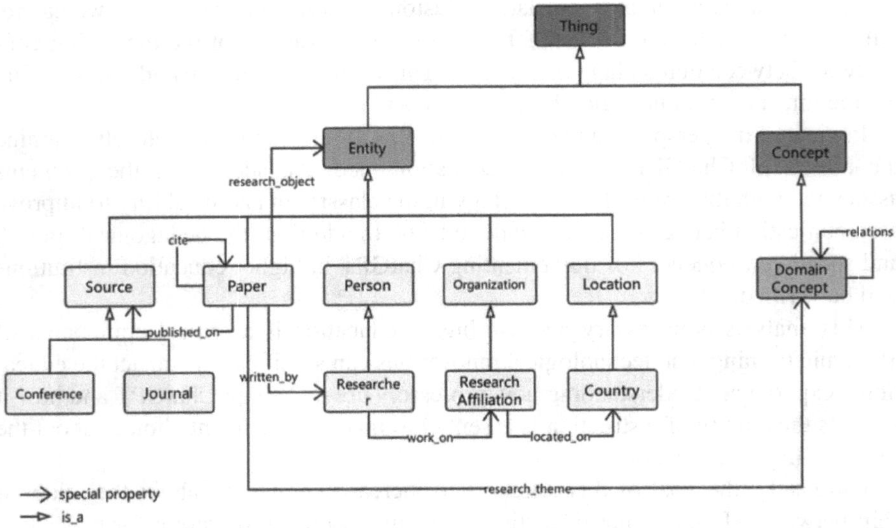

FIGURE 1.1 Schematic diagram using scientific database to analyze scientific articles.

We hope to find concrete results from scientific articles that can demonstrate the importance of the topic addressed in this article.

1.3 DISCUSSION

The results of this research showed that ChatGPT has had a significant effect on the ways in which teachers work in higher education. According to the data collected, teachers and educators reported that they communicated more effectively with students, especially when it came to frequently asked questions and issues that were related to the course content (Tayan et al., 2023). In addition, it was observed that ChatGPT was widely used as a learning support tool, providing instant answers to student questions both inside and outside the virtual classroom (Javaid et al., 2023). However, some problems were also found: although ChatGPT has proven to be useful in certain tasks, its interaction capacity is inferior to that of a human teacher, especially in situations that require emotional understanding, empathy, or personalized feedback (Roumeliotis & Tselikas, 2023). This becomes more evident when we look at Figure 1.2:

It should be noted that Figure 1.2 shows the impact of the evolution of ChatGPT in the most distinctive areas of knowledge. We observed that areas more focused on human aspects were more impacted initially compared to areas focused on mathematics and exact sciences. However, we observed that over the years, with new versions of ChatGPT, progress is more significant in areas that were not very adherent in the past. This means that this tool has been providing an evolution in many areas of knowledge in human education, enabling a new way of thinking for teachers, students, and all entities related to education on a global scale.

One area where ChatGPT is sure to make a significant impact is education. For example, as of March 2023, ChatGPT 4 is capable of passing a wide range of AP and

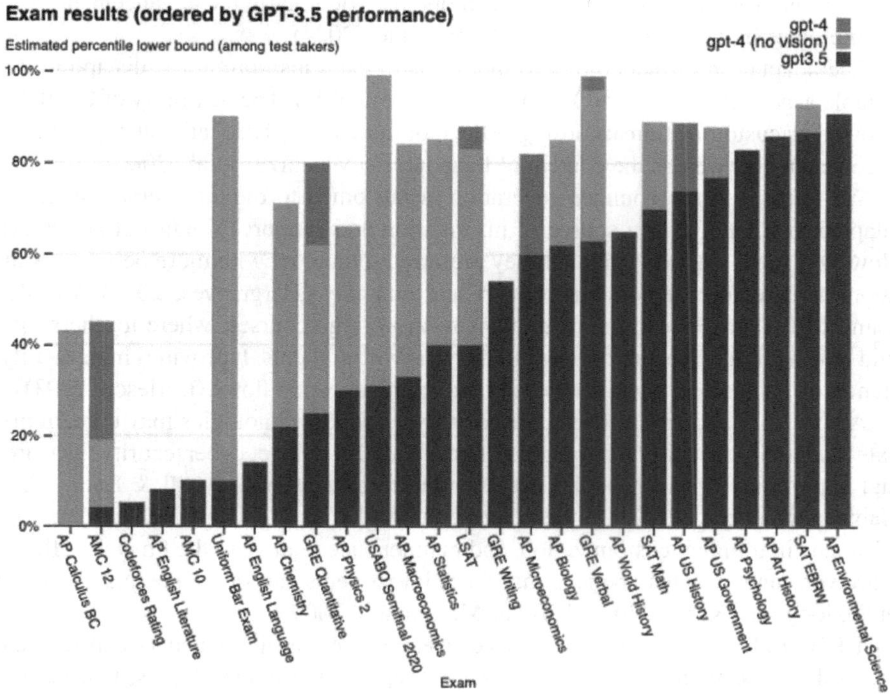

FIGURE 1.2 ChatGPT's education impact.

college-level tests. The challenge facing most educators in the near term is how to reduce the misuse of ChatGPT and, ideally, leverage it to improve students' learning experiences. Interestingly, in our opinion, the best way to reduce misuse is to incorporate it proactively as part of learning.

The data collected by this survey demonstrates that teachers report that ChatGPT is not a replacement for the need for a human instructor (Ahmed, 2023; Jeon & Lee, 2023). Instead, it helps instructors improve their skills, which allows them to focus on more complex and interactive tasks (Kasneci et al., 2023; Yu, 2023). The research also highlighted the importance of training teachers to use ChatGPT effectively. Many educators said they needed skills to use the technology in their respective teaching strategies; after all, each class or group has its own peculiarities. In addition, the ethics of using ChatGPT in higher education became an important issue. Transparency in the use of technology, student privacy, and bias in AI-generated responses were the main concerns when reading the researched articles (Fulgencio, 2024; Jeon & Lee, 2023).

1.4 ANALYSIS OF RESULTS

For all sectors that use AI, recent discussions about its impacts on education serve as a warning sign. Therefore, it is essential to engage in dialogue with society about the application of AI to avoid unfounded rumors that could harm the advancement of AI technologies in the field of education (Alves et al., 2024).

The adaptation of technology to the needs of each institution and discipline is a notable aspect of the results (Orlikowski & Barley, 2001). The flexibility of ChatGPT allowed its customization according to the requirements and educational objectives of each teacher, as well as the content of the courses (Majchrzak et al., 2000).

We found that the younger generation, accustomed to digital interaction, easily adapted to ChatGPT as a source of information and support (Yenduri et al., 2024). However, some students said that they preferred to interact with more people in situations that required emotional support and empathy (Hargreaves, 2000). We also found that ChatGPT can be especially useful in large courses, where teachers may find it difficult to have one-on-one interactions with students. It provides high-quality, standardized answers and also helps in managing question flows (Burleson, 2003).

A constant concern has been data security, as such technologies may contain private student information (B´elanger & Crossler, 2011). Strict cybersecurity measures and appropriate privacy policies are essential to mitigate risks (Gill & Kaur, 2023; Habibzadeh et al., 2019). On the other hand, implementing ChatGPT required a lot of money in training, customization, and maintenance relative to the financial effects. However, many teachers believe that these investments can save time and resources in the long run (Karam, 2023; Roxå & Mårtensson, 2009).

The introduction of ChatGPT raised questions about how teachers can reinvent themselves and focus on teaching more conceptual and human topics, that is, in a more philosophical and ethical context (Roxå & Mårtensson, 2009; Stahl & Eke, 2024). The research results showed that ChatGPT helps students to engage more in active learning when they seek information and answers outside the classroom (Rasul et al., 2023). However, the results demonstrate that ChatGPT performs best when it is used in conjunction with more conventional pedagogical approaches, such as classroom discussions and hands-on activities (Grassini, 2023).

The discussion also emphasized the need for ongoing monitoring and oversight of ChatGPT to ensure that responses are consistent with course content and quality standards. The data showed that students have varying opinions about ChatGPT: some see it as a useful tool, while others prefer to interact with their teachers (Tayan et al., 2023). Research shows that ChatGPT needs to be continuously improved, particularly with regard to understanding specific educational contexts and interaction skills (Grassini, 2023).

The discussion also discusses how ChatGPT can assist students with special needs, such as those who need translation or reading assistance, making higher education more accessible to everyone (Grassini, 2023; MacKenzie & Wajcman, 1999). Finally, the results of this research show that ChatGPT is changing the practices of teachers in higher education, offering significant advantages in terms of efficiency and accessibility. To maximize its potential, there are still ethical, training, and implementation challenges to be addressed (Floridi, 2014; Grassini, 2023). It is also of fundamental importance that educators, educational institutions, and developers collaborate to maximize ChatGPT as this technology continues to evolve, ensuring its use in teaching in an effective and, above all, moral and ethical manner (Grassini, 2023; Smith & Marx, 1994).

1.5 TECHNOLOGICAL EVOLUTION AND HUMAN DEVELOPMENT

Technological evolution has been an essential part of human history, shaping our society and influencing our interactions with the world. According to Schumpeter (1942), technological advancement drives economic and social progress and plays an important role in the dynamics of capitalism. The Industrial Revolution, one of the most important technological revolutions in history, brought significant changes in the organization of work and production. As Hobsbawm (1968) notes, this change was one of the few in history that can be compared to the Neolithic Revolution in terms of its influence on human existence (Schumpeter, 1942).

The Internet is considered a revolutionary technology in the modern world. Berners Lee (2000) describes the creation of the World Wide Web as a tool that radically transformed communication and access to knowledge, allowing the instantaneous dissemination of information around the world. Modern technological evolution, however, brings with it moral and social issues. AI is constantly developing, as Turing (1950) warned, sparking discussions about its ability to achieve singularity, which would mean that machines could have intelligence similar to that of humans (Berners Lee et al., 2000).

Bostrom (2014) states that although AI is considered one of humanity's most important inventions, it also raises important questions about its control and safety. Society must establish strong ethical guidelines to ensure that AI development is responsible (Roberts et al., 2014). In short, technological developments continue to change our society and our way of life. "We shape our tools, and then they shape us," McLuhan said in 1964. As a result, it is essential that humanity consider how to balance technological advances with moral concerns to ensure that technology contributes to the well-being and advancement of society, especially in the educational aspect, which is the foundation of any society (Roberts et al., 2014).

Finally, large databases have been created and used to successfully create complex AI models due to the expansion of computing capacity and the spread of Internet applications (Li et al., 2023). This is the case of ChatGPT, a natural language platform developed by OpenAI that was launched in November 2022 and stands out for its ability to write words in a way similar to that of a human being. Everyone was surprised by this technological advance, which can be expressed very well, especially in the English language, obtaining approvals in selection exams for doctors, lawyers, and postgraduate programs (Sun et al., 2020). ChatGPT is capable of processing and generating natural language responses in a variety of languages and dialects, including Spanish. It can also adjust its response to the tone and style of queries. Additionally, it is an essential tool for those seeking to extract knowledge from large data sets due to its ability to process large volumes of data (Sun et al., 2020).

1.6 AUTOMATED WORK AND COPYRIGHT

The increasing adoption of modern technologies is driving the automation of work, which is changing the economic and employment landscape. The rise of automation, which involves replacing manual tasks with automated systems, raises important questions regarding copyright because the results produced by these systems often raise questions about who the creator is.

The traditional concept of authorship and copyright has been based on originality and creativity, attributing legal rights to the human creator, as Boyle (2008) notes. However, the distinction between human authorship and automated authorship is becoming increasingly difficult as new technological advances are made.

The assignment of copyright to works created by algorithms or AI systems is the main issue in this context. According to Lessig (2008), when it comes to intellectual property in an automated environment, it is important to think not only about authorship but also about the impact that algorithms have on the creation of works.

The granting of copyright to non-human works is undefined in the Berne Convention, which deals with the protection of literary and artistic works. However, Gervais (2019) states that it is necessary to rethink copyright principles in relation to automation, emphasizing the importance of new legal frameworks that recognize the contribution of technology without disregarding human rights, ethics, and creative capacity.

Automation can result in the mass creation of content, which makes it difficult to apply conventional principles of authorship control fairly. Therefore, as automation becomes more prevalent across industries, copyright laws must be changed to address the complications of non-human authorship and ensure that human creators are recognized and protected in an increasingly automated world. To ensure fair and effective management of intellectual property and to foster creativity and innovation in a rapidly changing technological world, regulatory frameworks and ethical deliberation must be revised.

Recent research, such as Adeshola and Adepoju (2023), Elbanna and Armstrong (2024), Fütterer et al. (2023), Halaweh (2023), Lo (2023), Memarian and Doleck (2023), Montenegro-Rueda et al. (2023), Tan (2023), and Zhu et al. (2023), indicates the positive impact of the use of ChatGPT in education since it can break countless gaps in several problems in education nowadays. However, it is worth noting that, as the researchers and authors themselves reported in the aforementioned research, it is not known what the outcome will be. Therefore, it is extremely important to be cautious about the use of this tool in order to maintain a quality of excellence in education.

1.7 CONCLUSION

World history has shown that banning technology is not the best way to do things. Examples include the creation of drugs that have advanced medicine, the invention of food production methods that have reduced scarcity, and the construction of transportation and communication systems that connect people from all over the world into a single society. Attempting to prevent the development of new technologies is futile. Whether their use is beneficial or not, they will always be exploited by those who hold them. The use of the latest technologies, with regard to information and communication, will be no different. Most of them have already been created by companies as powerful as nations, with their respective global businesses. Their rapid development has occupied the discussions, especially in the school and academic spheres. New natural language models, such as ChatGPT, have particularly affected teaching and research.

Technological developments and the introduction of innovative tools such as ChatGPT have significantly changed the way teachers work in higher education. This chapter explores the complexities of this impact, identifying the benefits and potential drawbacks of technology. From this perspective, we conclude that ChatGPT is making a significant difference in higher education. However, it must be carefully evaluated and managed before it can be used.

The results of this research demonstrated that ChatGPT facilitates communication and learning support and eases the workload of teachers with quick responses to frequently asked questions. As a result, teachers can spend more time on more complex teaching activities, such as encouraging students to think critically and fostering classroom discussions. Managing large courses can be difficult, but with the advent of ChatGPT, it is possible to bridge the gap and help increase accessibility in higher education by supporting students with special needs, such as personalized readings or content translation. It is worth noting that the parameters were chosen empirically, based on vast literature and databases originating from scientific encyclopedias, such as Science Direct, Scopus, and Web of Science.

However, there are ethical, pedagogical, and implementation issues that need to be considered as we move into the AI era. Students have varying opinions about the use of ChatGPT, with some appreciating it as an enabling tool, while others appreciate the genuine human interaction and emotional support. The ethics of using ChatGPT must also be considered. Privacy issues, transparency in the use of AI, and bias in the responses generated by ChatGPT must be carefully addressed. To ensure that the technology is used ethically and effectively, teachers need to be properly trained to integrate such technology into teaching practices.

The impact of ChatGPT on higher education teaching practices is broad and varied, with significant benefits and challenges. It is critical that developers, educational institutions, and the academic community work together as we continue to explore and use this technology to ensure that it is used ethically and effectively in the educational context. By effectively integrating ChatGPT into daily practices, provided the challenges are carefully considered and managed, faculty have the ability to improve the quality of higher education and foster a more inclusive learning experience that this technological advancement offers.

Ultimately, none of these questions has a simple answer, and others will certainly emerge that are even more complex than currently imagined. Social networks emerged at the beginning of the century, and the debate still persists about whether or not regulation is necessary and in what form. However, since there is no record of legislation or brute force that has hindered the advancement of a technology, it is crucial to strive to always make the most of what they have to offer, always focusing on the ethical dimension that permeates all of science and society.

1.8 FUTURE WORKS

This article addressed a preliminary analysis point of view, using an empirical method, through world-renowned scientific databases. The perspective is that future works will emerge with a focus on discussing possible potential risks to education in general, mainly in the ethical framework, such as data security or long-term dependence on AI.

REFERENCES

Adeshola, I., & Adepoju, A. P. (2023). The opportunities and challenges of ChatGPT in education. *Interactive Learning Environments*, 1–14. DOI:10.1080/10494820.2023.2253858

Ahmed, M. A. (2023). ChatGPT and the EFL classroom: Supplement or substitute in Saudi Arabia's Eastern Region. *Information Sciences Letters*, *12* (7), 2727–2734.

Alves, A. M., Chuma, E. L., Vaz, G. C., De Oliveira, G. G., Roger, L. L. B., & Lano, Y. (2024). AI ethics in IoT: Involving society in the discussion. In *2024 international conference on electrical, computer and energy technologies (ICECET)* (pp. 1–4).

B´elanger, F., & Crossler, R. E. (2011). Privacy in the digital age: A review of information privacy research in information systems. *MIS Quarterly*, *35*, 1017–1041.

Berners Lee, T., Fischetti, M., Dertouzos, M., Sol´a, J., & Rubio Fern´andez, M. (2000). *Tejiendo la red: el inventor del world wide web nos descubre su origen*. Siglo Veintiuno.

Burleson, B. R. (2003). The experience and effects of emotional support: What the study of cultural and gender differences can tell us about close relationships, emotion, and interpersonal communication. *Personal Relationships*, *10* (1), 1–23.

Edwards, M. (2009). *Civil society*. Polity.

Elbanna, S., & Armstrong, L. (2024). Exploring the integration of ChatGPT in education: Adapting for the future. *Management & Sustainability: An Arab Review*, *3* (1), 16–29.

Floridi, L. (2014). *The fourth revolution: How the infosphere is reshaping human reality*. OUP Oxford.

Fulgencio, S.-V. (2024). Developing effective educational Chatbots with GPT: Insights from a pilot study in a university subject. *Trends in Higher Education*, *3* (1), 155–168.

Fütterer, T., Fischer, C., Alekseeva, A., Chen, X., Tate, T., Warschauer, M., & Gerjets, P. (2023). ChatGPT in education: Global reactions to AI innovations. *Scientific Reports*, *13* (1), 15310.

Gill, S. S., & Kaur, R. (2023). ChatGPT: Vision and challenges. *Internet of Things and Cyber-Physical Systems*, *3*, 262–271.

Grassini, S. (2023). Shaping the future of education: Exploring the potential and consequences of AI and ChatGPT in educational settings. *Education Sciences*, *13* (7), 692.

Grübler, A. (2003). *Technology and global change*. Cambridge University Press.

Habibzadeh, H., Nussbaum, B. H., Anjomshoa, F., Kantarci, B., & Soyata, T. (2019). A survey on cybersecurity, data privacy, and policy issues in cyber-physical system deployments in smart cities. *Sustainable Cities and Society*, *50*, 101660.

Halaweh, M. (2023). ChatGPT in education: Strategies for responsible implementation. *Contemporary Educational Technology*, *15* (2). https://eric.ed.gov/?id=EJ1385551

Hargreaves, A. (2000). Mixed emotions: Teachers' perceptions of their interactions with students. *Teaching and Teacher Education*, *16* (8), 811–826.

Javaid, M., Haleem, A., Singh, R. P., Khan, S., & Khan, I. H. (2023). Unlocking the opportunities through ChatGPT tool towards ameliorating the education system. *BenchCouncil Transactions on Benchmarks, Standards and Evaluations*, *3* (2), 100115.

Jeon, J., & Lee, S. (2023). Large language models in education: A focus on the complementary relationship between human teachers and ChatGPT. *Education and Information Technologies*, *28* (12), 15873–15892.

Karam, J. (2023). Reforming higher education through AI. In *Governance in higher education: Global reform and trends in the Mena region* (pp. 275–306). Springer.

Kasneci, E., Seßler, K., Ku¨chemann, S., Bannert, M., Dementieva, D., Fischer, F., …, & Hu¨llermeier, E. (2023). ChatGPT for good? on opportunities and challenges of large language models for education. *Learning and Individual Differences*, *103*, 102274.

Li, J., Herdem, M. S., Nathwani, J., & Wen, J. Z. (2023). Methods and applications for artificial intelligence, big data, internet of things, and blockchain in smart energy management. *Energy and AI*, *11*, 100208.

Lo, C. K. (2023). What is the impact of ChatGPT on education? a rapid review of the literature. *Education Sciences*, *13* (4), 410.

Lundstad, E., Brugnara, Y., Pappert, D., Kopp, J., Samakinwa, E., Hu¨rzeler, A., …, & Demar´ee, G. (2023). The global historical climate database HCLIM. *Scientific Data*, *10* (1), 44.

MacKenzie, D., & Wajcman, J. (1999). *The social shaping of technology*. Open University.

Majchrzak, A., Rice, R. E., Malhotra, A., King, N., & Ba, S. (2000). Technology adaptation: The case of a computer-supported inter-organizational virtual team. *MIS Quarterly*, 569–600. www.jstor.org/stable/3250948?casa_token=tCJIonElqWQAAAAA:iGa1Yk7bmZjp3k9yZJJPIberkpM2ZYr2NVebg7nAE6dMvEgCkAsSdEGyJp5lONZoGgIukr8P4MOIZybIo7znLBFhJiyO7kp1m6uhEsw1pbiSgtAgHNJpWg

Meghji, A. (2024). From public sociology to sociological publics: The importance of reverse tutelage to social theory. *Sociological Theory*, *42* (2), 114–136.

Memarian, B., & Doleck, T. (2023). ChatGPT in education: Methods, potentials and limitations. *Computers in Human Behavior: Artificial Humans*, 100022. www.sciencedirect.com/science/article/pii/S2949882123000221

Montenegro-Rueda, M., Fern´andez-Cerero, J., Fern´andez-Batanero, J. M., & L´opez-Meneses, E. (2023). Impact of the implementation of ChatGPT in education: A systematic review. *Computers*, *12* (8), 153.

Orlikowski, W. J., & Barley, S. R. (2001). Technology and institutions: What can research on information technology and research on organizations learn from each other? *MIS Quarterly*, 145–165. www.jstor.org/stable/3250927?casa_token=vc-6Exn4rowAAAAA:Ntibm3CN10DXlv4kOPK8xE41c-jW-jvczMXLib9plJQ1gcx7XrRsimdQu53lhAU59fdyq_s2sROjTcRKHtz0e_fBipEGokWH5jNSHMNyn1WYEaTVUsl0pw

Rasul, T., Nair, S., Kalendra, D., Robin, M., de Oliveira Santini, F., Ladeira, W. J., …, & Heathcote, L. (2023). The role of ChatGPT in higher education: Benefits, challenges, and future research directions. *Journal of Applied Learning and Teaching*, *6* (1), 41–56.

Roberts, L. D., Bostr¨om, P., O'Sullivan, J. F., Schinzel, R. T., Lewis, G. D., Dejam, A., …, & Georgiadi, A. (2014). β-aminoisobutyric acid induces browning of white fat and hepatic β-oxidation and is inversely correlated with cardiometabolic risk factors. *Cell Metabolism*, *19* (1), 96–108.

Roumeliotis, K. I., & Tselikas, N. D. (2023). ChatGPT and open-AI models: A preliminary review. *Future Internet*, *15* (6), 192.

Roxå, T., & Mårtensson, K. (2009). Significant conversations and significant networks – Exploring the backstage of the teaching arena. *Studies in Higher Education*, *34* (5), 547–559.

Schumpeter, J. A. (1942). *Capital, socialism, and democracy*. Harper.

Smith, M. R., & Marx, L. (1994). *Does technology drive history?: The dilemma of technological determinism*. MIT Press.

Stahl, B. C., & Eke, D. (2024). The ethics of ChatGPT – exploring the ethical issues of an emerging technology. *International Journal of Information Management*, *74*, 102700.

Sun, L., Jiang, X., Ren, H., & Guo, Y. (2020). Edge-cloud computing and artificial intelligence in internet of medical things: Architecture, technology and application. *IEEE Access*, *8*, 101079–101092.

Tan, X. (2023). The impact of ChatGPT on education and future prospects. *Highlights in Science, Engineering and Technology*, *61*, 138–143.

Tayan, O., Hassan, A., Khankan, K., & Askool, S. (2023). Considerations for adapting higher education technology courses for AI large language models: A critical review of the impact of ChatGPT. *Machine Learning with Applications*, 100513. www.sciencedirect.com/science/article/pii/S266682702300066X

Trithara, D. (2024). Agents of platform governance: Analyzing us civil society's role in contesting online content moderation. *Telecommunications Policy*, *48* (4), 102735.

Waller, I., & Anderson, A. (2021). Quantifying social organization and political polarization in online platforms. *Nature*, *600* (7888), 264–268.

West, J., & Mace, M. (2010). Browsing as the killer app: Explaining the rapid success of apple's iPhone. *Telecommunications Policy*, *34* (5–6), 270–286.

Yenduri, G., Ramalingam, M., Selvi, G. C., Supriya, Y., Srivastava, G., Maddikunta, P. K. R., …, & Wang, W. (2024). GPT (generative pre-trained transformer) – A comprehensive review on enabling technologies, potential applications, emerging challenges, and future directions. *IEEE Access*. https://ieeexplore.ieee.org/abstract/document/10500411

Yu, H. (2023). Reflection on whether ChatGPT should be banned by academia from the perspective of education and teaching. *Frontiers in Psychology*, *14*, 1181712. https://ieeexplore.ieee.org/abstract/document/10500411

Zhu, C., Sun, M., Luo, J., Li, T., & Wang, M. (2023). How to harness the potential of ChatGPT in education? *Knowledge Management & E-Learning*, *15* (2), 133–152.

2 The Gloomy Aspect of AI *Negative Impact of AI on Society*

Lokesh Jain

2.1 INTRODUCTION OF AI

Artificial Intelligence (AI) has become not just a vision of the future, but also a tool and an element that shapes the present and future of global society. This technology has gone through a process of transforming from an idea into a technological gadget that is vital in the current society, i.e., the innovation cuts across all domains of life, including using smart gadgets and managing vital utilities. Thus, over time, the use of integration and utilization of AI strengthens and expands throughout different areas, including healthcare, finance, law enforcement, transportation, education, as well as possesses the capability to dictate decisions. However, the advancement of AI calls for society to accept the opportunities that come with this technology as well as the dangers that are associated with it. Still, it is a very challenging task, which has to be solved by policymakers and other technology experts, as well as ethicists, to reach responsible development and application of AI systems.

This section focuses on both the opportunities of using AI and the threats that may negatively affect people's lives. On the positive side, AI technologies can increase effectiveness and optimize various organizational and research processes in addition to opening up new opportunities for innovations and developments, but on the downside, AI technologies might reproduce and even deepen social injustices, raise issues concerning privacy and confidentiality, and challenge democratic systems. However, these negative impacts will be discussed under critical aspects regarding issues like bias and discrimination, autonomous decision-making, manipulation, and fraudulent activities fostered by AI (Krishna, 2024). Thus, the primary objectives of the chapter are as follows:

- Address an introduction and a brief history of AI, along with its advantages and disadvantages, to depict the societal effects of AI.
- Examine job displacement, income destruction, and inequality caused by AI while suggesting strategies for workforce adjustment and economic fairness.
- Examine the role of AI in controlling and examining threats to data privacy.

DOI: 10.1201/9781003560463-3

- Interrogate the applications of AI in cyber defense, cybersecurity, and defense operations while addressing ethical and regulatory factors.
- Evaluate the influence of AI on mental well-being, cognitive function, and independence while proposing strategies to reduce reliance and encourage analytical thinking.

2.1.1 Brief History of AI

AI is a subset of computer science and a methodology that offers the ability of a computer-based system to learn and make the right decisions. However, despite present advancements in AI technology, this idea has been proposed for several decades, even dating back to the middle of the twentieth century.

In 1956, John McCarthy used the term AI as the start of innovative intelligence at the Dartmouth Conference. The early AI techniques aimed at developing mathematical models that could strictly follow directions to solve a problem or make a decision. However, these first forms of systems lacked the features of the enhancement of the acquired experience or changes in the systems depending on the circumstances. AI got its breakthrough with the evolution of Machine Learning (ML) in the 1980s and 1990s. One of the key differences between ML and rule-based systems is that the former can adapt and enhance its capabilities in processing data without being hard-coded into the desired behavior. This advancement was further boosted in the early 2000s with the emergence of big data, the large volumes of which ensured the operators of the mass of new tools and platforms, and supplied ML algorithms with fuel for achieving the highest levels of accuracy and execution. More recently, deep learning, a particular type of ML that uses artificial neural networks to mimic complex data patterns, has greatly progressed AI (Haenlein & Kaplan, 2019).

Another important factor in the development of AI is the advancement of technology, as it is growing exponentially due to recent improvements in computational power, the relative availability of big data, and the continual refinement of algorithms. The advancements in the current and upcoming hardware technology, such as Graphics Processing Unit (GPU), and chips dedicated to AI processing have made it possible to process large volumes of data within a shorter span, which has significantly boosted other fields like voice recognition, facial recognition, and translation (Guerrero, 2023; McLean et al., 2023).

2.1.2 Dual Nature of AI: Benefits vs. Drawbacks

Today, AI is integrated with a variety of applications that influence almost all domains of society. Table 2.1 highlights the way AI impacts different sectors with its power and efficiency and provides better personalized services in the real world (Smith & Eckroth, 2017).

As AI systems are used to automate different processes, there are also chances of threats if the system is hacked. In addition, there has been a question regarding their effectiveness due to the increasing use of AI in all sectors, which results in displacing human labor by automation (Marr, 2019). Table 2.2 outlines a few of the leading risks and challenges that are derived from the augmentation of AI in contemporary society.

TABLE 2.1
Role of AI in Various Domains

Sector	Role of AI	Examples
Healthcare	Diagnosis assistance, predictive analytics, personalized medicine, robotic surgery	AI in radiology, virtual health assistants
Finance	Fraud detection, algorithmic trading, risk management, personalized financial services	Credit scoring, robo-advisors
Transportation	Autonomous vehicles, route optimization, and traffic management	Self-driving cars, AI in logistics
Education	Personalized learning, AI-driven tutoring, administrative automation	Adaptive learning platforms, AI-powered grading systems
Manufacturing	Predictive maintenance, quality control, supply chain optimization	AI in smart factories, robotic process automation
Retail	Customer behavior analysis, personalized recommendations, and inventory management	Chatbots, recommendation engines, and dynamic pricing
Entertainment	Content recommendation, media creation, and gaming AI	Netflix recommendations, AI-generated music, and art
Security and Surveillance	Threat detection, anomaly recognition, facial recognition, behavioral analytics	AI in video analytics, smart surveillance systems
Agriculture	Precision farming, crop monitoring, and yield prediction	AI-driven drones, soil, and plant health monitoring
Energy	Energy optimization, predictive energy usage, grid management	AI for smart grids, predictive maintenance of power plants
Environment	Climate modeling, wildlife conservation, and pollution tracking	AI-driven climate models, deforestation tracking
Public Services	Smart cities, policy-making assistance, and disaster management	AI for traffic management, disaster response predictions
Human Resources	Recruitment automation, employee sentiment analysis, and performance management	AI-powered hiring platforms, employee engagement analytics
Legal	Document review, contract analysis, predictive case outcomes	AI in legal research, contract review automation
Marketing and Sales	Customer segmentation, targeted advertising, predictive sales analytics	AI-driven marketing campaigns, chatbots for customer support
Creative Arts	AI-generated content, tools for design and creativity	AI in graphic design, music composition, and video production

The duality of AI is embedded in its capability to bring positive and negative impacts of the same potential size. Thus, AI has both an opportunity and a potential threat for management and emphasizes that the development of AI should proceed along with the reflections of its benefits and risks (Zhao, 2023).

TABLE 2.2
Menaces Associated with AI in the Present Society

Category	Drawback/Risk	Description
Job Displacement	Unemployment	AI automation could replace jobs, especially in sectors like manufacturing, retail, and customer service.
Bias and Discrimination	Algorithmic Bias	AI systems can perpetuate or amplify societal biases due to biased training data, leading to unfair decisions in hiring, loans, law enforcement, etc.
Privacy Concerns	Data Exploitation	AI systems often rely on large datasets, risking misuse or mishandling of personal information, leading to privacy invasion or surveillance concerns.
Security Threats	Cybersecurity Risks	AI-driven systems can be vulnerable to hacking and malicious use, such as deepfakes or autonomous cyberattacks, increasing the risks of digital sabotage.
Concentration of Power	AI Monopoly	The development of AI is concentrated in a few large corporations, potentially leading to monopolies and unequal access to AI technology.
Lack of Accountability	Legal and Ethical Ambiguity	AI decision-making processes are often opaque ('black box'), making it difficult to assign accountability in case of errors or harmful outcomes.
Weaponization of AI	Autonomous Weapons	AI could be used to develop autonomous weapons, which may act unpredictably or lead to an arms race between nations, raising ethical and global risks.
Economic Inequality	Wealth Gap	AI's economic benefits may be unevenly distributed, increasing inequality between AI-rich corporations and the rest of society.
Environmental Impact	High Energy Consumption	Large AI models, particularly in ML, consume vast amounts of energy, contributing to environmental degradation and increasing carbon footprint.
Over-reliance on AI	Dehumanization of Decision-Making	Over-reliance on AI for decision-making in critical sectors (e.g., healthcare and judiciary) could reduce human judgment and ethical considerations.
Erosion of Human Skills	Decline in Human Expertise	Over time, heavy reliance on AI might result in diminished human skills and expertise in areas where AI systems dominate.
Mass Surveillance	Surveillance States	AI enhances surveillance capabilities, potentially leading to authoritarian control, tracking, and violation of civil liberties.
Misinformation	Spread of Fake Content	AI tools like deepfakes can generate realistic yet fake content, which can be used to spread misinformation, harm reputations, or manipulate opinions.
Ethical Dilemmas	Misalignment with Human Values	AI systems may be programmed with objectives that are misaligned with human values, leading to ethically questionable outcomes in social and moral contexts.

2.1.3 Theoretical or Conceptual Framework to Depict AI Impact

The impact of AI on the economy, employment, and surveillance is among the emerging hot issues in scholarly discussions. As for economic disparities, the context of the Skill-Biased Technological Change (SBTC) theory indicates that AI is primarily beneficial to highly skilled employees only since it removes low-skilled repetitive jobs, thus increasing the wage difference between highly skilled and low-skilled workers. Acemoglu and Autor argue and show how automation successfully eliminates low-wage employees' offers in the market while at the same time increasing the wages of those highly skilled workers' offer that can effectively support AI technologies (Acemoglu & Autor, 2011). Furthermore, capital-biased technological change claims that investments in AI result in higher average returns to capital instead of labor, which results in increased wealth disparities as AI investments produce more value for AI-driven firms or intellectual property owners. Another account of capital and income disparity similar to technology is brought forward by Thomas Piketty (Piketty, 2015). On an international level, AI technologies being distributed between developed and developing countries have contributed to a new world division of labor where developed nations reap the benefits of new technological advancements while the less endowed nations remain in the periphery and, at times, cannot even adopt new advancements (UNCTAD, 2021).

Considering that AI is capable of automating a considerable number of jobs, job displacement is analyzed via various economic models. Joseph Schumpeter's theory of creative destruction states that while automation or AI leads to unemployment, it also leads to the birth of new sectors and subsequent economic growth. Nevertheless, change from old to new industries can be disruptive and lead to substantial short-term problems for workers rendered jobless by AI (Schumpeter, 2017). After this, the routine-biased technological change theory expands on the information and reveals that with the advent of AI, middle-skill, routine positions like those in clerical or manufacturing are likely to disappear, creating a higher demand for high-skill, high-wage positions. As a result, some have suggested universal basic income (UBI) as a way to overcome the loss of income resulting from job disruption caused by AI. UBI is to provide a baseline income that will enable workers to move from one occupation to another or retrain for a new one (Van Parijs & Vanderborght, 2020).

2.1.4 Chapter Outline

This section of the chapter deals with the basic concept of AI, the history of AI, the motivation behind the development of AI, the pros and cons of AI, applications of AI in various sectors, and many theoretical or conceptual frameworks that impact the economy, employment, and surveillance. As there is an economic disruption due to the evolution of AI and automation, the next section covers a comprehensive analysis of the economic interferences due to AI that includes various factors for job displacement, long-term unemployment, income destruction, and inequality and wealth gap due to AI. The next section focuses on another adverse aspect of AI that deals with data privacy and mass surveillance, which involves various real-world use cases and policies implemented in India. The next section covers multiple security risks and threats from AI, including cyberattacks, ransomware, and other cybercrimes. This section also deals with how AI can be used in warfare and defense as a threat. The final section of the chapter emphasizes the social and psychological effects of AI, which include various issues like mental health, such as addiction and substance abuse, cyberbullying, and cyber harassment. Thus, the chapter attempts to cover most of the low-spirited aspects of AI in detail.

2.1.5 Section Summary

The summary of the section is as follows:

- Cover the introduction, history, and origin of AI, as well as the advantages and disadvantages of AI in various fields, and find solutions to solve global challenges while raising ethical and privacy concerns.
- Discuss the introduction of AI's negative aspects, including job loss, privacy invasion, bias, and environmental impact.
- Describe the need for the required policies for strong regulations, transparency, and equitable distribution of AI's benefits.

2.2 ECONOMIC DISRUPTION DUE TO AI: A COMPREHENSIVE ANALYSIS

AI and automation technologies are advancing in such a manner so fast that they are transforming the world economy rapidly. Consequently, economic growth and the promotion of a high level of technology through AI have also brought unprecedented risks to employment, income inequality, and social justice. The social consequences of AI consist of communities and persons losing their jobs, besides having their income wiped out in cases where AI is adopted in workplaces as well as firms rendering their services and products, which increase economic rifts that promise prospects of chaos in employment, income, and sociopolitical order.

2.2.1 Job Displacement and Automation

One of the main issues that can be observed in the past few decades is the role of AI in the job market. Automated intelligence is the use of software tools to make machines and algorithms do the work that was formerly accomplished by people, or what people may have done over and over again, especially where clerical, repetitive, or designing work is involved. Automation technologies are implemented to enhance efficiency, minimize costs, and enhance productivity. Therefore, most organizations are embracing AI to reduce employees, especially in industries that involve routine and predictable physical work (Shahvaroughi Farahani & Ghasemi, 2024). Various studies show the number of jobs lost due to automation in different countries, as shown in Figure 2.1 (McCarthy, 2017).

The job automation is visible in multiple industries in various fields, as shown in Figure 2.2. Some of the examples are as follows:

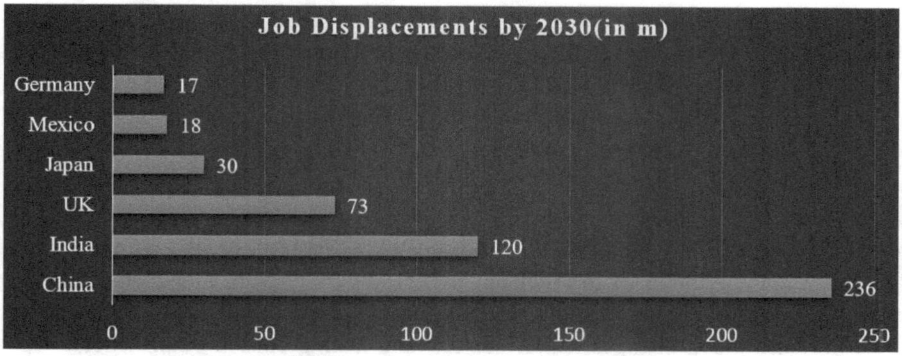

FIGURE 2.1 Job displacement due to automation by 2030.

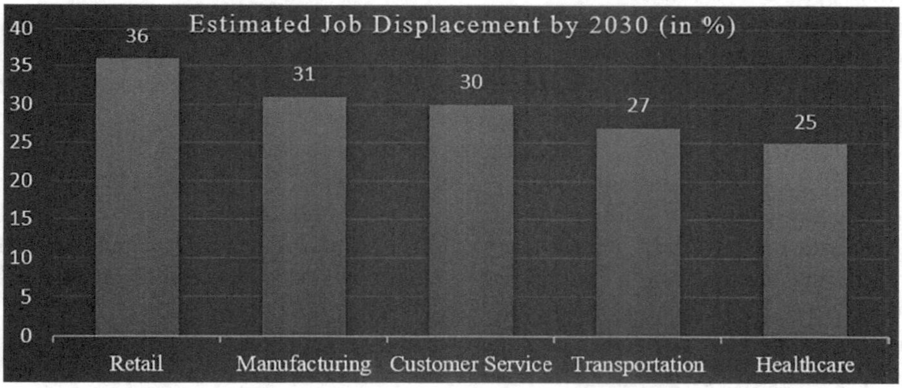

FIGURE 2.2 Industry-wise job displacement due to automation by 2030.

- **Manufacturing:** Factories have been one of the primary sectors in which automation started, as robots have begun to substitute assembly line staff. The International Federation of Robotics (IFR) stated that the installation of industrial robots has become a trend in the last ten years. It is estimated that approximately 4 million industrial robots will be used globally in 2024.
- **Retail:** Retail already has automated checkout technologies and smart inventory systems due to the application of AI. For instance, Amazon Go stores use AI and ML to design shops without checkouts, i.e., customers can simply go in, take the products they need, and leave without the assistance of a cashier.
- **Customer service:** Automated solutions such as AI chatbots and virtual assistants are taking over human agents in service-based sectors such as banking, telecommunication, and e-commerce. Such systems include the capacity to respond to inquiries from customers, identify problems, and offer solutions to them.
- **Transportation and logistics:** Technological advancements in the operational efficiency of logistics companies through the use of self-driving vehicles and AI to carry out delivery systems are changing the face of the logistics industry.

- **Healthcare:** In the current context, AI is applied to diagnoses, data or information processing, and administrative functions in the healthcare industry. For instance, Watson Health of IBM applies AI to diagnose patients and recommend diagnoses and treatments, while robotic surgery systems support surgeons in delicate operations.

2.2.1.1 Explanation of Long-Term Unemployment Due to AI

Although automation leads to efficiency and productivity gains and cost savings for firms, automation-driven technology can also put people out of jobs for the longest time possible. Such changes usually entail long-lasting cuts in employment, especially where jobs are highly routine. While robotic automation can reduce labor costs for some employers, displaced employees might struggle to find new employment or training for a new kind of work.

The influence of AI-driven automation can also be observed in history. For instance, during the Industrial Revolution, many people lost their job opportunities due to the invention of machines and robotic devices, which resulted in high unemployment and unrest. But over the period, as the new industries developed, they provided new employment opportunities, hence changing the economy. According to research done by McKinsey Global Institute, up to 375 million workers globally could be forced to change occupations or learn new skills as a result of automation by 2030. The World Economic Forum also anticipates that by 2025, AI and automation will come up with 97 million new job positions but displace 85 million workers, thus displacing workers (Çağda et al., 2021).

2.2.2 Income Destruction

AI's influence is directly proportional to the concept of job displacement and the changing structure of the workforce. On the one hand, the use of AI leads to the improvement of performance and productivity; on the other hand, it also destroys incomes, especially in industries where companies use AI extensively. On one hand, when companies implement efficient methods of working, they can decrease expenses for hiring employees and increase their profits, which, in turn, leads to additional layers of profit margins, but the positive outcomes of this are not transferred to the employees (Korinek & Stiglitz, 2019).

As per the Kuznets Curve model which was earlier applied to depict the relationship between economic growth and income disparities, it forms an inverted 'U' shape as shown in Figure 2.3. Inequality or environmental degradation increases during the early stages of development, but it is moderate as the economy evolves and policies are enhanced.

Similarly, in AI, there might be the same trend in the future, with a smaller number of unique works being produced. The various problems associated with the initial periods of AI definition and application are in the risks they create, such as job losses, ethical issues, such as bias, and inequality. However, with advancement, the right policies, and the right rules, the future of AI has the potential to fall from socially

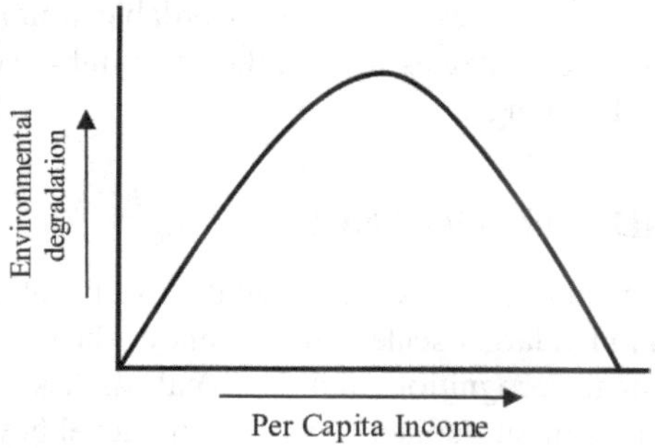

FIGURE 2.3 Kuznets Curve.

disadvantageous to socially constructive, meaning a tool for creating equal opportunities, leading to more positive change. Awareness of this balance enables one to manage the present stage of AI effects, which seeks to avoid AI's potential short-term harms while setting the conditions for its long-term benefits.

2.2.2.1 Analysis of Lower Income due to AI

AI creates a negative impact on workers since only a handful of employees will be required, reducing their wages. When jobs are mechanized, the workers who were involved in such jobs have the option of being rotated into low-paying jobs or even experiencing job loss, low wages, and job insecurity. In addition, automation takes most of the profits from the business people and investors, instead of the employees who lose their jobs.

The 'race to the bottom' effect is another negative concern since only a handful of employees will be required, reducing their wages. When jobs are mechanized, the workers who were involved in such jobs have the option of being rotated into low-paying jobs or even experiencing as much as AI minimizes the operational costs through cutting down the number of employees required, organizations may resort to forcing the remaining workers to offer their services cheaply or without benefits. This can create a vicious cycle of wage and working conditions deterioration, particularly in industries characterized by the most extensive use of AI.

2.2.3 Section Summary

The summary of this section is as follows:

- AI enhances throughput but concurrently risks job loss, income inequality, and social imbalance.
- Automation eliminates wages and negatively impacts lower and middle-class employees while escalating wealth in the hands of highly skilled personnel and organizations.
- AI helps large companies while increasing the gap and economic inequalities, leading to social stratification.

- Need for proper guidelines to ensure affordable training, equal opportunities, and preventive strategies against the potential adverse effects of AI on employment and earnings.

2.3 PRIVACY AND SURVEILLANCE

Modern surveillance has introduced AI in the gradients of tools and methods to conduct surveillance on a much larger scale and efficiency. The constant development of innovations in ML, image recognition, and data analysis has led AI to enhance the function and performance of surveillance systems to a level beyond human capacity by providing real-time analysis of data retrieved. If AI systems connect with large databases, the system can also track, evaluate, and modify with new information. While these capabilities present possibilities for crime prevention, public safety, and regulation, they also create avenues for scalable privacy violations. Such AI tools not only collect terabytes of personal data but may also forecast people's actions and choices, which raises ethical issues of automated surveillance. Analyzing mass surveillance and data privacy, it is possible to predict the further development of AI surveillance, which poses threats to individual rights and a lack of trust (Belanche et al., 2024).

2.3.1 Mass Surveillance

Surveillance is the monitoring and collection of a large amount of information about the activities of populations by governments or other private organizations. This process typically involves the application of the most advanced technologies like facial recognition, video analysis, social media surveillance, and others to collect, assess, and store a large amount of data regarding people's behaviors, locations, and communications (Königs, 2022).

2.3.1.1 Analysis of AI Technologies Applied in Surveillance

This type of analysis is made possible by the AI surveillance system, which comprises numerous technologies that are integrated to deliver a high-powered surveillance system that can track people and develop databases. These technologies include facial recognition, a technology that has been implemented by many cities, airports, and even some modes of transport. Facial recognition algorithms target special parts of the face, and the technology is highly accurate in recognizing people, even in crowded and changing situations. For instance, in China, intelligent surveillance cameras in public spaces can scan an individual's face and compare it to a database in real time, meaning the state can monitor a citizen's movements and actions in real time. It frequently works in tandem with gesture and gait recognition, where algorithms analyze human movement patterns, thereby making it another layer of identification.

Other primary mass surveillance techniques include behavioral analysis, in which AI systems analyze gestures, facial expressions, and mind–human interactions. Such systems can identify abnormal behavior patterns, which will notify authorities of impending danger. For instance, some airports employ the use of AI to scan people's behaviors, such as walking paces, to capture those who exhibit risky behaviors. Furthermore, AI is used in policing to conduct predictive policing, where the police would pinpoint areas that are likely to experience much activity in crime. However, these technologies based on AI also include new ethical dilemmas.

2.3.1.2 Real-World Use Cases of Government and Corporate Surveillance Programs in India

In India, surveillance programs put in place by governments and companies have evolved and spread to areas such as security and organizing spaces, and borders. One such successful implementation is the Crime and Criminal Tracking Network and System (CCTNS), a system of interconnectivity between the police station right from the local level through a central server for faster investigative purposes and crime analysis across the country. The CCTNS system helps the police track criminals and get their criminal histories and also assists law enforcement agencies in crime fighting. Another example is the deployment of Facial Recognition Technology (FRT) for crowd analysis and to identify people in real time across cities such as Delhi, Hyderabad, and Ahmedabad. In particular, this FRT system has been useful in crowd management as well as in finding missing individuals; however, it has been criticized too for violating the privacy rights of the individuals. In the corporate world, CCTV is used by businesses to conduct surveillance on their premises and secure their employees, especially in the retail and banking industries. Businesses also use data analytics to assess the performance of the workers they hire, while some organizations implement the biometric system for security purposes.

Apart from these systems, India has been extending surveillance programs to cover the goals of national security, urban management, and reducing crime. The Centre for AI and Robotics has designed the Network Traffic Analysis program (NETRA) that the Indian intelligence agency uses for scanning any internet traffic containing words/topics specified by them in counter-terrorism measures.

The Smart Cities Mission is another large-scale program in which metro cities deploy massive amounts of video analytics for traffic, energy, and emergency services. In addition, applications such as surveillance services that are Aadhaar-linked give different identifiers across services; some recommendations have even proposed the integration of Aadhaar with public security cameras to monitor individuals, which aligns biometric data with surveillance technologies and raises further questions about privacy in India (Raghavan et al., 2019).

2.3.2 Data Privacy

Data privacy, also known as information privacy, is the ability to control and manage personal information to prevent unauthorized access, use, or disclosure of such information. It includes rules, recommendations, and rules instituted to protect sensitive information, especially in a data-intensive setting where organizations and technologies collect, store, and process large amounts of personal and behavioral data (Bertini, 2023).

2.3.2.1 Analysis of the Method of Collecting, Storing, and Consuming Data

The AI system uses various ways to gather and store user data through browsing history, social media activity, biometric data, and location tracking. Furthermore, this information is then kept in gigantic databases or data clouds where it serves as the feed to AI, which forecasts or analyzes patterns based on this corpus. For instance, browsing data enables organizations to create an accurate database of people and monitor their activities in achieving specific goals. In the same way, the web advertisement targets or recommends relevant materials that the individual might be interested in going through, as supported by the excerpt. Further, the location data obtained from a mobile device describes the extreme physical behaviors and activities, including areas of interest or considerable time spent in specific areas.

2.3.2.2 Risks of Data Breaches and Misuse of Personal Information

The threats that may originate from breaches of personal information and data security are becoming more extensive because of the higher volume of data transmitted through digital platforms in India. For instance, in mid-2024, a critical exposure in the MOVEit file transfer software affected an excessive number of global organizations. Hackers exploited the weakness in the program to gain access to people's sensitive information, including personal, financial, and health information from several companies. It impacted millions of citizens in different countries, some personal data appeared on the dark web, which created threats such as identity theft, fraud, and personalized scams. This not only poses a major threat to individuals but also exposes organizations to considerable risks in terms of financial repercussions and litigation. This meant that victims incurred financial losses apart from stress due to the actualization of their information being used by third parties, which is a clear indication of the implications of these breaches on society. In 2024, an Indian public-sector bank was involved in customer data leakage and suffered fraud and identity theft at the hands of hackers. Hence, data breaches in the government sector also impact the security of a nation. In recent times, there was an instance where unauthorized parties were able to breach the Aadhaar database. It has, however, been enhanced, and there is always doubt concerning the sufficiency of the safety of data (Manheim & Kaplan, 2019).

2.3.3 Section Summary

The summary of this section is as follows:

- AI increases productivity but contains violations of privacy and ethical concerns.
- Mass surveillance, including initiatives like facial recognition and predictive policing, falls in the bias category.
- In India, some security systems like the CCTNS or the Aadhaar help in the matter but are highly criticized for privacy issues.

2.4 SECURITY RISKS AND THREATS

AI has become one of the most significant technologies transforming many industries and sectors within society. This aspect of designing AI as a system that is capable of both defending and attacking is a cause for concern in the future (Truong et al., 2020).

2.4.1 AI as a Tool for Cyber Defense

AI is a powerful tool to combat threats and a novel and effective weapon for cybercriminals in cybersecurity. Because of the growing complexity of digital infrastructure and expanding data volume, AI-based solutions are used as tools for identifying, preventing, and countering cybersecurity threats. Most legacy security solutions, which are based on rules and patterns, can no longer cope with constantly changing threats in today's complex environment. AI, particularly the learning type that can read big data and learn to look for outliers that could be an indication of fraud, has been revolutionary in this regard. Cybersecurity systems can be powered with AI technologies to handle large networks and identify anomalous occurrences that might indicate an attack. The application of AI can be used for detecting phishing attempts, malicious software, and unauthorized attempts at accessing confidential information in real time. These systems can incorporate new forms of cyberattacks since hack efforts continue to grow in complexity.

2.4.2 AI as a Tool for Cyber Offense

While using AI in protecting computer systems bears its advantages, the technology also creates new opportunities for attackers. AI is also a powerful tool in using it to plan and execute far more complex cyberattacks that are far beyond those that many organizations could envision or develop themselves. Cybercriminals can use AI to probe for vulnerabilities in the systems, launch attacks autonomously, and even develop self-evolving viruses.

Another worrying factor when it comes to AI-implemented attacks is the possibility of making the attacks more automated. Classic cyberattacks typically involve input and supervision, which makes it difficult to conduct them on a large scale. But with the use of AI, these cyberattacks can be carried out at an enormous scale, targeting many systems at a single instance, and keep on evolving to avoid being identified. For instance, AI can be employed to create phishing emails as well as to make them customized for certain persons, for those emails may be opened with interest by the target person, as well as contain a loaded virus. Again, such botnets can be applied to perform Distributed Denial-of-Service (DDoS) attacks where networks are flooded with traffic at a faster rate and on a larger scale than what a human can perform.

2.4.3 AI in Warfare

One of the more challenging and serious threats involves the use of AI in state-sponsored cyber war. Superior AI development may be harnessed by nations to perform cyberattacks on profound structures of competing nations, including electricity distribution systems, telecommunication networks, or the financial sector. This is capable of deepening conflicts in the international system because fully automated cyber operations with minimal human intervention are not traceable and can be easily denied compared to traditional military attacks.

AI also enhances the level of difficulty of cyberattacks as they get to be advanced. Thus, as opposed to AI-based malware, the use of traditional signature-based detection mechanisms is bound to prove less and less effective. As the threat continues to evolve and intensify, organizations must turn to technologies such as behavioral analysis and ML-based anomaly detection to boost their security measures (Hageback & Hedblom, 2021).

2.4.3.1 Ethical and Moral Concerns of AI in Defense Applications

The engagement of AI in defense has been deemed to pose deeper ethical and moral issues. One of the most critical issues discussed is the expansion of autonomous decision-making regarding human lives. In conventional war, human beings are the ones who deliberate on when and how the force will be applied and are answerable for every action that they take. In the case of autonomous weapons, this becomes the responsibility of machines, but there is a problem of responsibility and possible unmeritorious killings.

2.4.3.2 Role of International Regulation

Due to the heightened danger posed by the application of AI in modern warfare, global governance structures are increasingly needed to ensure supervision of the deployment of autonomous weapon systems. Large international groups such as the United Nations and many individual nations have urged the prohibition of autonomous weapons systems that can select and engage targets on their own. Critics believe that such a ban will help reduce the likelihood of the emergence of one dangerous technology or another and maintain moral norms in the use of armed forces.

Nevertheless, the goal of reaching a consensus on this matter on the international level is not easy. Most countries, especially those with highly developed military potential, are afraid to ban AI weapons because they can be beneficial for them. Therefore, the international community also needs to agree on the use of AI in cyber warfare in the same way it seeks to control the use of autonomous weapons.

2.4.4 Section Summary

The summary of this section is as follows:

- AI is a great beneficial tool for human lives and brings cybersecurity and warfare implications.
- AI in warfare poses risks to important assets and increases pressure on the accumulation.
- Sovereign weapons raise problematic concerns about command and/or moral responsibility.

- Specific guidelines and regulations that would govern efforts to implement AI in warfare and cyberattacks are critical but elusive.

2.5 SOCIAL AND PSYCHOLOGICAL EFFECTS OF AI

AI is probably one of the most revolutionary creations of this century, as it has revolutionized several sectors of human existence, such as communication, business, health, and education, among others. While the use of the radical form has effectively introduced these aforementioned advantages, the social and psychological impacts, especially on the mental state and the growing obsession with devices, ought not to be overlooked. This segment explores two important areas related to the impacts of AI on society and individuals, including its contribution to the development of mental illness and the fact that individuals of the new generation are more dependent on technology than on themselves. It ensures that the resultant effects of the introduction of these AI technologies on mental health issues such as addiction and substance abuse, cyberbullying, and cyber harassment, as well as how dependence on AI robs one of analytical minds and competence in handling problems, are given adequate and thorough scrutiny.

2.5.1 Impact of AI on Welfare

AI brings great change to mental health since it provides different solutions for diagnostics, treatment, and assistance. In the form of chatbots and virtual therapists, everyone can gain instant and convenient access to the care they need if they suffer from anxiety, depression, or any mental health issues. It improves individualized treatment with the help of data evaluation, flow patterns of behavior, and appropriate intervention. Physicians' dependence on AI results in privacy violations, a lack of compassion in patient care, as well as the risk of an incorrect diagnosis. Hence, AI is an effective tool in mental health but must be supervised and managed by people to provide the best results and services in this crucial sector.

2.5.1.1 Dependence on Supporting AI-Driven Technologies

One of the most striking examples of AI's negative effect on a population's mental health is digital dependencies. Today, many applications through which people spend time daily, such as social networks Facebook and Instagram, e-commerce portals like Amazon, Flipkart, and Snapdeal, streaming services such as Netflix, Amazon Prime, Disney+, and many more, contain AI that adapts content to users. From this perspective, AI systems select and promote content to make users remain active on the platform for a longer period.

One of the relatively recent problems that people experience is social media addiction. These platforms, regulated by AI, give users frequent notifications, likes, comments, and appropriate content. The negative consequences of addiction to AI-driven platforms include the following:

- **Anxiety and depression:** Several studies have established that increased use of social media leads to increased anxiety, depression, and loneliness. A daily feed of related material, which can, in turn, include posts and pictures of perfect, glamorous lifestyles, can cause feelings of inferiority, envy, and low self-esteem.

- **Sleep deprivation:** One of the common effects of developing a dependency on AI-advanced technologies is a lack of adequate sleep due to spending many hours online. These long hours of work lead to inadequate sleep, which in turn worsens mental health disorders, including anxiety and depression, to make it a cycle of unhealthy mental health.
- **Decreased attention span:** The current design of the AI-centered platforms leads to capturing people's attention with content that does not require much effort to process. In the case of continuous scrolling and fast proliferation of the content, it becomes challenging to engage in comprehensive and demanding efforts, which deteriorates the attention span and concentration.

2.5.1.2 Contribution of AI in Cyberstalking and Cyberbullying

Another important aspect concerning mental health is how AI contributes to or fuels cyberbullying and harassment. Social media and open-source forums powered by AI promote the type of content that generates more interaction, excluding the presentation of healthy interaction processes. Thus, contracted AI systems may contribute to the proliferation of toxic content or be insufficient in regulating it in their efforts to ensure users stay connected.

2.5.2 Dependence on Technology: Destruction of Critical Thinking

The integration of AI into almost every sphere of modern existence has led to a gradual delegation of decision-making, work, and data processing to AI systems. While the usage of AI helps to increase the efficiency and accuracy in various spheres of life, the tendencies toward the augmentation of reliance on it provoke concerns related to cognitive offloading, which implies the shift of human mental tasks to machines, and the potential degradation of human higher-order thinking skills. Some of the AI technologies that have influenced the daily lives of individuals include the Google search engine, Global Positioning Systems (GPS) navigation, Alexa from Amazon, and Siri from Apple. The availability and ease of use of AI templates reduce the skills of independent critical thinking, problem-solving, and decision-making within that person. For example, people are increasingly relying on GPS to guide and inform them that they do not need to bother developing orienteering skills. While using the GPS is comfortable, relying on it can reduce people's ability to find their way in new territories without using the technology. Similarly, Google and other search engines help users find answers to any query they might have. Also, recommendation systems such as Netflix, YouTube, and Amazon make users rely on algorithmically suggested content rather than having to take the time to look for it themselves, thus restricting exposure to different opinions and reducing the pursuit of novelty (Gabbiadini et al., 2024).

2.5.3 Section Summary

The summary of this section is as follows:

- AI interference has a positive impact on industries but a negative impact on the population's mental establishment and dependence.
- AI can help with care as well as harm with addiction, anxiety, and abusive experiences from apps and social media.
- AI decisions may thus erode autonomy, confound responsibility, and marginalize human judgment.

2.6 CONCLUSION

AI is one of the groundbreaking novelties of the modern world, with tremendous advantages and opportunities as well as threats. AI has progressed from abstract ideas to revolutionary tools and products across industries and global societies that can be associated with opportunities and risks. In its benefits, it creates economic opportunities through automation; in its risks, it unleashes unemployment caused by AI, social inequality, and further expansion of the income gap. In the context of privacy and security, intentional and unintentional deployments of AI in surveillance trigger ethical questions, whereas data vulnerability and unethical uses aggravate existing worries about personal data. Moreover, the practice of AI as a tool of both defense and attack in cyberspace, as well as the use of AI during warfare, also presents concerns as to why international control and ethical standards are needed. In social life, AI affects the health of individuals, reduces cognition, and raises some ethical questions about the actualization of autonomy and responsibility. In the meantime, there are a lot of opportunities that AI can provide for society, but it is also very important to address and balance the risks that technology might bring and regulate it to make it possible for all the participants of society to benefit from it.

REFERENCES

Acemoglu, D., & Autor, D. (2011). Skills, tasks, and technologies: Implications for employment and earnings. In *Handbook of Labor Economics* (Vol. 4, Issue Part B). https://doi.org/10.1016/S0169-7218(11)02410-5

Belanche, D., Belk, R. W., Casaló, L. V., & Flavián, C. (2024). The Dark Side of Artificial Intelligence in Services. *Service Industries Journal*, *44*(3–4), 149–172. https://doi.org/10.1080/02642069.2024.2305451

Bertini, F. (2023). Artificial Intelligence and data privacy. *Sistemi Intelligenti*, *35*(2), 477–484. https://doi.org/10.1422/108143

Çağda, Y., Bir, E., Endüstri, S., Yapay, Y., & Etkileri, Z. İ. (2021). An Old Problem in the New Era: Effects of Artificial Intelligence on Unemployment on the Way to Industry 5.0. *Journal of Yasar University*, 16, 77–94.

Gabbiadini, A., Durante, F., Baldissarri, C., & Andrighetto, L. (2024). Artificial Intelligence in the Eyes of Society: Assessing Social Risk and Social Value Perception in a Novel Classification. *Human Behavior and Emerging Technologies*, *2024*. https://doi.org/10.1155/2024/7008056

Guerrero, J. M. (2023). The history of modern Artificial Intelligence. In Mind Mapping and Artificial Intelligence (pp. 129–158). https://doi.org/10.1016/b978-0-12-820119-0.00007-8

Haenlein, M., & Kaplan, A. (2019). A Brief History of Artificial Intelligence: On the Past, Present, and Future of Artificial Intelligence. *California Management Review*, *61*(4), 5–14. https://doi.org/10.1177/0008125619864925

Hageback, N., & Hedblom, D. (2021). AI for Digital Warfare. In AI for Digital Warfare. https://doi.org/10.1201/9781003194965

Königs, P. (2022). Government Surveillance, Privacy, and Legitimacy. *Philosophy and Technology*, *35*(1). https://doi.org/10.1007/s13347-022-00503-9

Korinek, A., & Stiglitz, J. E. (2019). Artificial Intelligence and Its Implications for Income Distribution and Unemployment. In The Economics of Artificial Intelligence (pp. 349–390). https://doi.org/10.7208/chicago/9780226613475.003.0014

Krishna, V. V. (2024). A I and Contemporary Challenges: The Good, Bad and the Scary. *Journal of Open Innovation: Technology, Market, and Complexity*, *10*(1). https://doi.org/10.1016/j.joitmc.2023.100178

Manheim, K., & Kaplan, L. (2019). Artificial Intelligence: Risks to Privacy and Democracy. *Yale Journal of Law & Technology*, *21*(106), 106–188.

Marr, B. (2019). What Are the Negative Impacts of Artificial Intelligence (AI)? *Forbes*, 3–5. https://bernardmarr.com/default.asp?contentID=1827

McCarthy, N. (2017). Automation Could Eliminate 73 Million US Jobs by 2030. Retrieved December 31, 2024, from www.forbes.com/sites/niallmccarthy/2017/11/30/automation-could-eliminate-73-million-u-s-jobs-by-2030-infographic/

McLean, S., Read, G. J. M., Thompson, J., Baber, C., Stanton, N. A., & Salmon, P. M. (2023). The Risks Associated with Artificial General Intelligence: A Systematic Review. *Journal of Experimental and Theoretical Artificial Intelligence*, *35*(5), 649–663. https://doi.org/10.1080/0952813X.2021.1964003

Piketty, T. (2015). Putting Distribution Back at the Center of Economics: Reflections on Capital in the Twenty-First Century. *Journal of Economic Perspectives*, *29*(1), 67–88. https://doi.org/10.1257/jep.29.1.67

Raghavan, V., Jain, S., & Varma, P. (2019). India Stack-Digital Infrastructure as Public Good. *Communications of the ACM*, *62*(11), 76–81. https://doi.org/10.1145/3355625

Schumpeter, J. A. (2017). Capitalism, Socialism and Democracy. In *Modern Economic Classics-Evaluations Through Time* (pp. 242–269). https://doi.org/10.4324/9780203857090

Shahvaroughi Farahani, M., & Ghasemi, G. (2024). Artificial Intelligence and Inequality: Challenges and Opportunities. *Qeios*. https://doi.org/10.32388/7hwuz2

Smith, R. G., & Eckroth, J. (2017). Building AI Applications: Yesterday, Today, and Tomorrow. *AI Magazine*, *38*(1), 6–22. https://doi.org/10.1609/aimag.v38i1.2709

Truong, T. C., Diep, Q. B., & Zelinka, I. (2020). Artificial Intelligence in the Cyber Domain: Offense and Defense. *Symmetry*, *12*(3). https://doi.org/10.3390/sym12030410

UNCTAD. (2021). Catching Technological Waves. In *United Nations Conference on Trade and Development* (pp. 1–14). https://doi.org/10.18356/9789210056588c006

Van Parijs, P., & Vanderborght, Y. (2020). Basic Income: A Radical Proposal for a Free Society and a Sane Economy (Excerpts). *Ekonomicheskaya Sotsiologiya*, *21*(1), 44–59. https://doi.org/10.17323/1726-3247-2020-1-44-59

Zhao, B. (2023). Analysis on the Negative Impact of AI Development on Employment and Its Countermeasures. *SHS Web of Conferences*, *154*, 03022. https://doi.org/10.1051/shsconf/202315403022

3 AI-Powered Academic Knowledge Management

Transforming Academic Research and Teaching

Shruti Singh and Ragi T. S.

3.1 INTRODUCTION

The advent of advanced technologies has emerged as a transformative force in academia, offering unprecedented opportunities to revolutionize knowledge management (KM) practices. Academic institutions at the forefront of knowledge generation and dissemination face increasing challenges in managing the sheer volume and complexity of data generated through research and teaching activities (Eager, 2024). Traditional KM systems require revision to meet these demands, necessitating innovative solutions. These technologies, with their capabilities for automation, data analysis, and predictive insights, present a powerful alternative that can reshape the landscape of academic KM.

From an academic perspective, enhanced KM systems enable the streamlining of research processes, facilitating a shift from manual, repetitive tasks to activities that require critical thinking and innovation (Dabbagh & Kitsantas, 2012). By automating data organization, retrieval, and dissemination, these technologies empower researchers to focus on generating new knowledge while improving efficiency in scholarly activities. Studies have highlighted the capacity of these systems to foster interdisciplinary collaboration by breaking down barriers between fields, enabling integrated and innovative approaches to research. However, challenges such as institutional resistance to change and the need for faculty training remain critical barriers to technology adoption in academia (Madhu et al., 2023).

The flowchart below illustrates the process of integrating advanced technologies into academic KM systems, showing how the introduction of these tools can enhance various academic functions, including research, teaching, and administration. It represents a systematic approach that identifies challenges within traditional KM frameworks and culminates in transforming knowledge-related processes through technological advancement.

Identification of KM Challenges: Academic institutions face the challenge of managing an ever-expanding body of knowledge. Traditional systems are insufficient for organizing and sharing vast amounts of academic data.

DOI: 10.1201/9781003560463-4

Introduction of Technological Tools: Advanced technologies such as predictive analytics, intelligent tutoring systems (ITS), data mining tools, and automated administrative tools are introduced to address these challenges. These tools enable more efficient data processing, knowledge retrieval, and decision-making.

Transformation of Knowledge Creation, Sharing, and Application: These tools automate data collection, organization, and sharing processes. This transformation leads to faster knowledge creation, enhanced collaboration, and the application of insights across disciplines.

Enhanced Research Processes: By integrating these technologies, research becomes more streamlined with automated data analysis and predictive analytics, allowing researchers to focus more on innovation and higher-level problem-solving.

Improved Teaching and Learning Experiences: Enhanced learning systems, such as personalized tutoring and adaptive learning technologies, cater to individual student needs, improving educational outcomes and teaching effectiveness.

Streamlined Administrative Functions: Advanced technologies also optimize administrative processes, such as grading, performance tracking, and resource management, leading to greater operational efficiency within academic institutions (see Figure 3.1).

The application of advanced methodologies extends beyond research into teaching and learning. Enhanced KM systems enable data-driven personalization

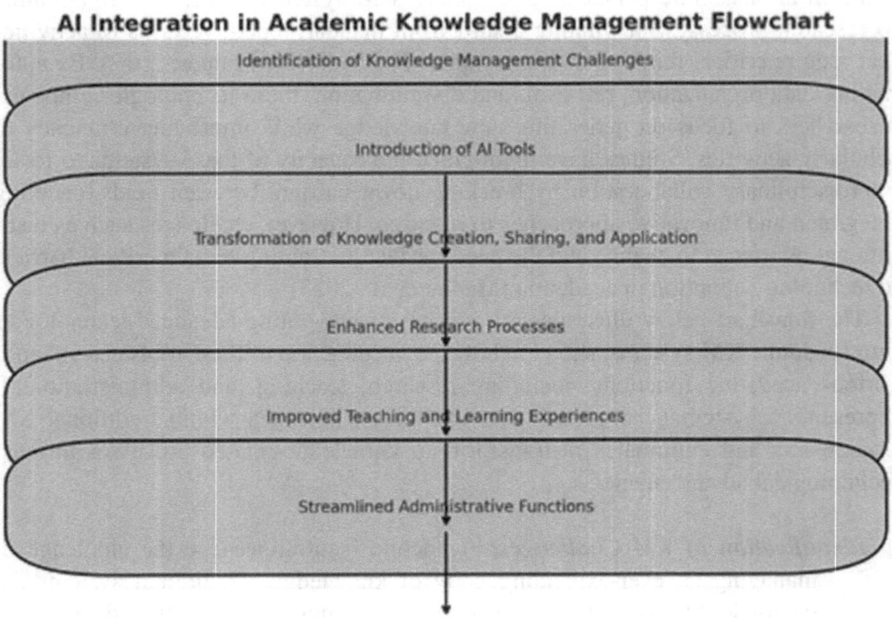

FIGURE 3.1 Technology integration in academic knowledge flowchart.

of educational experiences, tailoring learning resources and pedagogical strategies to meet individual student needs (Madabhushini, 2024). Research emphasizes that these platforms enhance student engagement by personalizing content and continuously improving teaching methods (Janaki, 2024). However, concerns about data usage and potential biases highlight the importance of transparency and accountability in deploying such systems. Additionally, studies have explored how these technologies facilitate knowledge sharing in digital learning environments, identifying tools like chatbots and virtual assistants as pivotal, while also noting scalability and accessibility concerns, particularly in resource-constrained institutions (West et al., 2018).

These transformative potentials extend to decision-making processes in academic institutions, where the ability to analyze vast datasets and identify patterns can significantly enhance institutional decision-making, guiding resource allocation and shaping research agendas (Patial, 2024). Nevertheless, the ethical and legal frameworks governing technology adoption still need exploration, raising questions about data privacy and the equitable application of these tools. This study investigates the theoretical and practical implications of integrating advanced technologies into academic KM (Aghaebe, 2024).

3.2 CONCEPTUAL FOUNDATIONS OF AI-DRIVEN KNOWLEDGE MANAGEMENT

Integrating advanced technologies into KM within academic institutions has become a significant development, transforming how knowledge is organized, managed, and shared. Traditionally, KM practices have focused on systematic approaches for creating, storing, and distributing knowledge. However, the rapid expansion of data and the growing complexity of research demand more sophisticated solutions. Systems that harness these advanced technologies offer a flexible and responsive approach, utilizing automation, analytical methods, and tailored experiences to meet these challenges and improve both research and educational outcomes (Holmes et al., 2019).

The conceptual framework of artificial intelligence (AI)-enhanced KM emphasizes the importance of knowledge sharing, collaboration, and innovation within organizations. AI tools enhance the dissemination of knowledge among various departments or teams and promote uninterrupted communication (Kumar et al., 2023). Thus, AI-driven KM systems are deemed crucial for promoting collaboration and creating competitive advantages in knowledge-intensive sectors. The implementation of AI in KM poses challenges, especially regarding ethical issues, data privacy, and the requirement for strong infrastructure and expertise to effectively deploy and sustain these systems, notwithstanding their potential benefits.

A key aspect of these modern KM systems is the automation of knowledge tasks. By employing algorithms, these systems can take over routine activities like data organization, retrieval, and distribution, markedly improving efficiency and precision (Aghaebe, 2024). This shift allows academic professionals to focus on more advanced intellectual endeavors, such as critical thinking, problem-solving, and innovation, ultimately enhancing overall productivity.

Another crucial element of such KM systems is the representation and interpretation of knowledge. Techniques like natural language processing and knowledge frameworks help in structuring complex academic information in an understandable manner (Jagtap & Epilli, 2024). These tools enable the identification of relationships and trends within extensive datasets, promoting collaboration across disciplines and transcending traditional boundaries. This collaborative research promotes connections across different fields, facilitating the resolution of intricate global issues.

Personalization and adaptability are also essential features of modern KM frameworks. By analyzing user behavior and preferences, these systems can deliver tailored knowledge to meet the specific needs of different users (Pamisetty, 2022). This approach fosters personalized educational experiences, enhancing student engagement and success (Williamson & Piattoeva, 2021). Adaptive learning environments allow educators to address various learning styles, thereby improving the quality of teaching. Similarly, personalized tools assist researchers in accessing pertinent information more efficiently, contributing to effective knowledge creation and sharing.

Incorporating real-time insights and predictive analysis is another fundamental aspect of contemporary KM systems. By examining extensive datasets, these systems can derive actionable insights that guide strategic decisions regarding resource allocation, research priorities, and institutional planning (Kumar et al., 2023). The role of predictive analytics is crucial in recognizing emerging trends and aligning academic goals with future possibilities, allowing institutions to operate effectively in a fast-changing educational environment.

Despite the many benefits, implementing these advanced KM systems poses significant challenges. Issues such as data privacy, bias in algorithms, and the readiness of institutions to adopt and manage these technologies are important considerations (Rayarao, 2024). Ethical aspects, particularly concerning transparency, accountability, and equitable access to these systems, remain central to the discussion about their use in academic settings (Neil, 2016). Addressing these challenges involves developing comprehensive frameworks that align technological innovations with institutional principles and practices. These guiding principles highlight the significant impact that technology can have in academia, underlining its role in boosting research productivity, fostering collaboration across disciplines, and enhancing educational outcomes (Reddy, 2022). As institutions work through the complexities of implementing these systems, these foundational concepts will be key to developing effective and sustainable KM practices.

Incorporating relevant theories into the framework allows for a deeper understanding of how new technologies in KM are reshaping academic institutions. Each theoretical perspective contributes valuable insights into the influence of technology on knowledge creation, sharing, and application in higher education. KM Theory and Nonaka SECI Model provide crucial perspectives for analyzing the role of technology in knowledge processes (Ardiabah, 2024). The SECI Model emphasizes the dynamic interplay between different types of knowledge and helps in understanding how these technologies can facilitate the transformation of tacit knowledge into explicit forms, which are essential for effective KM in academic settings.

The diagram illustrates the relationships between AI technologies, KM theories, and their impact on key academic functions such as research, teaching, and administration. This visual framework clearly shows how different theories (like Nonaka SECI Model, Technology Acceptance Model (TAM), Innovation Diffusion Theory (IDT), and organizational learning interact with AI tools to enhance various KM processes, ultimately benefiting academic functions. Each theory and component has been strategically placed to demonstrate its connection to the central concept of AI-driven KM.

3.3 AI TECHNOLOGIES AND TOOLS FOR KNOWLEDGE MANAGEMENT

KM within academic institutions is evolving to address the increasing complexities associated with managing extensive volumes of data and facilitating interdisciplinary collaboration (Holmes et al., 2019). The advancements in KM practices enable more effective organization, retrieval, and utilization of information, ultimately fostering innovation in both research and teaching methodologies. The integration of sophisticated management systems not only streamlines routine tasks but also enhances decision-making processes, promotes collaboration among scholars, and tailors user experiences to meet individual needs, as highlighted in earlier studies (see Figure 3.2).

One of the most significant advancements in KM involves the incorporation of natural language processing techniques, which empower systems to analyze, interpret, and generate human language (Reddy, 2025). These techniques have found diverse applications within academic settings, particularly in conducting automated literature reviews, executing semantic searches, and summarizing extensive documents. For example, advanced search engines aid researchers in identifying interdisciplinary connections, thus enabling more comprehensive approaches to knowledge generation across various fields (West et al., 2018).

Additionally, machine learning algorithms play a crucial role in transforming KM practices. These algorithms enable systems to learn and improve from data inputs, proving highly effective in categorizing and recommending academic resources. Personalized recommendation systems within digital libraries and academic databases employ these algorithms to predict user preferences, thereby streamlining access to pertinent information and reducing the time spent on manual searches. Furthermore, these tools are instrumental in analyzing large datasets, uncovering trends, and identifying gaps in research, which in turn drives strategic academic initiatives.

Another vital aspect of modern KM involves the utilization of knowledge graphs. These structures link concepts and entities, allowing for intuitive visualization of relationships among different pieces of information. Knowledge graphs have proven particularly useful in mapping complex academic research landscapes and fostering collaboration by highlighting connections among disparate fields (Ikeh, 2024). This structured representation supports hypothesis generation and exploration of novel research areas.

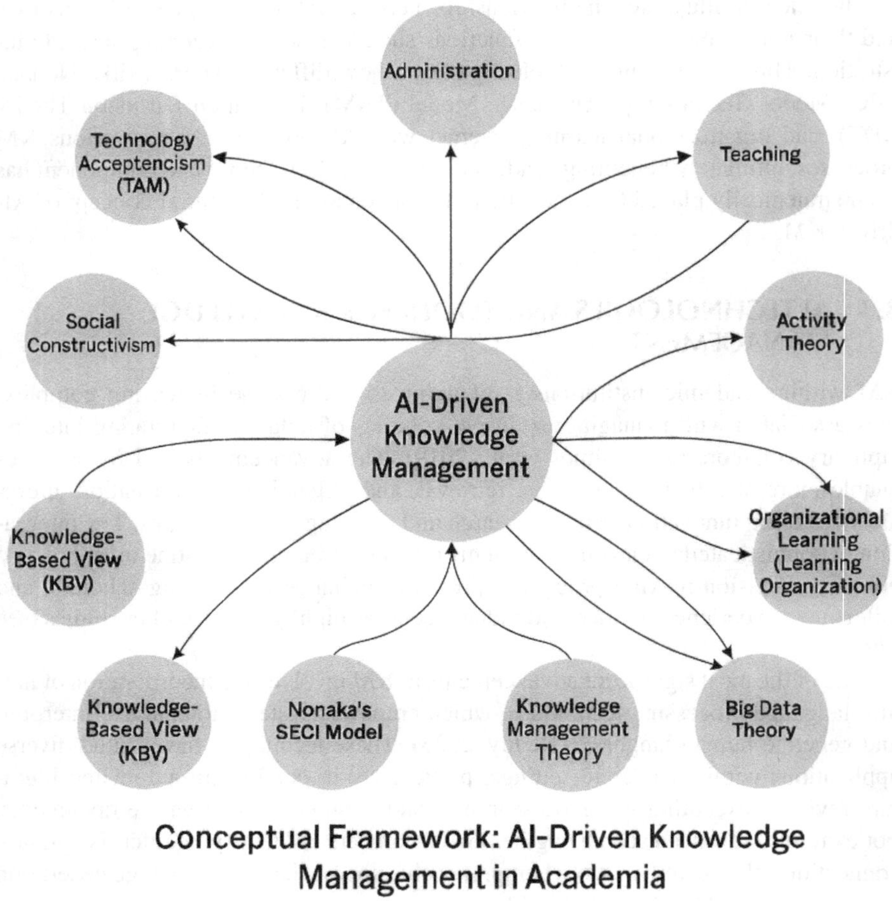

FIGURE 3.2 Conceptual framework on AI driven knowledge management academia.

Moreover, the prevalence of conversational technologies such as chatbots and virtual assistants is increasing within academic KM. These tools provide real-time assistance, answer user inquiries, guide researchers, and streamline administrative processes. For instance, chatbots can assist students in navigating online learning platforms, accessing essential resources, and resolving common queries, thereby relieving institutional support teams of certain workloads and enhancing user engagement.

Finally, the application of predictive analytics tools has emerged as an essential component in the realm of academic KM. These tools analyze both historical and real-time data to forecast trends and support informed decision-making processes. In an academic context, predictive analytics can be utilized to anticipate research impact, identify emerging fields of study, and optimize the allocation of resources. By generating actionable insights in real time, academic institutions can align their strategic initiatives with evolving priorities and challenges.

While these advancements in KM offer significant benefits, they also present challenges that warrant careful consideration. Issues surrounding data security, ethical use of technologies, and transparency in decision-making processes remain paramount (Madhu et al., 2023). Furthermore, ensuring equitable access to these technologies across institutions with varying capacities is essential to maximize their benefits. Addressing these challenges will necessitate interdisciplinary collaboration and the development of robust governance frameworks to guide the ethical and sustainable integration of KM systems.

Academic institutions can significantly enhance their knowledge ecosystems by leveraging advanced KM practices (Dahake, 2024). These systems improve efficiency and empower researchers, educators, and students to engage with knowledge more innovatively and meaningfully, thereby driving progress in research and education (see Figure 3.3).

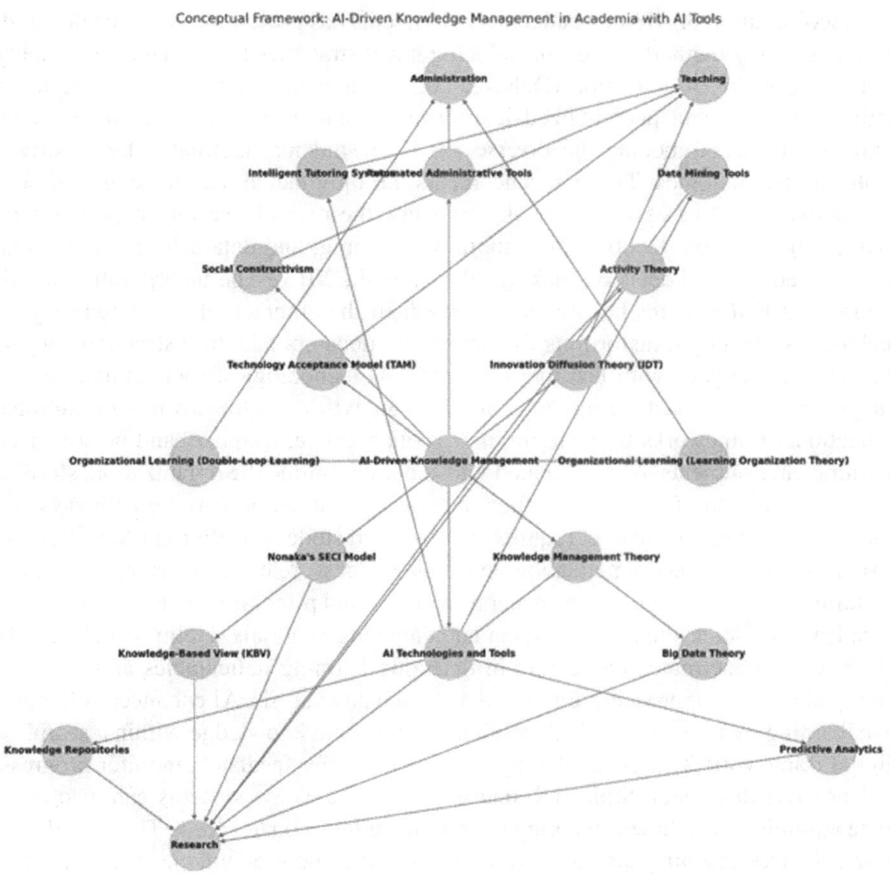

FIGURE 3.3 Conceptual framework: AI driven knowledge management in academia with AI tools.

The conceptual framework presented illustrates the interconnectedness of various KM elements, demonstrating how diverse tools and theoretical concepts enhance academic functions related to research, teaching, and administration. This model highlights the synergies among different KM systems, ultimately contributing to the evolution of educational methodologies that improve the quality and accessibility of academic resources.

This model integrates various AI tools, such as Knowledge Repositories, ITS, Predictive Analytics, Data Mining Tools, and Automated Administrative Tools, and shows how they are connected with both the theoretical concepts and academic functions (research, teaching, and administration). The model highlights the interlinkages between AI tools and theories like Nonaka SECI Model, TAM, and KM Theory, demonstrating how these technologies enhance academic KM processes.

3.4 AI-ENHANCED INNOVATIVE TEACHING METHODS IN KNOWLEDGE MANAGEMENT

The incorporation of AI into academic KM systems has profoundly transformed traditional teaching methodologies, introducing novel strategies that enhance the quality and accessibility of education (Dahake, 2024). Educators can foster a more interactive, adaptive, and personalized learning environment by tailoring instructional strategies to accommodate the diverse needs of students, facilitated by AI-driven tools and technologies. These advancements not only tackle the challenges of content delivery and assessment but also enhance the overall learning experience by facilitating real-time feedback, collaborative learning, and data-driven insights that support pedagogical decision-making (Woolf et al., 2013). The incorporation of AI-augmented KM systems has markedly expedited the embrace of novel pedagogical techniques, thereby transforming the approach educators take to instruction and the way students engage with learning resources. AI technologies, such as natural language processing, machine learning, and data analytics, are transforming traditional educational frameworks by fostering more collaborative, adaptive, and personalized learning environments. AI-augmented KM tools streamline the organization, storage, and dissemination of knowledge, allowing educators to tailor learning pathways and content to address the distinct requirements of each student (Heffernan & Heffernan, 2014). AI-driven recommendation systems suggest relevant learning resources customized to a student performance, preferences, and prior knowledge, ensuring that each learner obtains the most pertinent and engaging materials (Holmes et al., 2019). Moreover, AI facilitates instructors in detecting learning deficiencies and adapting their pedagogical approaches through real-time data analysis. AI enhances collaborative learning by allowing students to interact and share knowledge within a seamless digital framework. This is enabled by AI tools that offer feedback, monitor progress, and propose discussion topics (Kumar et al., 2023). These systems can also automate administrative duties, enabling instructors to focus on teaching. Thus, AI-driven KM enhances teaching efficacy while promoting a more adaptable, inclusive, and engaging learning environment. The theoretical foundations of AI-driven KM lie at the convergence of organizational theory, information systems, and AI. AI technologies are employed to enhance the acquisition, organization, retrieval, and distribution of knowledge. AI-driven KM focuses on employing data analytics, natural language processing, and machine learning to examine extensive data sets and generate insights that promote innovation, problem-solving, and decision-making. A principal objective of AI in KM is to create intelligent systems that automate the categorization, storage, and retrieval of knowledge, thus enabling organizations to access relevant information as needed. These systems aim to deliver more personalized and context-sensitive recommendations by emulating human cognitive abilities, such as contextual understanding, language interpretation, and learning from interactions. Moreover, AI-driven KM platforms utilize advanced algorithms to continuously refine their knowledge curation processes, ensuring that the information provided is relevant and appropriate for user needs (West et al., 2018). One of the most transformative applications of AI in education is the development of *adaptive learning systems*. These systems leverage machine learning algorithms to personalize the learning process, dynamically adjusting the content, pace, and complexity of lessons based on individual student performance. Such systems, including platforms like DreamBox and Knewton, have demonstrated the ability to deliver a customized learning experience, ensuring that students receive appropriate challenges and support. By analyzing student data in real time, adaptive learning systems help optimize instructional strategies, contributing to more effective learning outcomes (Siemens, 2013). The ability to adapt to individual learning styles and needs makes these systems particularly valuable in addressing diverse learner populations and promoting student engagement.

In addition to adaptive learning, AI has fostered the development of *ITS*, which simulate one-on-one tutoring experiences for students. ITS provides immediate, personalized feedback and guidance, enabling students to master concepts at their own pace while receiving targeted assistance. These systems have been effectively employed in subjects such as mathematics and language learning, where they support students through interactive problem-solving exercises and offer corrective feedback as needed. The role of ITS in promoting active learning is significant, as they encourage students to engage deeply with content, fostering the development of critical thinking skills and enhancing retention of complex material (Woolf et al., 2013).

Furthermore, AI technologies have facilitated the emergence of *collaborative learning tools*, which enable students to work together in more effective and structured ways. These tools, which include platforms such as Microsoft Teams and Slack, leverage AI to enhance communication, manage group tasks, and provide relevant resources for collaboration. AI can analyze group interactions, offering real-time suggestions to improve teamwork and ensuring that all participants contribute meaningfully. These tools are particularly useful in fostering interdisciplinary collaboration, allowing students from diverse academic backgrounds to engage with one another and solve complex problems collectively (Dabbagh & Kitsantas, 2012). By supporting group dynamics and offering structured support, AI-driven collaborative platforms help students develop key competencies in teamwork, communication, and problem-solving.

Another notable contribution of AI to educational practices is the development of *automated assessment and feedback systems*. Traditional forms of assessment often fail to provide timely feedback, hindering students ability to improve their performance continuously. AI-based assessment tools, such as automated essay scoring and peer review platforms, offer faster and more personalized feedback, enabling instructors to provide detailed insights on student progress. These systems assess a wide range of assignments, from essays to presentations, and provide real-time evaluations, helping students identify areas for improvement. By reducing the grading burden on instructors, AI also allows for more frequent assessments, ensuring that students receive ongoing guidance throughout their learning journey (Heffernan & Heffernan, 2014).

Additionally, AI-powered *learning analytics* are increasingly used to inform teaching and decision-making. Learning analytics tools analyze large volumes of student data to uncover patterns in learning behaviors, engagement, and performance (Kalra, 2024). These insights enable educators to identify students at risk of underperforming and implement targeted interventions to support their success. By leveraging predictive analytics, institutions can forecast trends in student outcomes, assess the effectiveness of different teaching methods, and allocate resources more effectively. The data-driven nature of AI learning analytics allows for evidence-based decisions that enhance both teaching effectiveness and student outcomes (Siemens, 2013).

While AI presents significant opportunities for enhancing teaching and learning, its integration into educational contexts requires careful consideration of ethical implications, such as privacy concerns and algorithmic biases (Neil, 2016). It is essential that AI systems be transparent and that data usage comply with ethical standards to ensure that students rights are protected. Moreover, AI should complement, rather than replace, the human elements of teaching, emphasizing the importance of critical thinking, creativity, and emotional intelligence, skills that AI, in its current form, cannot replicate (Woolf et al., 2013). The implementation of adaptive learning systems, intelligent tutoring, collaborative platforms, automated assessment tools, and learning analytics enhances both teaching and learning processes. As AI technologies continue to evolve, their integration into educational practices will undoubtedly further transform the academic landscape, offering new opportunities for improving educational outcomes and fostering deeper learning experiences.

3.4.1 Leveraging AI for Innovative Teaching: Enhancing Learning Experiences and Pedagogical Practices

The advent of AI is fundamentally reshaping the landscape of education by introducing new, innovative teaching methods that enhance KM processes within academic institutions. AI-enabled tools and technologies are not only improving how knowledge is shared and accessed but also how it is created, processed, and applied in teaching and learning environments (Cheng & Feng, 2024). This section critically examines how AI-driven KM is influencing educational practices and transforming traditional pedagogies into more dynamic, personalized, and effective learning experiences.

- **Personalization of Learning Experiences**

AI ability to analyze large datasets allows for the customization of educational content according to individual learning styles, preferences, and progress. AI-powered learning management systems (LMS), such as ITS, adapt the delivery of content to match the unique needs of each student. For example, systems can monitor student interactions, assess performance in real time, and provide tailored feedback or learning materials that target specific areas of weakness. This level of personalization not only improves student engagement but also optimizes learning outcomes by catering to the diverse abilities of students. The application of AI in this regard is supported by the principles of constructivist learning theories, which emphasize student-centered approaches where learners build knowledge through their own experiences and interactions with the content (Piaget, 1950; Vygotsky, 1978).

However, the adoption of AI in personalized learning raises concerns regarding data privacy and the ethical implications of algorithmic decisions. Critics argue that while AI offers tremendous potential for customizing education, the reliance on data-driven models could lead to biases or reinforce existing educational inequalities. These concerns highlight the importance of establishing clear ethical guidelines and frameworks for the responsible use of AI in education.

- **Collaborative Learning and AI Facilitation**

Another significant contribution of AI to innovative teaching is its capacity to foster collaborative learning environments. AI tools, such as chatbots, virtual assistants, and collaborative platforms, facilitate seamless communication and teamwork among students, enabling them to engage in real-time discussions, problem-solving activities, and shared learning experiences. For instance, AI-powered platforms like Google Classroom and Microsoft Teams allow students from different geographical locations to collaborate on projects, share resources, and provide peer feedback. These platforms, underpinned by AI, make learning more interactive and ensure that students develop key collaboration and communication skills, essential for future professional success (Cheng, 2024).

Social constructivist theories, which advocate for the importance of social interactions in the learning process (Vygotsky, 1978), align well with AI-enhanced collaborative learning. AI tools can bridge the gap between isolated learning experiences and create opportunities for group learning, even in virtual environments. AI systems can also track student interactions within groups, analyze collaborative patterns, and provide insights into team dynamics and individual contributions. This allows instructors to better support collaborative activities and ensure equal participation from all students.

Despite the promise of AI-driven collaborative learning, there are challenges related to maintaining meaningful interaction and preventing over-reliance on AI tools. Critics argue that while AI can facilitate communication, it may not necessarily promote the depth of interaction required for effective collaborative learning. Moreover, over-dependence on AI for collaboration could limit students ability to develop interpersonal skills and critical thinking in face-to-face interactions.

- **AI-Driven Content Delivery and Interactive Learning**

AI is transforming content delivery by providing educators with tools that can automate repetitive tasks, such as grading, feedback, and content creation, thereby allowing them to focus more on interactive and student-centered teaching methods. AI-driven platforms enable instructors to design interactive learning environments that go beyond traditional lectures and textbooks. For example, virtual reality (VR) and augmented reality (AR) powered by AI provide immersive, experiential learning opportunities in fields such as medicine, engineering, and history, enabling students to visualize and interact with complex concepts in a more intuitive way.

Furthermore, AI facilitates the creation of dynamic, multimedia-rich content, allowing students to engage with materials in various formats, such as videos, simulations, and gamified experiences. This aligns with multimedia learning theories, which propose that students learn more effectively when information is presented through both visual and auditory channels (V., 2024). The integration of AI in content delivery thus enhances students ability to engage with material and retain knowledge through more interactive and personalized means.

However, there are challenges in ensuring that AI-based content delivery remains pedagogically sound. Educators must ensure that AI-generated content aligns with curricular objectives and fosters deep learning rather than superficial engagement. Moreover, there is a risk that AI-driven tools could further widen the digital divide, leaving students in under-resourced areas at a disadvantage when it comes to access to advanced learning technologies.

- **The Role of AI in Assessment and Feedback**

AI impact on assessment and feedback represents another area where teaching methods are being radically transformed. Traditional assessment methods, such as exams and quizzes, are being augmented or even replaced by AI-powered tools that provide continuous, formative feedback (BankoleFalaye, 2024). These AI systems can evaluate student responses, offer real-time corrections, and suggest further resources or practice materials tailored to the student progress. This not only reduces the time instructors spend on grading but also enables more frequent and actionable feedback, which has been shown to enhance learning outcomes.

Moreover, AI-driven assessments offer the potential to assess a wider range of skills beyond what traditional exams can measure, such as creativity, problem-solving, and collaboration (Eager, 2024). These capabilities align with more contemporary models of assessment that focus on student holistic development rather than rote memorization.

However, the use of AI in assessment raises important questions about the accuracy and fairness of algorithmic evaluations. Concerns about the potential for AI systems to penalize students based on incomplete or biased data, or to misinterpret complex human responses, must be addressed (Taktak-Kallel, 2024). This highlights the need for transparent AI systems and ongoing monitoring to ensure that assessments remain equitable and fair.

The integration of AI into academic teaching methods represents a significant shift toward more personalized, collaborative, and interactive learning environments. While AI enhances teaching through customization, real-time feedback, and collaborative learning opportunities, it is essential to critically assess its limitations, particularly in terms of data ethics, algorithmic biases, and equitable access. The future of AI-enhanced education will depend on how these challenges are addressed and how

AI tools are integrated into pedagogical practices that prioritize deep learning and critical thinking (Yu, 2024).

Table 3.1 synthesizes the key AI tools, their roles in enhancing teaching methods, the positive impacts on student learning, and the challenges that need to be addressed for effective implementation.

TABLE 3.1
Roles of AI Tools in Enhancing Teaching Methods

AI Tool	Role in Teaching and Learning	Illustrative Applications
Intelligent Tutoring Systems (ITS)	Facilitate personalized instruction by adapting learning pathways and providing real-time feedback based on learner performance.	Adaptive quizzes, guided problem-solving platforms, and step-by-step instructional support.
Natural Language Processing (NLP)	Enhance interaction between instructors and learners, automate assessment tasks, and support language acquisition.	AI-powered chatbots for academic queries, automated essay scoring, and grammar correction tools.
Predictive Analytics	Analyze learner data to identify at-risk students, predict performance trends, and recommend targeted interventions.	Early warning systems, predictive dashboards, and personalized study plan recommendations.
AI-driven Content Development Tools	Enable automated creation of instructional materials tailored to learning objectives and student needs.	Generation of lecture slides, video summaries, adaptive simulations, and interactive learning modules.
Automated Assessment Systems	Ensure consistent, objective, and rapid evaluation of student work, reducing instructor workload.	Multiple-choice question grading, plagiarism detection, and automated evaluation of coding or mathematical assignments.
Speech Recognition Technologies	Improve accessibility and engagement through voice-enabled interactions and automated transcription.	Voice-command-enabled presentations, lecture transcriptions, and real-time language translation.
AI-integrated Virtual and Augmented Reality	Create immersive and adaptive environments to facilitate experiential and practical learning.	Virtual laboratories, AI-enhanced simulations, and interactive 3D educational models.
AI-powered Learning Management Systems	Optimize course delivery by tracking learner engagement and recommending adaptive content pathways.	Personalized course sequencing, AI-curated learning resources, and real-time progress tracking.
AI-based Recommender Systems	Provide tailored educational resources aligned with learner preferences and performance data.	Suggested readings, multimedia content recommendations, and adaptive learning playlists.
AI-enabled Gamification Platforms	Increase learner motivation and engagement through adaptive game-based learning strategies.	Personalized challenges, adaptive scoring systems, and AI-driven educational games.

3.5 CHALLENGES AND CONSIDERATIONS

The integration of AI into academic KM systems presents several challenges that need to be addressed for successful implementation.

1. **Data Privacy and Security:** AI systems in academia rely heavily on personal, academic, and behavioral data, raising significant concerns about data privacy and security. Institutions must implement stringent data protection measures to comply with regulations like GDPR (General Data Protection Regulation) and FERPA (Family Educational Rights and Privacy Act) to safeguard sensitive information (Madhu et al., 2023). As note, the rapid adoption of AI often outpaces the development of adequate security protocols, risking data breaches and misuse.
2. **Algorithmic Bias and Fairness:** AI systems can inherit biases from historical data, potentially leading to unfair outcomes, such as disadvantaged marginalized student groups. Aghaebe (2024) warns that biased AI algorithms can perpetuate inequalities, particularly in predictive systems that inform decisions about student success or resource allocation. To mitigate bias, it is essential to ensure that AI systems are transparent, regularly audited, and trained on diverse datasets.
3. **Faculty and Staff Resistance to AI Integration:** Resistance from faculty and staff can hinder AI adoption. Concerns about job security, loss of autonomy, and the potential dehumanization of teaching are prevalent. Dahake (2024) emphasize the need for universities to involve educators in the planning process and provide training on how AI can enhance, rather than replace, teaching and research tasks. Addressing these concerns is crucial for fostering acceptance and support for AI technologies.
4. **Equity and Access:** AI integration risks exacerbating the digital divide, particularly for students from lower socio-economic backgrounds who may lack access to necessary technology. As Aghaebe (2024) highlights, without equitable access to infrastructure like high-speed internet and personal devices, AI tools could further entrench educational inequalities (). Institutions must ensure that all students, regardless of their background, have access to the technologies needed to benefit from AI-enhanced education.
5. **Ethical Considerations in AI Deployment:** The ethical implications of AI use in education, such as transparency, accountability, and consent, are critical. AI decisions, especially in assessments and personalized learning, must be transparent and explainable to students. Binns (2018) stresses the importance of ethical guidelines to ensure AI aligns with educational values, promoting fairness and academic integrity. Institutions must develop frameworks to address potential ethical concerns and protect students rights.

3.6 CONCLUSION

The integration of AI into academic KM holds the promise of reshaping the educational landscape, offering new opportunities for personalized learning, enhanced research capabilities, and more efficient administrative processes. However, the path to fully realizing these benefits is not without its challenges. From ensuring data privacy and security to addressing algorithmic biases, overcoming faculty resistance, and promoting equitable access, these hurdles must be met with thoughtful, strategic approaches.

As AI continues to evolve, its potential to democratize education, bridge knowledge gaps, and streamline research processes is immense. Yet, its success hinges not only on technological advancements but on how institutions confront these complexities with foresight and responsibility. The ethical deployment of AI focused on fairness, transparency, and inclusivity, will be critical in maintaining the integrity of academic environments while reaping the rewards of AI transformative power.

Ultimately, the future of AI-driven KM in academia is a balancing act: navigating the promise of innovation while safeguarding the core values of education. As we stand on the precipice of this technological revolution, academic institutions must commit to a vision where AI complements human expertise, enriches learning experiences, and fosters an inclusive, fair, and transparent academic ecosystem for all. The potential for a more intelligent, dynamic, and equitable academic future is within reach if we can navigate the complexities of AI integration with care, purpose, and an unwavering commitment to ethical standards.

3.6.1 Implications of the Study

Practical Implications: The practical implications of this study are profound, especially for academic institutions seeking to modernize their KM practices. AI technologies offer significant advantages by automating data-intensive tasks like data retrieval, content curation, and information dissemination. This allows researchers, educators, and administrative personnel to focus their time and effort on more creative and intellectually demanding activities. For example, ITS powered by AI can provide personalized learning experiences for students, addressing individual learning needs and boosting educational outcomes. Additionally, predictive analytics and data mining tools can facilitate the identification of emerging research trends and inform strategic planning in academic institutions. AI-based administrative tools can also help streamline routine processes such as student enrollment, grading, and performance tracking, ultimately leading to increased operational efficiency.

Managerial Implications: From a managerial standpoint, the integration of AI technologies into KM systems requires academic leaders and administrators to shift their approach to strategic management. Managers must play a critical role in ensuring that faculty and staff are equipped with the necessary digital literacy and training to effectively adopt and use AI tools. Moreover, it is crucial for managers to facilitate a culture of innovation, collaboration, and continuous learning to ensure the successful integration of AI. This means fostering an environment where AI technologies are not seen as replacements for human roles but rather as complementary tools that can enhance human decision-making and creativity. Furthermore, managing the organizational change associated with AI adoption is vital. Leaders must address potential resistance, provide adequate support systems, and promote an understanding of AI transformative potential across all levels of the institution.

Theoretical Implications: This study contributes to the theoretical body of knowledge in the domain of AI-driven KM by merging traditional KM theories with modern AI capabilities. The integration of frameworks like KM Theory and Nonaka SECI Model with AI technologies provides a nuanced understanding of how AI can enhance knowledge creation, sharing, and application in academic settings. Furthermore, theories such as the TAM and IDT provide insights into how faculty, students, and administrators perceive, accept, and adopt AI tools. The incorporation of Organizational Learning theories also expands the understanding of how AI can facilitate organizational learning and adaptation in academia. By bridging these diverse theoretical perspectives, this study lays the groundwork for future research that explores how AI impacts organizational behavior, knowledge-sharing practices, and decision-making processes in educational institutions.

Societal Implications: On a broader societal level, AI-enhanced KM in academia has the potential to democratize education and contribute to greater societal equity. By personalizing learning experiences, AI can help level the playing field for students from diverse backgrounds, ensuring that they receive the support they need to succeed. AI-powered KM systems can make educational content more accessible, creating opportunities for lifelong learning and skill development. Moreover, AI tools can enable collaboration across borders, fostering global networks of knowledge exchange and contributing to the development of a more interconnected, informed global society. However, it is also essential to consider the ethical implications of AI in academia. Issues such as data privacy, algorithmic bias, and the digital divide must be addressed to ensure that AI benefits are equitably distributed and do not exacerbate existing inequalities.

3.6.2 Limitations and Future Scope of the Study

While this study provides a comprehensive framework for understanding the integration of AI in academic KM, several limitations must be acknowledged. First, the research primarily relies on secondary data and theoretical analysis, which limits the ability to empirically validate the proposed framework in real-world academic settings. Future studies could benefit from empirical research, such as case studies or surveys, to explore how AI tools are implemented and utilized in specific academic institutions. This would allow for a more nuanced understanding of the practical challenges and benefits of AI integration.

Additionally, this study does not fully address the ethical concerns surrounding the use of AI in academic environments. Issues such as the privacy of student data, potential bias in AI algorithms, and the consequences of AI replacing traditional academic roles warrant further investigation. These concerns are critical to the responsible adoption of AI technologies in education and should be explored in future research.

Another limitation is the focus on AI tools within traditional academic frameworks, without fully exploring the potential of AI in non-traditional learning environments, such as online learning platforms or remote education settings. The impact of AI on these evolving forms of education is an important area for future research.

Looking ahead, future research should investigate the long-term effects of AI adoption on academic institutions, particularly in terms of its impact on faculty-student interactions, knowledge-sharing practices, and institutional culture. Studies could explore how AI tools reshape the roles of educators, researchers, and administrators, and examine how they adapt to these changes. Additionally, as AI technologies continue to evolve, future research could explore the potential of emerging AI tools such as natural language processing, machine learning algorithms, and AI-driven content generation in further enhancing KM and educational practices.

Finally, given the rapid pace of technological advancements, future studies should explore the implications of AI on interdisciplinary research, international collaborations, and cross-institutional knowledge sharing. By focusing on these areas, researchers can ensure that AI-driven KM systems are developed and implemented in ways that foster innovation, inclusivity, and global collaboration.

REFERENCES

Aggarwal, R., & Woolliscroft, J. O. (2020). Artificial intelligence in medical education: A review of the current state of the art and future directions. *SSRN Electronic Journal*. https://doi.org/10.2139/ssrn.5161242

Ardiabah, D. A. (2024). Implementing knowledge graphs and ontologies for AI-powered recommendation system. *African Journal of Biomedical Research*, 27(4s). https://doi.org/10.53555/ajbr.v27i4s.6687

BankoleFalaye, A. (2024). AI-powered knowledge systems for SME financial management: A conceptual approach. *International Journal of Science and Research (IJSR)*. https://doi.org/10.21275/sr241117060640

Binns, R. (2018). *On the Ethics of Artificial Intelligence*. The Atlantic.

Cheng, J., & Feng, R. (2024). A comparative study of AI-powered workforce development via forensic analytics, blockchain, and metaverse. *Journal of Knowledge Management Practice*, 25(3). https://doi.org/10.62477/jkmp.v25i3.525

Dabbagh, N., & Kitsantas, A. (2012). Personal Learning Environments, social media, and self-regulated learning: A natural formula for connecting formal and informal learning. *The Internet and higher education*, 15(1), 3–8. https://doi.org/10.1016/j.iheduc.2011.06.002

Eager, B. (2024). The gift of AI. In AI-Powered Scholar (Chap. 1). *Routledge*. https://doi.org/10.4324/9781032665276-1

Heffernan, N. T., & Heffernan, C. L. (2014). The Assistments ecosystem: Building a platform that brings scientists and teachers together for minimally invasive research on human learning and teaching. *International Journal of Artificial Intelligence in Education*, 24(4), 470–497. https://doi.org/10.1007/s40593-014-0024-x

Holmes, W., Bialik, M., & Fadel, C. (2019). *Artificial Intelligence in Education: Promises and Implications for Teaching and Learning*. Center for Curriculum Redesign.

Ikeh, C. (2024). AI-powered market segmentation and personalization strategies for enhancing digital product lifecycle management. *International Journal of Research Publication and Reviews*, 6(6), 21107. https://doi.org/10.55248/gengpi.6.0625.21107

Jagtap, V. & Epilli, R. (2024). *AI-Powered HR Finance: Transforming Workforce Management and Financial Strategies in the Digital Age, IIP Series, Volume 3,* pp. 115–127, e-ISBN: 978-93-6252-594-9. www.doi.org/10.58532/nbennurptch11

Janaki, G. (2024). Exploring the Influence of AI-Powered Learning Tools on Student Understanding and Academic Performance: A Comprehensive Analysis. *Shanlax International Journal of Management*, 11(S1-May), 1–12. https://doi.org/10.34293/management.v11is1-may.7834

Kalra, V. (2024). Coupling NLP for Intelligent Knowledge Management in Organizations: A Framework for AI-Powered Decision Support. *Journal of Information Systems Engineering and Management*, 10(10s). https://doi.org/10.52783/jisem.v10i10s.1337

Kumar, S., Gupta, U., Singh, A.K. and Singh, A.K. (2023) Artificial Intelligence. *Journal of Computers, Mechanical and Management*, 2, 31–42. https://doi.org/10.57159/gadl.jcmm.2.3.23064

Madabhushini, A. (2024). AI-powered knowledge graphs: The future of enterprise business intelligence. *International Journal of Information Technology and Management Information Systems*, 16(2), Article 50. https://doi.org/10.34218/ijitmis_16_02_050

Madhu, M., Manoj Kumar, K., Pratyaksha, B., Sushmita, S. & Javed, G. S. (2023). Striking Ethical Balance in AI-TAI: Promoting Academic Integrity through AI-Powered Tools. *Proceedings of the 2023 IEEE TEMSCON-ASPAC*. https://doi.org/10.1109/temscon-aspac59527.2023.10531521

O'Neil, C. (2016). *Weapons of Math Destruction: How Big Data Increases Inequality and Threatens Democracy*. Crown Publishing Group.

Pamisetty, K. (2022). AI-powered decision support systems for enhancing tax compliance and public revenue management. *International Journal of Science and Research (IJSR)*, 11(12), 1490–1504. https://doi.org/10.21275/ms2212143253

Patial, S. (2024). AI-powered banking: From customer service to risk management. *OSF Preprints*. https://doi.org/10.31219/osf.io/tmvns

Rayarao, V. (2024). AI-powered agents transforming 401(k) compliance and risk management. *TechRxiv*. https://doi.org/10.36227/techrxiv.173611324.44788458/v1

Reddy, G. K. (2025). SmartKitch: AI-powered kitchen inventory management and personalized culinary assistant. *In Proceedings of the 2025 International Conference on Knowledge Engineering and Communication Systems (ICKECS)*. https://doi.org/10.1109/ickecs65700.2025.11034836

Reddy, S. S. (2022). AI-powered content management systems: Automating workflows and enhancing user experience. *International Journal of Science and Research (IJSR),* 11(8), 1566–1569. https://doi.org/10.21275/sr220811132343

Siemens, G. (2013). Learning Analytics: The Emergence of a Discipline. *American Behavioral Scientist, 57*(10), 1380–1400. https://doi.org/10.1177/0002764213498851

Taktak-Kallel (2024). Advancing Entrepreneurship Knowledge with Artificial Intelligence (AI) Methods: Insights from Scoping Review of Emerging AI-Powered Research on Student Entrepreneurial Intentions. *Journal of Innovation Economics & Management*, (1), 175. https://doi.org/10.3917/jie.pr1.0175

V., Kamalesakuma. (2024). BigBrain: A Unified AI-Powered Personal Assistant for Intelligent Personal Knowledge Management and Digital Content Organization. *Authorea Preprints*. https://doi.org/10.22541/au.174491151.15381490/v1

Vygotsky, L. (1978). *Mind in Society: The Development of Higher Psychological Processes*. Harvard University Press.

West, D. M., Eronen, H., & Duvall, D. (2018). *The Future of Artificial Intelligence in Education: Privacy, Ethics, and Equity*. Brookings Institution.

Williamson, B., & Piattoeva, N. (2021). *Education Governance and Datafication: Exploring the Politics of Data in Education*. Routledge.

Woolf, B.P., Lane, H.C., Chaudhri, V.K. & Kolodner, J.L. (2013), AI Grand Challenges for Education. *AI Magazine*, 34: 66–84. https://doi.org/10.1609/aimag.v34i4.2490

Yu. (2024). InfoSecPilot: Navigating the Complex Landscape of Information Security with an AI-Powered Knowledge Management Chatbot. *Journal of Knowledge Management Practice*, 25(4). https://doi.org/10.62477/jkmp.v25i4.536

4 AI and Storytelling

Reimagining Learning in Film Institutes of India

Priyanka Yadav and Sachin Bharti

ABBREVIATIONS

AI	Artificial Intelligence
FTII	Film and Television Institute of India
SRFTI	Satyajit Ray Film and Television Institute
HEI	Higher Educational Institution
MOOC	Massive Open Online Course
NEP	National Education Policy
OTT	Over-The-Top
SDG	Sustainable Development Goal
UGC	University Grants Commission
UNESCO	United Nations Educational, Scientific and Cultural Organization

4.1 INTRODUCTION

The Indian film industry has been creating magic on screen for more than 100 years now. The Industry has come a long way since its humble beginnings with Dada Saheb Phalke's *Raja Harishchandra* (1931). The Indian film industry has welcomed the post-partition filmmakers[1] from Lahore with as much support as they could provide to create an emotion that would go on to bind the nation. It has always tried to put up with the times, be it the lesser means of production, the studio, and the star system or the changing technologies (Kumar, 2017). A country that produces the largest number of films in the world deserves special attention right from the primary stage of educating youth while they are at film schools.[2]

Artificial intelligence (AI) seems to be rapidly invading the world! From simple problem-solving tasks to heavy data analysis, this technology is helping automate tasks in medicine, arts, education, accounts, and many other fields. It is a technology that is capable of doing tasks that would ideally require human intelligence. This chapter however argues that despite mimicking human intelligence, AI does require human supervision to see if the final output has been correctly generated. The future will call for a balanced relationship between AI and human intelligence. AI has immense potential to help students by adapting to their needs and creating a personalized environment (Sari et al., 2024). It has the potential to help educators by

 DOI: 10.1201/9781003560463-5

making teaching and learning more accessible. This chapter aims to uncover the role of AI in education, particularly film education. A lot of implications and suggestions proposed here do however apply to other media-related fields as well (albeit with minor alterations).

Preparing the future filmmakers is quite a task and the educators are all up for it (Petrie & Stoneman, 2014) With film institutes like the Film and Television Institute of India (FTII) boasting of envious alumni, an updated educational pedagogy will come in very handy. Enhanced curriculums, practical applications, and creative freedom in film schools will ensure that they produce students who are ready to take on the world, quite literally. Mehboob Khan's *Mother India* in 1957 marked the first ever official entry for Indian Cinema at the Oscars. From the times of Mehboob Khan to that of present times Kartiki Gonsalves gracing India with the Oscars in 2025 with *The Elephant Whisperers* (2022), the industry continues to thrive!

With the intervention of technology like AI, this chapter attempts to reimagine learning in film schools to be able to put up with the changing times. The pressing need to realize the Sustainable Development Goals (SDGs) laid down by the United Nations will make this work a possible contributor to the literature surrounding the development of India. The Government also seeks to enhance the education system to create a vision for a developed India by 2047 by ushering in the NEP (2020).

This chapter also suggests various ways in which the students can be made ready for employability. Additionally, it also suggests a roadmap for higher educational institutions (HEI)[3] to alter their teaching practices to be able to produce film graduates equipped with better skills for the industry (NSDC, GOI). The intervention of AI models has called for changes in the curricula of various courses and this chapter highlights the importance of adopting the same.

The education model of India can help reshape the future by resorting to a slew of measures that can integrate emerging technology with education at various levels. Access to such advanced tools is the need of the hour. This access will gradually increase as India aims to attain the agenda of 2030 for Quality Education and further the vision of a Viksit Bharat by 2047. The tools can be used to provide personalized experiences for the students. This will enable the filmmaking students to increase the scope for creativity by curating the resources according to their needs. Students will be able to get much advanced training in learning processes of pre- and post-production which will help bridge the gap between the technologically adept institutes and otherwise. It shall also motivate other film institutes to adopt newer technologies like AI models into their infrastructure and consequently the curriculum. This will further promote critical thinking and other problem-solving skills.

Filmmaking will be a much more interdisciplinary field as a result of the automation of processes in the industry. This possibility has arisen because of the exposure of students to various fields like big data, analytics and machine learning along with the fields of film and media. This further aligns with SDG 4 of Quality Education in making students more skilled apart from mere teaching and learning processes.

Therefore, this chapter shall help in uncovering insights from the world of AI and Films in India and explore the possibilities that surround the higher education system in the country.

4.1.1 Realization of the SDG of Quality Education in India

The United Nations 2030 Agenda for Sustainable Development has been adopted by all its member states. It provides a future roadmap for the countries to ensure prosperity for all. This chapter focuses on the fourth SDG of quality education which intends to guarantee an inclusive and equitable education for all.

Every broad SDG has been broken down into small actionable targets for the countries to ensure that they are on the right track toward the achievement of those goals. Some of the targets centered on the usage of technology in education are:

Target 4.3, which talks about affordable technical and higher education, and

Target 4.4, which lays emphasis on expanding and enriching skills which will be of relevance to financial stability (see Section 4.4).

In the context of India, the issue of the Digital Divide as a roadblock to achieving the SDGs still looms large. This calls for intensive mapping of the trajectories of such a developing country in the pre- and post-pandemic era. Despite that, recent assessments highlighted in SDG Report (2024) have identified several alternate factors apart from Covid-19 which have impacted the global education model. However, internet availability for pedagogical purposes in the Indian educational landscape has been satisfactorily progressive as highlighted by the United Nations Educational, Scientific and Cultural Organization (UNESCO) Report that monitors the SDG 4 global indicators.

The Government of India has been taking various initiatives toward the realization of the goal of Quality Education, namely, the National Skill Development Corporation. Launched in 2009, it provides vocational training to the youth to increase their prospects of employability in an otherwise fast-paced Indian economy.

Another initiative of the Government of India is that of Digital India. Launched in 2015, it lays emphasis on Digital Literacy and provides skills training to people to help them with employability and contribute toward the bigger vision of a digitally empowered economy.

NEP (2020) has been a path-breaking initiative of the Government of India. Launched amid the pandemic, this policy aims to bring about a reform in the education sector of the country. It is a robust agreement that lays stress on the holistic development and curriculum enrichment of the educational institutes of India. Implementation of NEP (2020) by the educational system in the country is still in its nascent stages and it will be interesting to see how the institutions align their learning systems with the technological shifts (Bharti & Yadav, 2024).

4.1.2 Multiliteracies

There are many film institutions in India that are aligning themselves with the NEP (2020) guidelines that throw light on integrating technology into teaching. This further aligns with the 2030 vision of Quality Education as an SDG, thus making India ready for development on a more global scale.

The method of teaching a subject is essentially referred to as Pedagogy. Talking about pedagogy, enhancing the curriculum becomes imperative if the above is planned to be included in classroom teachings. This will mean that the educators will have to update their teaching curricula accordingly. Integration of AI shall enable the teachers to be able to develop lectures that adopt multiple ways of communicating with the class. These ways comprise what is referred to as '*Multiliteracy Pedagogy*'. An alignment with the multiliteracy pedagogies will make such institutions create a comprehensive learning environment and produce graduates who are better equipped for what the industry holds. Since there can be myriad ways of teaching a subject, therefore there are different pedagogies involved in making up a course for teaching. This approach can be adopted by encouraging the students to use different modes of communication which will mean their learnings get combined with various technologies. This approach is in two ways: teaching and learning. Therefore, any study on educational models is incomplete without examining its learning environments. Back in the 1990s, multiliteracies as a concept evolved in response to some major developments happening in the globalized world. One of them was the takeover of technologies like the internet and digital media.

Figure 4.1 highlights the key aspects of what an ideally balanced classroom design looks like. In this design, the first key component is that of Situated Practice which stresses the connection laid down between what is taught in classrooms compared to what is learned by practical exposure. This focuses on the importance of the practical application of what film students learn theoretically in classrooms.

The second component focuses on the need for overt instruction. You would have dealt with a situation wherein your teacher taught you how to do a certain assignment and gradually let the students take charge of the process. This is called the process of scaffolding in terms of teaching. It is the process of providing support to students while they are developing new skills and gradually withdrawing to let them practice by themselves. This practice is proving to be of significance as students learn the application of AI technologies and gradually tread the path of becoming AI literate. Learning graphic designing, illustrations, scripting, editing, etc. in film school

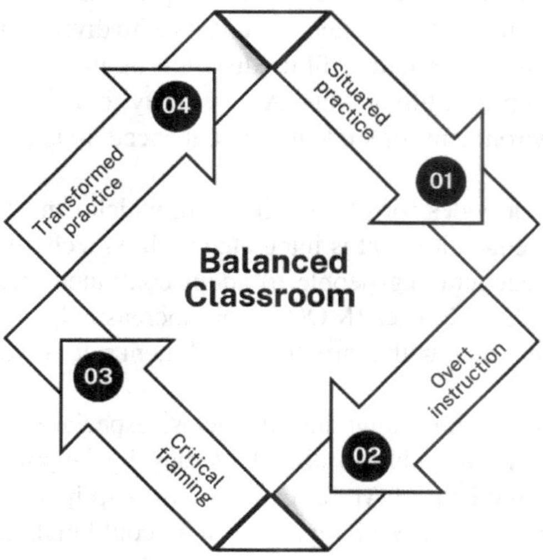

FIGURE 4.1 Balanced classroom design.

provides the students with hands-on training in what the industry will expect as they enter the world of films.

The third aspect of critical framing holds an increased importance in classrooms in contemporary times as the United Nations focuses on enhanced inclusion and equity among people. As a part of critical framing, students learn how to assess the social and cultural contexts of their learning. This is crucial for film school graduates as they go on to create films for the viewers, especially in India which accounts for one of the oldest and largest film industries in the world.

The final component is that of Transformed Practice. It is rightly the final one as it combines all the previous components to reflect upon and apply in the context of a newer practice.

Talking about inclusiveness, AI has the power to assist filmmakers with access to wider language translations. This shall mean multiple things. One, it could mean a greater inclusion of the workforce in the filmmaking industry. It could prove helpful for the lower strata that migrate from different cities in search of a job in the bustling city of Bombay, now Mumbai (the heart of the Bollywood Film Industry). It could also prove to be helpful in the day-to-day workings with diction and foreign language experts on film sets.

Two, it could mean a wider distribution among the box offices as the dialogs could be better curated for multiple languages by the individual film studio itself rather than mere mechanical dubbing of the film content. This way the viewers would be able to get access to the film in close proximity to the vision of the director and minimize losing the essence of the story which would have ideally happened due to a language barrier. Third, on an institutional level, students will be able to understand the importance of accessing multiple languages to curate content.

Lastly, students from diverse states would be able to gain access to major film schools like that of *FTII*, irrespective of the languages they know. This, however, is a potential possibility in the near future. Institutions are yet to align themselves on the lines of NEP (2020) to become better equipped for providing such access to students.

Inclusiveness shall not just be limited to students with diverse cultural backgrounds and languages. It shall mean letting film education be more accessible to disabled students willing to enroll in filmmaking. Accessibility tools based on AI shall mean a more inclusive environment for students and enhanced assistance with better pedagogical mechanisms.

NEP (2020) also provides for lifelong learning which can be facilitated through online mediums. An education that is integrated with AI technologies has the power to promote this by encouraging people to adopt continuous learning if they wish to. Massive Open Online Courses (MOOCs) are increasingly gaining traction in the Indian Education landscape with a growing enrollment rate of students from diverse age groups.

AI can also help in enhancing multiliteracies, especially in a field like Film Education in India. It can enhance digital literacy by helping students automate repetitive tasks with the help of AI Tools. This will help in saving time and enable the digital fluency of students. Such repetitive tasks could include automated editing software, VFX, and computer-generated images (CGI) software. Consequently, this could help in the visual literacy of students by using technologies for tasks like color grading. Analyzing the trends, audience behavior, and feedback for particular content and learning to analyze big chunks of data can help students become more media literate. Educators can play an important role in teaching students about the ethical considerations of using such tools to become critically media literate (see Section 4.3.2).

4.2 THEORETICAL BACKGROUND

The technology of AI seems to be shifting the focus from a human-driven way to a data-driven one. This reminds me of the short yet powerful phrase of *The Medium is the Message* (Mcluhan, 2001). In the backdrop of his concept, it would be interesting to witness how AI changes the media landscape. This technology has the potential to change the way stories are told which can shape the way the content will be perceived. McLuhan viewed every form of media as something that could reconfigure an entire society by changing the perceptions of people. As a result, this could potentially alter the film culture of India.

The chapter lays its theoretical foundations by emphasizing the role of technology in fostering change in society. It builds upon the theory of Technological Determinism to highlight the inevitable impact of AI on society at large. Despite various criticisms, this technology has the power to shape societies. Although harnessing that power in our favor will always rest with us humans, hopefully, this is because the technology can cause potential threats to the systems if not used with caution. Hereon, this chapter takes forward the perspectives of this theory in society, particularly the education system. It discusses the role that educators can play in classroom teaching. It also predicts the contributions of AI-literate graduates to the economy of a country. The chapter further elaborates on the need for media research on the integration of AI technology and film education. It concludes with suggesting further areas of research and recommends future directions of this convergence.

4.3 HUMAN-CENTERED APPROACH TO AI

AI models should be designed in a way that prioritizes human needs. This has immense potential to help the education model if it is able to account for ethical value additions as well. Unlike typical machines, it is essential to keep human rights in consideration rather than just focus on optimization for performance. This approach in an education system will ensure that AI models are designed to be human-centric and focus on making the content understandable by the layman. The content provided by such models will ensure better engagement by students and teachers alike.

Adoption of this technology has the potential to foster inclusivity by being human-centric. This will enable a fair approach to the users of such models enabling higher engagement for the students from various social groups. Frequent updates to AI models will provide users with enhanced tools and more power over the kind of information they desire. Such models will further be able to give suggestions to students to execute their creative strategies in comparatively newer and more innovative ways. With a focus on ethical considerations, the organizations of AI models will be more careful about the actions they will be taking in their development mechanisms and have a greater sense of accountability. These actions can ensure that this technology can act as a boon for society and minimize the potential risks that can occur if the ethical considerations of human beings are not taken care of.

4.3.1 Role of Film Educators in Teaching Ethical Usage of AI

The human-centered approach is a two-way street, quite literally! While it is important to monitor ethical considerations in AI Models, it is equally important to teach students about the implications of using such technologies (Zhai et al., 2021). Viewing the heavy integration of AI into filmmaking, it is of utmost importance that students are guided by educators about the ethical ways of using this technology. In filmmaking courses particularly, students need to be guided by such educators at every step. Considering that AI integration into filmmaking can start as early as the pre-planning stage of the film, it means that the educators need to be cautious about intervening with student activities at every step of the way.

There are several ways in which film educators can guide and monitor students in the ethical use of such technologies. This can range from creative rights to transparency. Frequent classroom discussions about data privacy shall lead to a better understanding of what it takes to be a socially responsible filmmaker, especially in a developing country like India. AI models have the potential to create creative content that simply depends on the prompts that are generated by the user. That is where the educators step in! They can teach students about the significance of respecting the intellectual property of someone. That is because the ownership of the content can be heavily misused while using AI which makes the human-centric approach imperative to AI.

Another concern is that AI-generated content can heavily favor the kind of answers one is seeking. The algorithms in such models can influence how stories are told because they tend to be biased toward the prompt of the user. Therefore, it necessitates that educators guide the students in the appropriate usage of this technology. This has completely different implications for the filmmaking students as they are taught to present various stories on screen. AI prompts can start from the very first step of ideation. At such a stage, it can become difficult for students to understand the novelty of their stories for the film (if any) because the AI prompts will make sure to adhere to the requirements of the user in the best way possible. That is where the educators can help the students in forming a concrete foundation for their film- a foundation that is unperturbed by the intervention of a *biased technology*. This intervention by educators will ensure that the students are able to use these models sparingly. Heavy dependence on AI models has a chance of disrupting the natural ideation process of filmmaking students. Once the ideation process starts, it moves into scripting and henceforth the final stages of making a film. Educators can teach the need for maintaining transparency with the audience. This can start from as early as the assignment stages of students during their courses. This will help the students in fostering a sense of authenticity in their works which will further reflect with the audiences. It will help in building trust with the audiences on a larger level and help in maintaining loyalty to their work.

This also raises the question of authorship and copyright. AI-generated content can raise such concerns over the ownership of creative content and educators need to teach students how to attribute such authorships. Though the ownership of such content is continuously debatable, this loophole will pose a pressing need for clarification on the part of content creators.

A growing concern in the contemporary use of this technology is misinformation and the rise of deepfakes. Educators will need to teach the ethical considerations of using such content for film production in the near future. Critical media literacy and digital literacy as a form of multiliteracies mentioned in previous sections will require improvisation and update to the teaching curricula in film schools and other media courses alike.

4.3.2 Contributions of Film Educators in Teaching Data-Driven Insights

With the gradual penetration of AI into films, the need for a greater awareness of this technology while studying in film schools has been felt. Therefore, an integration of the same in the curriculum would be a welcome step by most institutions. Teaching in the 21st century will require educators to alter their curriculum in order to produce more industry-ready graduates (Perry, 2018). The educators can teach students how to do data analytics and monitor audience responses to enhance engagement in future projects. Students can learn this at the educational level by learning how to analyze various data-driven insights like audience demographics. They can be taught to analyze the data so they learn how to trace different patterns that can help them in curating projects accordingly.

Tracing patterns is a subset of assessing trends in the film markets for such students. The industry witnessed turbulent times in Covid-19 by trying various ways that could work for the success of their content. The pandemic demanded newer formulae for success and the gradual rise of OTT platforms (Akser, 2020). With OTT platforms forming an indispensable part of the Indian Viewership, it is not just limited to the box office anymore. The educators can guide them in assessing the box office performance of contemporary films. Therefore, it will be interesting to analyze the streaming statistics of such content. This coupled with the metrics of social media will help students in assessing the factors that have influenced the film's performance. Educators can play an active role in guiding students to get better at determining such factors and teach them how to understand the distribution and marketing strategies of a film. Thus, analyzing trends in Indian film markets can be an important value addition to the course curriculum for filmmaking students. Additionally, this can help students in making decisions about their creative work thereby maintaining a higher probability of meeting audience expectations on the big screen. This will also help the students kick start their film careers with confidence and gradually shape the path they want to take.

Educators have been incorporating real-world data insights into their teaching approaches of late. Using data-driven approaches to teach filmmaking students can provide the students with an understanding of how such data is shaping the filmmaking industry. Using case studies like that of OTT streaming platforms and their algorithm analysis has the potential to make classroom teaching more engaging.

Teaching students what inferences they can make from the distribution strategies of various films can help students in viewing their content from the point of view of other stakeholders as well.

Organizing timely workshops with AI Experts can help the students gain better insights into the use of this technology. Workshops that are specially tailored for film school students shall help students understand the use of this technology in the most efficient way possible. Such workshops coupled with the insights of the educators in teaching about the ethical considerations of this technology shall lead to a more robust understanding for the students before stepping into the industry. This shall ensure that filmmakers are not just creative artists but are data literate and ethically aware of their industry bearing a social responsibility toward them.

Film schools can consider implementing the usage of AI Tools in their course curricula. A generative AI tool like *Adobe Firefly*,[4] for example, could be incorporated to aid students in curating videos that let their imagination overflow.

Premier film institutions in India like the *FTII* offer certificate courses in Animation and Visual Effects Design. Integration of Generative AI in curricula could enhance the creative outputs of students. Educators can start by teaching AI for simple design-related tasks that minimize the time required for a particular work that can ideally be automated. For example, automating repetitive tasks like color grading can help save time and utilize it for more creatively demanding work of a student.

In addition to standard software for learning Video Editing like *Adobe Premiere Pro* and *Final Cut Pro*, film schools can consider introducing AI Tools into the curriculum to enhance the practical exposure of students.

Institutes like *Satyajit Ray Film and Television Institute (SRFTI)* already offer courses in animation and VFX. The students could benefit from the integration of tools like versions of AI-driven software for animation. This will help students be more creative while doing VFX techniques like CGI animation. CGI technique is primarily used to create moving scenes and effects. This will help students generate myriad outputs enhancing healthy competition in a creative field like this.

Specific case studies pertaining to the integration of AI technologies in film schools in India are still emerging and the higher education sector is yet to witness this paradigm shift in film education. Implementation of NEP (2020) guidelines shall hopefully ensure that film education in India witnesses some technological shifts.

4.4 INTEGRATING AI INTO FILM EDUCATION FOR EMPLOYABILITY

The *Jobs of Tomorrow* Report of the World Economic Forum (2023) states that AI may not be able to completely take over those tasks that involve some critical thinking, but it does state that AI can always lend a helping hand to make such tasks easier. This is the underlying thought behind the intervention of AI technologies in the Film Education Sector in Indian higher educational institutes. Instead of being intimidated by a new technology, educators will have to teach students to embrace the technology and use it to their advantage.

Students of filmmaking can incorporate various AI models in their work to increase their chances of employability. Apart from curriculum incorporation, educators can ensure that the technological use is tailored according to the needs of every student (Anderson & Jefferson, 2020). That means students will have options for learning technology integration in scriptwriting, storyboarding, digital illustrations, graphic designing, VFX, editing, etc. Apart from the obvious tasks for filmmaking students that can be assisted by AI, it can also help in saving time and effort in doing repetitive tasks during the filmmaking process. As a result, it can help the filmmaker in cost-cutting by automating tasks that would have otherwise required an extra crew member. This enhanced skill set has a chance of ensuring better career placements for students graduating from film schools. An industry that is continuously evolving with the times will require people who are willing to adapt to newer technologies to save time and effort. In hindsight, this shall ensure that the filmmaker can focus on content thereby delivering a heightened visual experience for the audience.

Increase in employability also refers to the expansion of job opportunities (India Employment Report 2024). By learning skills in both filmmaking and AI, students can ensure an interdisciplinary knowledge that shall expand their job prospects and not limit them to filmmaking alone. This means they will be able to gain multi-faceted skills that can ensure jobs in other areas of media like animation, editing videos for social media, gaming, creating brand posters, making marketing videos for varied brands, and shooting advertisements and other promotional videos. With an increasing demand for AI-literate filmmakers in the current industry dynamics, such people will have an edge over others in the coming years. AI integration will become a part of most of the production workflows which shall mean the demand for AI-literate filmmakers will be on the rise. This competitive edge will potentially enhance the employability of students thereby increasing enrollments in film schools in the first place. Thus, it will turn out to be a win-win situation for the students as well as the higher educational institutions (HEIs) that teach filmmaking.

4.4.1 Contribution to the Economy of India

Film graduates equipped with AI knowledge will help in extensive cost-cutting thereby producing greater films on limited budgets (see Section 4.4). This will enhance the economy of the country. AI-literate filmmakers have the potential to ensure the generation of better-quality films, thus increasing their chances of attracting larger audiences. As mentioned before, educators can teach students about analyzing data-driven insights to gauge market trends. This will ensure that such filmmaking graduates are better equipped for the industry thereby contributing to the economy of the country. A flourishing work shall also ensure that there is a greater labor utilization which will help optimize efficiency. This way the industry has better potential to ensure employability at all levels of the hierarchy. AI literacy will ensure that film graduates will be able to streamline the basic processes of production and generate an enhanced output. This way they will be able to give undivided attention to the creative process of the film and help in inducing better national and international revenues by attracting larger audiences.

AI will help in the creation of new job opportunities in the forms of roles like that of data analysts which will become imperative to gauge the market trends. This shall help the directors in better analysis of their output and in turn help producers to get a more robust idea of what they are betting upon! This learning will potentially start from the classrooms and in turn ensure better placement opportunities. The fresh graduates will have wider arenas to contribute to the economy of the country thus enabling the need for better campus placements by the Higher Educational Institutes (HEI). This shall possibly lead to a growth in the employment sector especially in the entertainment sector. This is also discussed in the AnnualMedia and Entertainment report by EY and FICCI (2025). Another avenue of placement for AI-literate graduates will open up in the domain of streaming services. By fuelling the technology-based streaming services with creativity, the newer workforce shall be able to ensure that the digital economy rises. This collaboration of technology and creativity will also uplift the rising content creation sector. Talking about content creation, we are witnessing an increase in innovative startups (Damasio & Bicacro, 2017). The blend of AI models into the creative space of the filmmakers shall ensure that there is a rise in video production startups which will not be limited to filmmaking alone. This shall mean a rise in startups that are delivering services of short video production, trailers, teaser poster making, other forms of graphic design, and various forms of marketing. Accessibility to a technology like AI has enabled the creation of such services at lower costs thereby liberating them from the confines of the big-budget production houses that conventionally dominated the market. This has the potential to contribute to the startup model of the country and the generation of more diverse income streams for the people.

Apart from increasing the economy of a developing country like India, this shall even provide a chance to enhance its reputation in the global film market. A better-equipped workforce in the film industry shall ensure the creation of finer and quicker content. This will potentially lead to the parallel filmmakers being able to focus on generating content that is able to compete for the international film festivals. This shall push the recognition of the country globally enhancing its vision for a Viksit Bharat by 2047. This competitiveness and recognition in the global arena have chances of attracting more tourism and foreign investment in hindsight. The dream of Viksit Bharat can also be integrated with the governmental initiative of *Incredible India* (2002) by showcasing the diverse cultural heritage of the country. The National Geographic channel documentary called *Ganga: The River from the Skies* (2022) is a small example of the point highlighted above. This documentary was featured on the channel in March of 2022 and the YouTube trailer itself garnered as much as 2.5 million views. The tourism ministry of the country has engaged with the advertising sector to create new campaigns for such initiatives time and again. Apart from the filmmaking industry, AI-literate graduates shall also have an opportunity to expand their income streams by venturing into different sectors of the entertainment and media industry.

4.5 CONTRIBUTION TO MEDIA RESEARCH ON AI PENETRATION INTO FILMS IN INDIA

This chapter makes its foray into the scanty literature available to explore the penetration of AI into films. With AI emerging as a widely debated topic of research, works in specific fields are still in the stages of infancy. To say that AI penetration in the media has not been researched would be an understatement. Although research in the Indian media landscape is continuously evolving, exploration of this technological integration into films in India and the subsequent impact is less studied. Filling the knowledge gap would mean that academic literature in this field is increased so that future scholars have a chance to build upon some underlying work.

These academic research can also help set the stage for the practical world of cinema to draw upon the findings of such studies and inculcate them in tweaking their work accordingly. Apart from the creative expression in the cinema, there is so much more that goes into the process of filmmaking. This could mean the processes of pre-production like scripting will benefit from this technology. The data analytics to determine the trajectory of future projects based on previous and current box office outputs will minimize risks to some extent. Determining content personalization as per the industry-specific needs which can be based on the data derived from OTT User Analytics will foster better content creation. Such studies will expand the role that AI as a technology will play in assisting a creator in the times to come.

Additionally, media scholars will be able to explore the ethical implications of the novelties that have penetrated this craft. Coupled with legal implications, it will help in drafting guidelines for what this new technology can be used for. It will help in determining the degree to which it can possibly impact society and create much more responsible filmmakers. Determinants like creative ownership will possibly pave the way for discourses on ethical AI usage by the industry.

This approach will manifest at multiple stages. First is the foundational stage where the students will be able to learn the theoretical usage of AI and ultimately learn about using it in the industry. Second is the practical stage wherein the filmmakers will learn the practical implications of using such a technology. Third is the regulatory stage wherein the policymakers will be able to draft guidelines and proposals regarding the usage of AI in films. With the government making provisions via the recent Cinematography (Amendment) Act 2024, a fresh slew of changes has already been proposed in the initial Cinematograph Act of 1952. This act was established to oversee processes like the exhibition, censorship, and certification of films. An act established almost 70 years ago does require additions to the evolving film industry. Therefore, AI penetration will require policymaking at various levels if the film industry plans to welcome the new technology for good.

A blend of AI into media education will also call for myriad interdisciplinary research projects that will foster newer discipline collaborations and align with the vision of NEP (2020) in hindsight. Ethics, digital education, cultural studies, films, and media are just some of the disciplines that will converge to form robust research in Indian Academia.

4.5.1 Strengthening Industry–Academia Collaborations

The Industry and Academia collaborations will help in bridging the gap between classroom education and the workforce of the 21st century (see Section 4.5). This is imperative for the upcoming film graduates to enhance their contributions to the economy.

There are some suggestions for fostering such collaborations:

1. Situating the Curriculum: Aligning the curriculum with the needs of the industry is an obvious step toward fostering the Industry and Academia collaborations.
2. Guest Lectures with Industry Professionals: This shall include workshops with industry professionals which can provide exposure to students within

the institutions themselves. This shall also let students make choices of what aspect of the industry interests them. Sufficient masterclasses, guest lectures, and workshops have the possibility of bringing clarity in the minds of a lot of students to choose the part of the filmmaking process that they are interested in. The alumni of various institutions will prove to be extremely helpful in luring in such contacts from the industry. Alumni working in the practical space will also be able to lead by example for the students pursuing the courses.

3. Making Internships Mandatory: The above point can have a chain reaction of motivating a student to take up an internship in the allied part so they know if they wish to pursue it further. For others, making the internships a mandatory part of the curriculum might ensure that students go to practice in the field even if that duration is for a couple of months (Damasio & Bicacro, 2017). Institutions can play an important role in making students aware of entrepreneurship initiatives. This has the scope of creating multiple income streams providing better options for employability.
4. Recruiting more Professors of Practice: University Grants Commission (UGC) has provided for the recruitment of professors of practice in educational institutions. The mission of NEP (2020) to better the quality of higher educational institutions by providing skill enhancement courses has led to the need for professors of practice. This means such educators who have been practicing in the industry are therefore able to bring their professional experience into the classrooms. As the commission rightly puts it, it is about making the learners think like the employers and vice versa.
5. Industry-Academia Collaborative Projects: The practice of letting students work for a day in an advertisement or film shoot in colleges is common. This lets students volunteer for tasks they would be interested in and provides a helping hand to the crew at a nominal cost. These projects can be increased to encourage the film crews to let more students assist with small projects to foster better collaborations between the industry and the academic world. Especially graphic designing projects where AI as a technology comes in handy, students of the newer generation can handle such projects better and provide excellent assistance for the same.
6. Remote Mentorships: Institutions can also provide remote mentorships for students which can be between multiple institutions that are separated by geographical distances. This could be an opportunity to engage with those institutions and know what is happening in the academic world outside. It shall also foster higher research collaborations on mutual fields of academic expertise.
7. Film Festivals: Automating tasks at Film Festivals via AI technologies will act as a boon for the organizers thereby cutting costs and dedicating time to bigger tasks. However, there will also be a lot of tasks that will require timely supervision despite being automated. Students can assist in tasks like these and contribute toward the smoother functioning of such Festivals.
8. Automating Content with AI: The filmmakers can consider AI for automating basic tasks that are repetitive in nature and do not require a creative streak.

AI can help assist filmmakers by carrying out tasks like color grading film scenes and creating a rough cut for a voiceover or a film scene at the edit table, thereby saving time for the editor in terms of applying the creative vision of the filmmaker. The beginnings of such automation can be traced to the students in film schools in contemporary times who are being technologically smart enough to know what tasks can be automated to free up time for other aspects of a project (albeit with ethical considerations).

Such automations also apply to scriptwriting. *Writer's Block* is for real and that should not stop a creator from creating something. This reminds one of *Salim-Javed* - hit writer duo of yesteryears which wrote films that went on to become blockbuster hits. Cue *Zanjeer, Deewaar, Sholay, Seeta aur Geeta, Haathi Mere Saathi*, and an endless list! In contemporary times, however, filmmakers and students alike seem to be resorting to AI technologies to be able to get suggestions for story plots. Having the ability to suggest a variety of prompts and be able to tweak them accordingly ensures that the story idea can be laid out as close to the vision of the maker as possible. Of course, refining those narratives and giving direction to the story will require the indispensable human touch in ensuring optimal usage of such technological models.

Assisting makers with AI-powered tools to generate script ideas, do graphic designing, use VFX and SFX will also help students carry out such tasks in whatever capacities they choose to work.

9. Audience Analytics: Leveraging the power of audience analytics with the help of AI will mean that the makers will be able to get an overview of the viewing trends and audience preferences. The students can be taught the power of storytelling keeping data analytics in mind to possibly enhance the output and minimize risk. However, filmmaking is an extremely creative place and a data-driven analysis cannot ensure success with complete surety. It will only mean minimizing the incurred loss if the film does not do well at the box office or online in case of a virtual release on an OTT Platform. By no means does this lead to the certainty of a zero-loss situation. After all, the audience is varied and viewer preferences fluctuate. Keeping this in mind, data can assist makers in content but not guarantee an utterly positive outcome. While the students are learning in film schools, they have the opportunity to try out various permutations and combinations in predicting the outcome of a certain film. It is easier because money is not at stake for students. This means they have the window to learn the applications of this technology in multiple ways to find out what works best for them.
10. Policymaking: A newer technology that requires immense ethical considerations will call for robust policy advocacy on the part of the learners, educators, and makers in the film world. A technology that has the potential to threaten the Intellectual Property Rights of creators needs a set of guidelines and policies to be able to ensure ethical usage. Pedagogical

interventions and creator responsibilities have already been talked about earlier in the chapter. What is left unsaid is the need for a policy framework on the part of the government to ensure the Indian film industry continues to thrive in an otherwise demanding world of newer content. This is important to maintain the sanctity of an industry that relies heavily on creativity and carries the responsibility of showcasing the craft of the filmmakers of India on multiple global platforms.

4.6 CONCLUSION

A trailer like that of the popular Bollywood Series talking about the camaraderie of film writers Salim-Javed as *Angry Young Men* garnered more than 9 million views alone. Sadly, Generation Z will not get to see the duo create magic in the Bollywood films of the 21st century. What they do have now are stories from the past that are preserved or recreated on various OTT platforms and sites like YouTube. With contemporary times facilitating AI penetration into films, the newer generation gets to create much-advanced content for the viewers to watch, provided AI is considered a boon rather than a bane for the creative industry. The technology calls for responsible and ethical usage, all of which can be taught at the institutional level when young minds are still making sense of this craft. Filling this knowledge gap would require curriculum updations on the part of institutions. Additionally, educators will have to step up to learn the appropriate usage of this technology and incorporate newer pedagogical approaches to impart teachings to the students.. The population gap will also need to be addressed to promote inclusivity while treading the path of sustainability. This inclusion and diversity shall also help realize the vision of a Viksit Bharat (Developed India) by 2047.

The love for Indian cinema brings the country together on so many occasions and makes us worship the craft of these creators. Therefore, this form of entertainment brings a greater responsibility on the shoulders of such people and calls for government, industry, and academic collaboration. With multiple films doing us proud at the Academy Awards and other global stages, it is important that we start somewhere. This 'somewhere' could be the film schools that light up the spark for creativity in the first place. It will foster greater representation back home and on a global stage to encourage higher enrollment rates in film institutions of the country. After all, the magic of cinema lies in creating films that will stay with us long after the credits have rolled. And the foundations of that magic are built at film schools!

4.6.1 Scope for Future Research

On the academic front, it will be interesting to see the development of case studies on different aspects of the film industry. This shall help build upon the previous examples to determine what can work in the future or where one can take inspiration from. This can also equip future filmmakers with what to restrain from in their works. Extensive interviews with film journalists articulated and thematically analyzed into long-form research studies shall help in forming a base for conversations around how the industry is evolving. Case in point is an interesting book called *First Day First Show* (Chopra, 2011) that talks about her experiences with the film industry. Such film journalists can possibly form the backbone of studies that shall help in tracing the journey of films from the post independent era to contemporary times. Additionally, releases on popular OTT Platforms can help in gauging what the audience wants to see. Future research could explore the different ways in which AI can influence the creative thinking of a student. Intensive surveys and open-ended interview questionnaires could be employed to investigate the implications of this technology on students pursuing such film courses.

Another area of research could be a detailed analysis of the changing pedagogical approaches to accommodate this technology. This can prove beneficial to chart out roadmaps for universities in areas like curriculum development.

Scholars interested in studying inclusion could investigate the access to such AI tools in various institutes. They could further build upon its implications for areas with a lack of a better education or development.

Future scholarly works can also uncover various ethical challenges of AI and pose suitable solutions. This could also result in community empowerment projects to raise awareness among creators about its ethical use. Institutions could harness the potential of these projects to educate students and teachers alike.

This chapter also proposes the possibility of carrying out Longitudinal Studies by the scholars. Such studies mean repeatedly studying the same individuals over a prolonged period (years) to observe and deduce any changes over time. A longitudinal study to investigate the potential of AI in shaping a better career graph could be conducted on AI-literate film graduates to explore the possibilities and suggest findings. Consequently, such studies could prove to be helpful for multiple stakeholders, namely, students, educational institutions, and policymakers.

NOTES

1 For the purpose of this chapter, the term *filmmakers* refers to all people who are presently working or plan to work within the film industry in India in any capacity.

2 Various points in the Working Group Report of 11th Five Year Plan (2007–12) on the Information and Broadcasting Sector released by the Govt. of India use the terms *school* and *institute* synonymously. Therefore, for the purpose of this chapter, the two terms shall be used interchangeably with each other.

3 For the purpose of this chapter, *Higher Educational Institutions* shall include those institutions that offer a film-related course in the country. The focus of this chapter is on the integration of AI into the filmmaking classrooms, irrespective of the institution that offers these courses.

4 This is just an example of the multitude of Gen AI tools available to explore. Stakeholders are advised to use such tools at their own discretion.

REFERENCES

Akser, M. (2020). Cinema, life and other viruses: The future of filmmaking, film education and film studies in the age of covid-19 pandemic. *CINEJ Cinema Journal*, *8*(2), 1 - 13. https://doi.org/10.5195/cinej.2020.351

Bharti, S., & Yadav, P. (2024). Three Years of NEP 2020: Assessing Some Opportunities and Challenges for Higher Education Institutions. In *Foundations of change: Analyzing the impact of National Education Policy, 2020 on India's educational landscape* (pp. 199 - 214). HP Hamilton Limited.

Chopra, A. (2011). *First Day first show: Writings from the Bollywood trenches*. Penguin Books India.

Damasio, M. J., & Bicacro, J. (2017). Entrepreneurship education for film and Media Arts. *Industry and Higher Education*, *31*(4), 253 - 266. https://doi.org/10.1177/0950422217713110

FICCI, E., Young. (2025, March 27). *Revolution in Indian Media & Entertainment Sector*. Annual 2025 Report on Indian Media and Entertainment by FICCI and EY. www.ey.com/en_in/insights/media-entertainment/shape-the-future-the-revolution-in-indian-media-and-entertainment-sector

Incredible India, Govt. of India.(2002) *Discover incredible India: Travel guide & tourism info*. Available at: www.incredibleindia.gov.in/en

Kumar, K. J. (2017). *Mass communication in India*. Jaico Publishing House.

Ministry of Human Resource Development. (n.d.). *National education policy 2020*. www.education.gov.in/en/nep/about-nep. www.education.gov.in/sites/upload_files/mhrd/files/NEP_Final_English_0.pdf

McLuhan, M. (2001). *Understanding media*. Routledge.

National Geographic. (2022, April 29). *The mighty river | ganga: River from the Skies | National Geographic*. YouTube. www.youtube.com/watch?v=Q0gYQrebGwY&t=2s

National Skills Development Corporation, Govt. of India. Available at: NSDC Technical Support I Home.

Perry, M. S. (2018). 21st century skills through film production in tertiary education: A transformative assessment in a literature and media course. *3L the Southeast Asian Journal of English Language Studies*, *24*(4), 214–232. https://doi.org/10.17576/3l-2018-2404-16

Press Information Bureau, Govt. of India. (2023, Winter). *PM launches Viksit Bharat @2047: Voice of Youth*. Available at: www.pib.gov.in/PressReleaseIframePage.aspx?PRID=1985077#:~:text=Viksit%20Bharat%20%402047%20is%20the,environmental%20sustainability%2C%20and%20good%20governance

Sari, T., Nayir, F., & Bozkurt, A. (2024). Reimagining education: Bridging Artificial Intelligence, transhumanism, and critical pedagogy. *Journal of Educational Technology and Online Learning*, *7*(1), 102–115. https://doi.org/10.31681/jetol.1308022

Statistics Division. (2024, June 28). *SDG indicators*. United Nations. The Sustainable Development Goals Report 2024 (United Nations).

World Economic Forum. (2023, September). *Jobs of tomorrow: Large language models and jobs report 2023* Available at: www.weforum.org/publications/jobs-of-tomorrow-large-language-models-and-jobs/

Zhai, X., Chu, X., Chai, C. S., Jong, M. S., Istenic, A., Spector, M., Liu, J.-B., Yuan, J., & Li, Y. (2021). A review of Artificial Intelligence (AI) in education from 2010 to 2020. *Complexity*, *2021*(1). https://doi.org/10.1155/2021/8812542

Additional Readings

Anderson, M., & Jefferson, M. (2020). *Teaching the screen: Film education for generation next*. Routledge, Taylor & Francis Group. https://doi.org/10.4324/9781003117827

Petrie, D. J., & Stoneman, R. (2014). *Educating film-makers: Past, present and future*. Intellect. https://doi.org/10.2307/j.ctv36xvhbb

5 Equity in AI-Driven Education

Addressing Gender Bias and Promoting Inclusive Learning Environment

Manisha Singh, Vinod Bhatt, Dev Brat Gupta, and Sonali Srivastava

5.1 INTRODUCTION

The development of more advanced systems driven by artificial intelligence (AI) is accelerating. These systems have the potential to significantly improve everyday life, including the quality of education for students at every level. As education continues to shift to digital platforms, the ability to capture data and use it to individualize learning experiences has become more feasible, and in some cases, essential.

In the context of education's evolution it is crucial to proactively foresee and address any potential adverse effects of AI. This is necessary to prevent a resurgence of inequality and to guarantee that AI contributes to a future where education is more fair and just. Now is a favourable time for collective self-reflection to tackle deep-rooted structural prejudices and disparities in education (Aiston & Walraven, 2024). This chapter will especially address the identification of potential gender biases that may arise from AI in education, as well as strategies to counteract these biases in order to foster inclusive learning environments.

5.1.1 The Study Has Three Primary Objectives

- To anticipate the potential impacts of AI in education on gender equity, identify how these impacts may result from structural biases in education, and consider ways in which AI could be leveraged to offset these impacts and promote equity.
- To identify current gaps and opportunities for educational research and practice aimed at leveraging AI to create more inclusive learning environments for the students.
- To address the challenges that girls and women face in education today, the technology shift, and the potential impact that AI-driven education can have on teaching and learning.

DOI: 10.1201/9781003560463-6

The concerns listed above mainly pertain to educational equity, in that AI technologies should be deployed so as to minimize disparities in educational outcomes and provide increased support for underprivileged students. Issues of equity are no more salient than they are now to widen disparities in students' educational outcomes. Hence, it is more important than ever to develop AI in ways that are aligned with equity goals. At the same time, the potential for AI in education has sparked apprehension among scholars of education. Their concerns often revolve around the unequal distribution of computing resources among schools and that the new technologies may widen gaps between rich and poor students (Alam, 2021).

AI is the theory and development of computer systems capable of performing tasks that typically require human intelligence, such as visual perception, speech recognition, decision-making, and translation between languages. It is a rapidly evolving field with the potential to greatly impact society (Dwivedi et al., 2021). Interest in AI has increased substantially over the past few years with a significant increase in investment and research. In primary and secondary education, one AI application is to provide intelligent tutoring and tailored instruction to students.

5.2 PURPOSE OF THE CHAPTER

The purpose of this chapter is to explore why gender equity in AI-driven education is an important issue to solve. It begins by addressing the challenges that girls and women face in education today, the technology shift, and the potential impact that AI-driven education can have on teaching and learning. This is followed by a deeper discussion of the potential and the reality of how AI systems learn gender bias. Unfortunately, most AI is already reinforcing and, in some cases, even creating new gender biases. The chapter argues that if left unchecked, AI-driven education will only serve to further educational gender inequity. Following that, evaluate the potential for AI-driven education to address gender bias and create more inclusive learning environments, and the risks associated with AI replicating or creating new biases. It is important to note that the focus is not on excluding boys and men from educational resources or creating separate resources for them and women, but rather on creating more equitable educational resources and outcomes for all.

This chapter is focused on critically examining equity in AI-driven education through a gender lens. The decision to focus on gender bias is rooted in a long-standing commitment to challenge all forms of inequity in education, but also pragmatic recognition that issues of gender bias are often deprioritized in education research and policy discussions around AI. It is also a response to the recent call from UNESCO, for gender equality in and through education, 'ensuring that every individual has equal access to the resources they need and equal opportunity to succeed in their educational path'. Our commitment to gender equity aligns with the Sustainable Development Goals (SDGs), recognition of the pivotal role that education plays in achieving these goals and broader social justice agendas.

5.3 SCOPE AND LIMITATIONS

To ensure that perception and implications of gender bias in educational AI was effectively captured, it was primarily on changes to students' performances in each of these tools, and educational outcomes as a result. This chapter will be exploring and analyzing bias in AI tutoring software, and its implications for online education. While problems with bias in AI education have been explored before, as well as it aims to address the lack of research on gender bias in virtual learning tools and its impact on the education system. In addition, increasing reliance on technology for education has created a new urgency to address these issues, as the system that these tools are biased toward is rapidly becoming the new norm of educational delivery. As such, this chapter has the potential to bring about a significant positive change for future students using these tools, as well as a greater understanding of the potential pitfalls of this system from educators and software developers.

5.4 UNDERSTANDING AI-DRIVEN EDUCATION

5.4.1 Definition of AI-Driven Education

AI has been put to use in a variety of educational settings, from small-group tutoring to massive open online courses (MOOCs). AI systems are being implemented in various ways, including intelligent tutoring systems, intelligent tools for learning, and AI-based learning environments. These are just a few examples and the list of possibilities grows each day (Akgun & Greenhow, 2022; Kabudi et al., 2021). AI has also been used to produce automated grading systems that enable instructors to provide students with instant feedback, freeing the instructors up to focus on other aspects of their teaching. Such systems are said to be useful in not only grading students, but also in providing students with additional feedback, as the more advanced of these systems can pinpoint areas where a student needs improvement.

AI as an efficient tool to scale education and teaching methodologies. There is and will continue to be a demand for online education which can be global in its reach. AI will be used to provide individualized education, help students to work at their own pace, and direct students in their learning. The deployment of intelligent tutoring systems helps to fulfill this need providing instruction and feedback to a student. These systems are interactive and can be very effective in the absence of a human instructor. AI which provides a virtual learning environment where students can explore and learn as they would in the real world could be put to use in training for complex real world tasks (Bhutoria, 2022; Maghsudi et al., 2021). The utilization and advantages of AI in the education sector are stupendous. It has transformed the way in which education is conveyed and delivered to the learners. It has been quite successful in eradicating the blocks which earlier hindered a typical learner with learning disabilities. The use of AI in special education is a booming area with a vast amount of potential. This area has been quite successful with children, especially those suffering from autism (Luan et al., 2020; Gocen & Aydemir, 2020).

5.4.2 Applications of AI in Education

When discussing the application of AI in education, it is useful to consider how it has been implemented so far and how it could be used in the future. Existing applications can be categorized as those employed by the institutions themselves to improve the management and delivery of education and those used to provide a better service to students. The former includes systems to automatically timetable classes and resources to a room and time, manage library resources, tutor students, or manage coursework, all with the aim of reducing the financial or staff resources required for a given task (Dwivedi et al., 2021). Whilst these systems have an indirect effect on students, the latter category will be the most contentious as it directly affects the quality of education the learner receives. AI tutors are likely to be the most influential future technology. Currently, they are limited to searching a database for solutions to student queries, but in the future, they may take the form of a virtual human which can converse with the student in natural language, interpreting the student's emotions and motivation as a human tutor would and adapt its teaching style and material to suit (Chen et al., 2020).

5.4.3 Benefits and Challenges of AI-Driven Education

The benefits of AI-driven education are numerous. As hinted at by its potential in the field of intelligent tutoring systems, AI provides the possibility of a truly individual education (Srinivasa et al., 2022). By detecting the needs of a student, an intelligent tutor can tailor education to the level and pace of each student, providing reinforcements and teaching while continually assessing the student's progress. This is a radical improvement over the current situation of human tutors, where the high cost of skilled professionals can only provide a low teacher-student ratio in exceptional situations. Outside of tutoring, intelligent agents can provide personalized advice to students in a huge range of areas such as subject and career choice. This is invaluable in societies which provide a vast array of choices to citizens, but are unable to counsel each one individually. AI also has the potential to revolutionize education for disabled students (Chen et al., 2022; Adıgüzel et al., 2023). Students who struggle with the social demands of traditional education or who learn better without them have an option in the form of intelligent agents, who don't care if their counterpart is a human. More importantly, intelligent agents can be considered the only fully successful means of educating students who are unable to leave their homes because they do not require any kind of physical connection.

5.5 GENDER BIAS IN AI-DRIVEN EDUCATION

Research on gender bias in AI education is still in its early phases, but it has already produced some alarming results. The depiction of gender roles in an instructional math game and its subsequent effect on children's math performance were examined by Sato et al. (2019). They discovered that the game demonstrated minor but observable instances of gender bias, such as giving the male character in the game more difficult mathematics problems and portraying him as more intelligent and capable than the female character. After playing the game, children were more likely to be unsatisfied with their arithmetic skills and to have a strong preference for math with their same gender peers. These findings have important consequences: although it is implausible to anticipate that gender bias in AI education would completely disappear, action must be taken to lessen its effects in order to reduce its impact on the emergence of larger social gender inequities (Putnam et al., 2022). It will take some years before researchers fully comprehend the complicated and sophisticated ways that gender bias creeps into AI education, as well as the best strategies to recognize and mitigate its effects, as research on this area is still in its early stages. Nonetheless, it is evident that comprehending and resolving this issue are crucial.

5.5.1 Overview of Gender Bias in AI

When considering AI in the field of education, it tends to envision it as an impartial instructor that fosters an impartial learning atmosphere. However, the effectiveness of AI is solely dependent on the quality of data it gathers. In the past, this data has been extensively documented to exhibit gender bias. This bias is rooted in societal bias that has been around for centuries. Some of it can be as simple as societal norms, such as what gender usually does what chores in the house, to more complex sexism, such as there being more male than female doctors in the world. All of these things can influence the outcome of an AI's learning.

One particular study by Latanya Sweeney found that Google's ad system was more likely to show ads for high-income jobs to men than it was to women. When both were shown an ad for a subscription of a background check for said jobs, Google asked the man if he wanted to do a criminal background check, and didn't ask the woman anything. This was not a choice made by a person, but by an AI interpreting which ads were more relevant to the two users. This didn't have any profound effect, seen that both users could still get to the same background check through various means, but it reflects this societal norm that men are more likely to be employed in more well-to-do professions. This is a biased reflection of how the world should be, and an AI using this information could treat students differently depending on their gender with the belief that it is treating them the same (Kashyap & Verkroost, 2021).

5.5.2 Gender Bias in Educational Technologies

Gender bias is a well-known issue in education. When it comes to utilization of AI in education, it may further reinforce the stereotypes and traditional roles, and may even worsen the situation in certain aspects. This is particularly an issue with single gender datasets, and an all too common occurrence in the software development industry. The majority of AI developers are men, developing tools on a subject (education) which they have never been directly affected by (Golendukhina et al., 2022). The inevitable result is an oversight of the needs of female students and teaching staff, and the implementation of systems which are inherently biased toward the needs of men. Simulation is one example of this, where in an attempt to use behaviorist methods to make learning more scientific, the complex and often more empathetic methods used by teachers with girls are overlooked, causing disengagement for female students. An Australian study on the implementation of an intelligent tutoring system showed that the system largely ignored the questions asked by girls, who were the lower academic achievers at the school, choosing to help the male students it identified as having learning difficulties(Tang et al., 2023). This resulted in the girls becoming even less involved with their schooling and furthering the gap in learning outcomes between the genders.

5.5.3 Impact of Gender Bias on Learning Outcomes

Gill (1989) mentions six psychological dimensions of assessment through which bias may operate. These are content, construct, context, presentation, response and affect. An example of bias in the dimension of content is seen in national testing where texts that are of greater interest to boys are chosen (Chávez & Mitchell, 2020). In the affective dimension, teacher feedback to boys is often of a directive nature aiming to correct mistakes, whereas for girls it is often more descriptive aiming to encourage further effort. It has been found that directive feedback is most effective in enhancing performance (McCaslin & Good, 1992) and may thus lead to different levels of achievement for the two pupils. Affect from teacher to student has been shown to correlate with student perceptions of ability and subsequent achievement (Eccles et al., 1983) and so differences in teacher behaviour toward students may affect their self-perceptions and motivation in specific subjects. Gender differences in assessment also occur due to the fact that many educational outcomes are defined as measures of masculinity or femininity, for example boys are often thought of as maths people and encouragement of non-traditional subject choices may be interpreted as discrimination against boys. Finally, bias may occur due to the lack of knowledge and training teachers have in the art of assessment and their lack of awareness of issues surrounding gender and assessment (Wong et al., 2023). The theories and findings reviewed above demonstrate that an educational system is pervasive with gender bias, which is often unintentional. Gender bias has the potential to inhibit women as well as men and it is therefore important to address and overcome. Gender bias in assessment has been shown to disadvantage girls (Napier et al., 2020).

5.6 PROMOTING INCLUSIVE LEARNING ENVIRONMENT

A variety of changes have occurred in the education sector. These changes have had differing effects on students. Some students were able to adjust to the change more readily, while others have struggled (Jussim et al., 1996). Students in special education and English learners are two groups that have had a particularly difficult time adjusting to the new environment. These groups of students require more individualized instruction and access to support services (Ortogero & Ray, 2021). The change to online instruction has created a situation in which many of these students do not have access to the instruction and services that they require. This is even truer for students in special education who are in separate instructional settings. These students were already experiencing marginalization and have likely had the most difficult time of any student group in adjusting to the new educational environment (Supratiwi et al., 2021).

Inclusive education has always been a challenge for schools and educators. Due to the COVID-19, education has shifted from the traditional classroom to a more virtual setting. This change has provided even more challenges and opportunities to provide a quality and inclusive education to all students (Kamruzzaman et al., 2023).

Discussion about inclusion in educational environments often falls short of its mark by conflating the idea of inclusion with equality. While they are conceptually similar, the implications of striving for an inclusive educational environment as opposed to an equal one are massive. Equality in education merely means that both male and female students receive identical treatment, resources, and protection from educational abuses such as harassment, exclusion, and discriminatory biases. While it is essential that educational environments are free from gender bias, an equal educational environment is not necessarily an inclusive one (Stenman & Pettersson, 2020). In addition to offering equal treatment and protection to all, inclusive educational environments acknowledge, address, and work to eliminate the disparate effects of bias on various groups.

5.6.1 Importance of Inclusivity in Education

Education should be inclusive for a number of reasons, including moral, social, and economic ones. All children and young people can receive a higher quality education from an inclusive education system, equipping them with the knowledge and abilities they need to succeed in life. Education has the power to mitigate poverty and act as a catalyst for constructive social change. Social cohesion and social marginalization can be lessened through inclusive education. According to De Schutter et al. (2023), it can also help end the generational cycle of poverty and disadvantage. Studies indicate that a parent's engagement in their child's education is essential to their academic achievement.

Effectively involving all parents in their children's education will contribute to an inclusive education system. This is good for society as a whole and it is relatively cost effective; making small changes to educational practice now can have a big impact on the way we live in the future.

5.6.2 Creating Gender-Neutral AI Tools and Platforms

One method for addressing gender bias is to involve more women in the development of AI education tools and platforms. Increasing the diversity of perspectives within a development team can help to identify potential sources of gender bias and develop strategies for avoiding them. Gender stereotypes about computer science and engineering being male-dominated fields may dissuade women from pursuing careers in AI/tech, and as a result, there are currently fewer women working in AI development (Enholm et al., 2022). Measures to counteract this and encourage women to pursue careers in AI development could have long-term benefits for the quality and inclusivity of educational AI.

Creating AI tools and platforms that are gender-neutral is important for mitigating the negative effects that biased gender stereotypes have on education. Gender biases can be implicit or explicit. Implicit in this case refers to AI systems that provide educational materials or lessons that reflect cultural stereotypes about gender. Explicit bias, on the other hand, might involve a system that discourages a girl from pursuing a career as an AI technician (Yu, 2024). Given the rapid uptake of AI in education, there is a growing need to ensure that these systems do not serve to perpetuate harmful stereotypes about gender and do not disadvantage students of a particular gender. Measures need to be taken to identify and remove gender biases from existing educational AI systems and prevent biases from being introduced into new systems.

5.6.3 Empowering Educators to Address Gender Bias

To build teacher awareness about bias, we first need to better understand what teachers already know and believe about how bias affects students. Although both qualitative and quantitative methods can provide useful information, this issue is so sensitive that teachers' honest reflections about their own experiences require approaches that ensure confidentiality.

When we consider all strategies for promoting equity in education, equipping teachers to recognize and address bias is perhaps the most powerful lever for change. Since educators have the most day-to-day contact with students, they are best positioned to identify bias when it occurs and help students to cope with its effects. At the same time, research has shown that teachers, like others, tend to be less aware of bias when it is implicit. Consequently, when incidents of bias are ambiguous or minor, educators may miss the importance of intervening. Given that much bias is subtle and difficult to address, helping teachers to recognize it and respond effectively is an essential step in creating more equitable educational experiences (Valdez et al., 2020; McCray, 2022; Mayfield, 2020).

5.7 ETHICAL CONSIDERATIONS IN AI-DRIVEN EDUCATION

There are many ethical considerations in AI about education. Despite the fact AI can provide personalized learning experiences and decrease the need for human intervention, it still develops a couple of dangers to certain human beings. This includes, however, is not confined to, loss of control, bias and discrimination, unemployment, cultural and lifestyle changes, and comfort.

5.7.1 Privacy and Data Protection

Privacy in educational settings is a significant concern due to the sensitive nature of information disclosed to those in the education system. AI systems might look to gather, examine, and use student data to inform forecasts or choices. This could become an issue if people's data is used for purposes that are not in their best interests or in ways that they do not anticipate. If a student is informed that an AI system has predicted that they would not succeed in a specific line of study, they can become discouraged from following their passion. AI systems may occasionally use data to make decisions (like course suggestions) that could significantly affect a person but are not disclosed to or include the person in the decision-making process. There are questions regarding the accountability and fairness of AI-driven educational systems because of this lack of openness and personal interaction.

Therefore, education and AI stakeholders should work together to ensure that the use of AI in education is in compliance with privacy and data protection legislation, and that best practices with respect to privacy are built into AI systems for education from the outset.

5.7.2 Transparency and Explainability of AI Algorithms

Transparency pertains to the openness of a party's attributes, where explainability refers to the capability of an AI system to provide background for human understanding. AI vendors, whether they be developing products for use in higher education, need to be transparent regarding how the product's algorithms work, what data sets the algorithms were trained on, and the expected implications of algorithm use. Without such information, it is difficult for consumers to assess the effectiveness of AI products and weigh the potential risks against the benefits. Additionally, vendors should explain to what degree their AI product mimics human cognitive processes and whether the AI's decision-making is dynamic or static. Static decision-making occurs when an AI system comes to a conclusion on an issue and the solution to the issue remains constant over place or time. Dynamic decision-making occurs when an AI system makes a decision that is subject to change given different circumstances. Static AI decision-making is desirable in situations involving diagnosis and problem resolution.

5.7.3 Ensuring Fairness and Accountability in AI-Driven Education

AI is driven by data that has the capabilities to replicate human decisions. However, sometimes these decisions can also be biased based on issues like gender, race, and the courtesy of coming from a privileged background. In AI education, the selection of resources can be influenced by said biases. AI educators must understand what specific groups within a culture find important for their children to learn. If not, resources could be shifted from rote memorization to conceptual teaching.

The large risk with AI in education is that it is built on an infrastructure that is meant to be impartial in regards to teacher quality or resources. This is especially important in public education. AI systems may change course and teacher recommendations based on what they perceive as effective or by comparing similar schools in differing income areas and their success. A biased system may suggest that a teacher change a course to an easier format for inner-city students opposed to staying on course and having high expectations. This process will perpetuate the education gap rather than closing it. These changes can be minute, and educators may not realize that they are coming from computer data rather than professional decisions.

5.8 CHALLENGES AND OPPORTUNITIES FOR INCLUSIVE EDUCATION

Students with disabilities faced significant obstacles during their education. The lack of resources and tools not only hindered their learning, but also affected their social and emotional development. Many of these students regressed, losing both academic and functional skills. Furthermore, the absence of necessary accommodations and modifications prevented some students from effectively participating in distance learning. This put both students and their families in a challenging position as they had to advocate for their rights in special education during an unprecedented situation. However, it offers an opportunity to transform education and make it more inclusive. Education leaders, administrators, teachers, and families can take this time to reflect on what has and hasn't worked for students and work toward creating a more inclusive learning environment. It is clear now, more than ever, that inclusive education should be prioritized and made a reality for all students, including those with disabilities. Achieving this goal requires providing the necessary tools, resources, professional development, training, and support to all educational stakeholders. By making these efforts, we can ensure that all students, regardless of their abilities, can effectively learn in virtual, blended, or traditional learning environments. It is crucial to recognize that high-quality inclusive education has a significant impact on the lives of students with disabilities, and this impact should not be underestimated (Brzezinska & Cromarty, 2022; Toquero, 2020; Easop, 2022).

5.8.1 Leveraging AI for Inclusive Remote Learning

AI helped to build inclusive remote learning environments and identify and provide intervention for students who are disengaged or struggling. AI tools can help teachers adapt instruction to the online environment and to individual student needs, like by automating the creation of alternative formats and adaptive learning paths from the same materials and by providing data dashboard visualizations of engagement and learning progression (Agarwal et.al. 2024). In implementing AI in remote learning settings, it will be important to consider the implications both for students with access (Fitria, 2021) to a lot of face-to-face support and specialist teaching and students who are already struggling in the current context with reduced teacher contact and less specialized learning support. It will also be important to ensure AI solutions are inclusive in design and that their use does not contribute to the widening of education disparities. Other AI innovations such as simulations and interactive learning environments and intelligent tutoring tools have potential to greatly enhance learning experiences for the diverse learners and address barriers presented by remote learning (Chen et al., 2020). However, there is a risk that during rapid ad hoc implementations of AI responses in remote learning, certain biases and inequalities may arise. These biases can disproportionately affect marginalized students, creating further barriers to equitable access and participation in education. Furthermore, leveraging AI in remote learning can provide personalized and tailored educational experiences for students, regardless of their background or circumstances.

5.8.2 Ensuring Access to Technology for All Students

Key Considerations:

a. Electronic devices and internet: To ensure that all students have universal access to educational opportunities, purchase or give tools and resources, such as gadgets and internet access.
b. Technological help: To enable students and teachers to fully engage in online learning, make sure they have access to prompt and continuous help for their technological needs and issues.
c. Pedagogical support: Provide teachers with the direction and tools they need to leverage technology to enhance instruction.
d. Digital learning resources: Give every student access to excellent digital learning tools. It takes more than just giving each student a device and an internet connection to ensure access to technological education. It all comes down to giving kids the instruments and materials they need to access and participate in top-notch education. As a result, while deciding how to give each student access to technology, districts and schools must take into account every element required for engaging learning opportunities.

5.9 STRATEGIES FOR ADDRESSING GENDER BIAS IN AI-DRIVEN EDUCATION

The first strategy is to retrain the AI in a clear and focused manner. Due in large part to the fact that AI is a byproduct of current societal views and prejudices, gender biases in the field currently exist. AI learns by being exposed to existing human-generated data and is trained to replicate human language and behaviour. Gender biases are common in this data; other studies have indicated that there may be up to 4.5 times as much information about men as women (Dworkin et al., 2020). Then, male experiences and perceptions are disproportionately represented by AI as being more frequent or usual than they actually are. For instance, it has been observed that Google's translation technology consistently interprets a doctor as a man and a nurse as a woman, translating the professions to pronouns in other languages appropriately. It takes deliberate effort to expose AI to positive examples of women in roles or engaging in activities that are currently automatically assumed to be male-only, as well as to identify the specific areas of bias in order to retrain AI to change these biases.

The above key strategies stand out as central to addressing gender bias on the part of AI tools and in the instructional approaches inspired by AI. All four strategies share in common a reduction in the extent to which automatic assumptions are made regarding girls and boys, and recognition of the diversity within each gender.

5.10 FUTURE DIRECTIONS AND RECOMMENDATIONS

Enriching software development practices to address gender biases can best be accomplished by having a diverse team. To prevent bias from occurring in the first place, we must seek to create a development environment that is diverse and inclusive. Having a gender-balanced team is the best way to ensure that the software does not exhibit any one group's biases. It brings a variety of perspectives to the table and makes it less likely that the group shares the same implicit biases. Gender-balanced teams also require fairer treatment of women, which can only benefit the organization. Therefore, it is strongly recommended that development teams aim to be gender balanced. This is no easy task and may take many years to accomplish. Organizations might consider setting an affirmative action goal or establishing special recruiting practices for women. One proposal is to use blind assessment at several points in the hiring process to prevent evaluator bias. An audit is the first step to discovering and understanding existing biases so that they may be mitigated.

In application stages, it is required to frequently check the decision boundaries of models in real-time scenarios. It would be very easy to slide back into a biased environment at these key stages and increase the gender gap. The factors have identified certain characteristics that may be misinterpreted, especially in the candidate assessment factors. The decision phase for AI could benefit from guidance or a checklist to ensure that decisions do not display gender bias. This could take the form of a question series to determine suitability for the role and guidelines of when it's appropriate to reject a candidate. To satisfy and align with diversity targets, the public sector and organizations with diversity goals may consider quotas to ensure the fair treatment and higher selection of female candidates in male-dominated sectors. An open and informed decision is morally superior to an unconscious decision driven by hidden bias.

There are several areas that need to be investigated further, relying on better equality ambitions for both men and women with respect to correcting the influence. The findings and factors that have been identified would apply equally to any future improved model or algorithm and should not rely on changes that would benefit from increased prediction accuracy for a male candidate over a female candidate.

REFERENCES

Adıgüzel, T., Kaya, M. H., & Cansu, F. K. (2023). Revolutionizing education with AI: Exploring the transformative potential of ChatGPT. *Contemporary Educational Technology*, 15(3), ep429.

Agarwal, P., Swami, S., & Malhotra, S. K. (2022). Artificial intelligence adoption in the post COVID-19 new-normal and role of smart technologies in transforming business: A review. *Journal of Science and Technology Policy Management*, 15 (3), 506–529. https://doi.org/10.1108/JSTPM-08-2021-0122

Aiston, S. J. & Walraven, G. (2024). A re-view of educational inequalities. *Educational Review*, 76(1), 1–12. https://doi.org/10.1080/00131911.2023.2286849.

Akgun, S. & Greenhow, C. (2022). Artificial intelligence in education: Addressing ethical challenges in K–12 settings. *AI and Ethics*, 2 (3), 431–440.

Alam, A. (2021, November). Possibilities and apprehensions in the landscape of artificial intelligence in education. In *2021 International Conference on Computational Intelligence and Computing Applications (ICCICA)* (pp. 1–8).

Bhutoria, A. (2022). Personalized education and artificial intelligence in the United States, China, and India: A systematic review using a human-in-the-loop model. *Computers and Education: Artificial Intelligence*, 3, Article 100068. https://doi.org/10.1016/j.caeai.2022.100068.

Brzezinska, M., & Cromarty, E. (2022, June). Emergency remote teaching in the university context: Responding to social and emotional needs during a sudden transition online. In *International Conference on Human-Computer Interaction* (pp. 30–47). Springer International Publishing.

Chávez, K. & Mitchell, K. M. W. (2020). Exploring bias in student evaluations: Gender, race, and ethnicity. *PS: Political Science & Politics*, 53(2), 270–274. https://doi.org/10.1017/S1049096519001744.

Chen, L., Chen, P., & Lin, Z. (2020). Artificial intelligence in education: A review. *IEEE Access*.

Chen, X., Zou, D., Xie, H., Cheng, G., & Liu, C. (2022). Two decades of artificial intelligence in education. *Educational Technology & Society*, 25(1), 28–47. https://www.jstor.org/stable/48647028.

De Schutter, O., Frazer, H., Guio, A. C., & Marlier, E. (2023). *The escape from poverty: Breaking the vicious cycles perpetuating disadvantage*. Policy Press Shorts Research. https://doi.org/10.51952/9781447370611

Dwivedi, Y. K., Hughes, L., Ismagilova, E., Aarts, G., Coombs, C., Crick, T., …, & Williams, M. D. (2021). Artificial intelligence (AI): Multidisciplinary perspectives on emerging challenges, opportunities, and agenda for research, practice and policy. *International Journal of Information Management*, 57, 101994.

Dworkin, J. D., Linn, K. A., Teich, E. G., Zurn, P., Shinohara, R. T., & Bassett, D. S. (2020). The extent and drivers of gender imbalance in neuroscience reference lists. *Nature Neuroscience*, 23(8), 918–926.

Easop, B. A. (2022). Education equity during COVID-19: Analyzing in-person priority policies for student with disabilities. *Stanford Law Review*, 74 (1), 223–275. http://dx.doi.org/10.2139/ssrn.3869935

Eccles, J. (1983). Expectancies, values and academic behaviors. In J. T. Spence (Ed.), *Achievement and achievement motives: Psychological and sociological approaches* (pp. 75–146). Free Man.

Enholm, I. M., Papagiannidis, E., Mikalef, P., & Krogstie, J. (2022). Artificial intelligence and business value: A literature review. *Information Systems Frontiers*, 24(5), 1709–1734.

Fitria, T. N. (2021). Artificial intelligence (AI) in education: Using AI tools for teaching and learning process. In *Prosiding Seminar Nasional & Call for Paper STIE AAS*.

Gill, H. S. (1989). The will in psychoanalytic theory and practice. *British Journal of Medical Psychology*, 62, 1–11. https://doi.org/10.1111/j.2044-8341.1989.tb02805.x

Gocen, A. & Aydemir, F. (2020). Artificial intelligence in education and schools. *Research on Education and Media*, 12(1), 13–21. https://doi.org/10.2478/rem-2020-0003.

Golendukhina, V., Lenarduzzi, V., & Felderer, M. (2022, May). What is software quality for AI engineers? Towards a thinning of the fog. In *Proceedings of the 1st International Conference on AI Engineering: Software Engineering for AI* (pp. 1–9).

Jussim, L., Eccles, J., & Madon, S. (1996). Social perception, social stereotypes, and teacher expectations: Accuracy and the quest for the powerful self-fulfilling prophecy. In *Advances in experimental social psychology* (Vol. 28, pp. 281–388). Academic Press. https://doi.org/10.1016/S0065-2601(08)60240-3.

Kabudi, T., Pappas, I., & Olsen, D. H. (2021). AI-enabled adaptive learning systems: A systematic mapping of the literature. *Computers and Education: Artificial Intelligence*, 2, 100017.

Kamruzzaman, M. M., Alanazi, S., Alruwaili, M., Alshammari, N., Elaiwat, S., Abu-Zanona, M., …, & Ahmed Alanazi, B. (2023). AI-and IoT-assisted sustainable education systems during pandemics, such as COVID-19, for smart cities. *Sustainability*, 15(10), 8354.

Kashyap, R. & Verkroost, F. C. J. (2021). Analysing global professional gender gaps using LinkedIn advertising data. *EPJ Data Science*, 10 (1). https://doi.org/10.1140/epjds/s13688-021-00294-7.

Luan, H., Geczy, P., Lai, H., Gobert, J., Yang, S. J., Ogata, H., …, & Tsai, C. C. (2020). Challenges and future directions of big data and artificial intelligence in education. *Frontiers in Psychology*, 11, 580820.

Maghsudi, S., Lan, A., Xu, J., & van Der Schaar, M. (2021). Personalized education in the artificial intelligence era: what to expect next. *IEEE Signal Processing Magazine*, 38(3), 37–50.

Mayfield, V. (2020). *Cultural competence now: 56 exercises to help educators understand and challenge bias, racism, and privilege*. ASCD.

McCaslin, M. & Good, T. S. L. (1992). Compliant cognition: The misalliance of management and instructional goals in current school reform. *Educational Researcher*, *21*(3), 4–17.

McCray, C. R. (2022). Equitable instruction, empowered students: A teacher's guide to inclusive and culturally competent classrooms. Solution Tree Press.

Napier, J. L., Suppes, A., & Bettinsoli, M. L. (2020). Denial of gender discrimination is associated with better subjective well-being among women: A system justification account. *European Journal of Social Psychology*, 50(6), 1191–1209.

Ortogero, S. P. & Ray, A. B. (2021). Overrepresentation of English learners in special education amid the COVID-19 pandemic. *Educational Media International*, 58(2), 161–180. https://doi.org/10.1080/09523987.2021.1930485.

Putnam, M. M., Ryan, R. M., Barba, E., Nissen, J., & Calvert, S. L. (2022). Young children's mathematics learning from same-gender and other-gender intelligent character prototypes. *Technology, Mind, and Behavior*, 3(2). https://doi.org/10.1037/tmb0000069

Sato, K., Hamamoto, K., Wright, D. & Nakagawa, H. (2019). A virtual reality-based language learning environment to teach sentence word order. *Journal of Advances in Humanities and Social Sciences*, 5(5), 196–206. https://doi.org/10.20474/jahss-5.5.1

Srinivasa, K. G., Kurni, M., & Saritha, K. (2022). Harnessing the power of AI to education. In *Learning, teaching, and assessment methods for contemporary learners: pedagogy for the digital generation* (pp. 311–342). Springer Nature Singapore.

Stenman, S. & Pettersson, F. (2020). Remote teaching for equal and inclusive education in rural areas? An analysis of teachers' perspectives on remote teaching. *The International Journal of Information and Learning Technology*, 37(3), 87–98.

Supratiwi, M., Yusuf, M., & Anggarani, F. K. (2021). Mapping the challenges in distance learning for students with disabilities during COVID-19 pandemic: Survey of special education teachers. *International Journal of Pedagogy and Teacher Education*, 5(1), 11–18.

Tang, K. Y., Chang, C. Y., & Hwang, G. J. (2023). Trends in artificial intelligence-supported e-learning: A systematic review and co-citation network analysis (1998–2019). *Interactive Learning Environments*, 31(4), 2134–2152.

Toquero, C. M. D. (2020). Inclusion of people with disabilities amid COVID-19: Laws, interventions, recommendations. *Multidisciplinary Journal of Educational Researchi*, 10(2), 158–177. https://doi.org/10.17583/remie.2020.5877.

Valdez, A., Takahashi, S., Krausen, K., Bowman, A., & Gurrola, E. (2020). Getting better at getting more equitable: Opportunities and barriers for using continuous improvement to advance educational equity. WestEd. Retrieved from www.wested.org/resource/getting-better-getting-more-equitable

Wong, J. T., Bui, N. N., Fields, D. T., & Hughes, B. S. (2023). A learning experience design approach to online professional development for teaching science through the arts: Evaluation of teacher content knowledge, self-efficacy and STEAM perceptions. *Journal of Science Teacher Education*, 34(6), 593–623.

Yu, C. (2024). Gender inequality in the age of AI: Predictions, perspectives, and policy recommendations. OSF Preprints. https://doi.org/10.31219/osf.io/5zrh9

6 Artificial Intelligence in English Literature

Close to Distant Reading

Jaydipkumar Devabhai Pandya

6.1 INTRODUCTION

Close reading – a consistent feature in literary studies – is a difficult feature to ignore, which demands a careful analysis of every detail of a single piece of text in its countless dimensions, themes and stylistics. However, the situation has completely changed with the emergence of digitalization and the advent of artificial intelligence (AI), which opened up new opportunities for both distant and focused reading. The gradual introduction of AI in literary studies may be considered a breakthrough since it broadens the scope of what we can interpret, aiming to be able to examine an almost limitless number of literary works. This also means that literature has to be accepted as a multi-faceted entity, where machine learning algorithms and natural language processing (NLP) technologies nowadays become necessary for scholars and push towards the age of computational criticism. This chapter considers the consequences of this development, providing insight into the synergies of AI with literature and illustrating the questions the new practice raises about the nature of authorship or creativity and indeed the act of interpretation.

6.2 DEFINITION OF ARTIFICIAL INTELLIGENCE IN THE CONTEXT OF LITERATURE

The intersection between technology and literature was introduced by AI which has been transforming the traditional approaches taken in literary studies. AI has enhanced both close and distant reading strategies wherein a scholar seeks and zooms in on detail patterns of the text or examines broader literary trends over the large corpus of literature. This dual capability disrupts the traditional ideas about authorship and creativity and therefore advances calls for re-thinking the concept of literary work in a world that is becoming more and more machine-based. (Underwood 72). Furthermore, the deployment of AI technologies for the analysis of literary texts raises certain ethical questions of a more fundamental nature – questions of bias, questions of authorship and questions of the meaning of a text as interpreted by a human being. Embracing these tools presents an opportunity to engage with literature untainted while emphasizing the need to embrace technological evolution which would orchestrate a new era in the discipline by integrating qualitative and quantitative lenses.

DOI: 10.1201/9781003560463-7

6.3 THE ROLE OF AI IN LITERARY STUDIES

AI technology, which has emerged in the modern world, has brought forth new opportunities across varied social and academic domains including the field of literary studies. AI has provided new resources that allow literature researchers to transcend the limitations inherent to human analysis – for example, improving the level and quality of literary text analysis and pattern recognition. This also includes methodologies of distant reading, which allow for the analysis of large-body literary works to identify genre, stylistic and theme change over time. The application of AI allows literary studies to tell not only narratologies and meta-analysis of interconnections systems within texts but also deeper concepts through works (Underwood 56).

In literary studies, AI has proven to be a game changer due to its capacity to perform distant reading where the machine is able to read and comprehend a large volume of text simultaneously. Distant reading employs a broader focus than close reading in that it employs many texts, enabling the identification of trends in literature across various genres, authors or periods. AI is helping scientists analyse vast amounts of data in short periods due to NLP and machine learning. This ability is a watershed moment in the field, broadening the scope of who can access literary data and how complex patterns may be perceived even without high computing skills (Jockers 12).

6.4 THEORETICAL BASIS AND CULTURAL IMPLICATIONS

The recent integration of AI into the literary field challenges the conventional approach to eBooks and authorship. It is easy to see how creating and analysing text shifts the focus away from the human aspect of literature as it demands the revaluation of originality. For instance, AI's Application in sentiment analysis has been utilized in the interpretation of complex metaphors in literature and it appears to be on the same level with humans. A similar case involved GPT-4 interpreting metaphors from a novel work of fiction and a jury of humans rated the AI's performance higher than that of college students, suggesting that AI has the capacity to provide different views of the literature in question (Underwood 68; Hammond 45).

In addition, the issue of AI's involvement in the process of translation has also been raised. The intention of a Dutch publishing company to deploy AI for the translation of its fiction books into English has raised the ire of translators and writers who contend that AI is incapable of grasping the greater context that is necessary for a good translation. This tendency brings the implications of the use of AI in literature into focus as it raises issues to do with the loss of fine details in the language and the authenticity of translation.

These trends point to the possibility that AI is not just a means to its critical end but quite an agent that alters the core beliefs in the field of literature. With the capacity to produce and interprete more texts, AI thus provides for a need to revisit the historical and social contexts of writing and reading literature, both with the potential to span great distances between previously unrelated ideas across space and time.

6.5 EMPIRICAL EVIDENCE AND PRACTICAL APPLICATIONS

Genre Classification: Luis Alberto Sanchez pointed out that AI may complement someone's writing style, but does not replace it. AI has indeed been proven effective at classifying, predicting and augmenting narrative arc structures and other genre influences. As seen in her studies of stories, Katherine Elkins was able to analyse how stories convey a deeper emotional meaning through the help of AI, which itself can generate a new layer of complexity in literature and serve as an additional factor in genre classification.

Topic Modelling in Historical Analysis: To make a clearer sentence, Persians used AI tools that allowed them to Leonid storm patterns within the literary works of Forough Farrokhzad which are important and interesting metrics that show different eras of literature and its transformations. This application indeed shows AI as the means of cultural and civilizational changes tracing over the history of literature.

These applications showcase how AI can improve traditional literary analysis by offering quantitative data that bolsters qualitative interpretations. By revealing patterns and trends across various periods and genres, AI aids in developing a deeper understanding of literary history and its evolution.

6.6 FROM CLOSE READING TO DISTANT READING

For decades, close reading has remained a fundamental practice in reader-response literary studies that seeks to connect a reader to the text. Thus, different scholars using this method have been able to grasp shades of meaning, figures of speech and patterns of work and put them together into an interpretation based on its details. However, while close reading warrants astonishing evidence, the practice itself is narrow, especially with regard to an extensive body of works. Didactic analysis of themes, style or structure of an entire genre or an era of literature is a possible but very exacting task without the computational assistance AI offers. This is where distant reading comes in, offering an opportunity to interpret the literature of the past on a larger scale than a narrow conception.

The concept of distant reading was first popularized by Franco Moretti who sought a perspective that did not limit scholars to the study of an individual text but encouraged scholars to focus on groups of texts (Moretti 46). This method has become more relevant with the development of AI since it has made it possible to analyse datasets that would have been too complex for manual analysis. Distant reading is not a substitute for close reading; rather it enhances it in terms of placing a particular piece of literature within the grander scheme. AI tools can trace themes, stylistic features and historical development throughout a variety of texts, which will deliver information that would not be possible from a close reading perspective.

6.7 COMPUTATIONAL METHODS IN AI-DRIVEN LITERARY STUDIES

The field of literary studies is beginning to incorporate certain technological features with the most salient ones seeming to be natural language processing and machine learning which are useful in both distant reading and even extending close reading to the next level. NLP allows machines to absorb and decode the intricacy of humankind's voice and thus makes it feasible to investigate large-scale text collection to reveal trends in discourse and in emotional and even stylistic aspects of the language. AI systems, on the other hand, can employ machine learning, which enables them to improve the precision of their analyses on the basis of the accumulated information. These approaches work well in combination and make it possible to conduct analysis at an entirely different level with unprecedented sophistication to literary studies.

For instance, NLP methods can analyse variations in a particular author focusing on specific themes and their variations over the years. Likewise, machine learning can agglomerate texts according to stylistic or genre features and this will provide some hidden connectivity in a dataset which are otherwise difficult to comprehend. AI allows scholars to construct statistical models that explain the relationship across different styles or motifs throughout history, putting everything into perspective. In this way, AI does not merely help in carrying out the analysis but alters the way literary texts are interpreted by providing means to understand literature in a more integrated manner.

6.8 DEMOCRATIZING ACCESS TO LITERARY ANALYSIS

The availability of AI as a promising and competent tool has thus almost revolutionized and expanded the horizons of numerous literary scholars and critics by allowing them to analyse literature without the need for enormous investment in time and resources. In the past, even the most thorough examinations of texts required significant funding and special knowledge, leaving the research for the most part to established scholars or wealthy institutions. This is however changing, as a large number of people, who are almost everyone involved in the education industry, be it a teacher, a student or an educator, are now able to study literary works in depth because of the use of AI-driven tools. Such barriers that have been removed broaden the scholarly community as they allow people from different categories and abilities to participate in the ongoing discourse in literature and literary criticism.

AI also enables researchers to work together across disciplines so that synergies between computer science, linguistics and literature can be achieved to address specific and complex research problems. Such an integration further enriches the study, and in this case, the study of literature and AI, as deep insights enable the formulation and answering of research questions on different scales, which would have been impossible with traditional methods. Different types of inclusivity in AI and literature research practices help to enhance the development in the area. More and more types of literary artefacts can be explored, leading to newer forms of interpretation with the addition of more cultural, historical and linguistic contexts.

6.9 IMPLICATIONS OF AI-DRIVEN ANALYSIS ON CREATIVITY AND AUTHORSHIP

As AI becomes increasingly integrated into the study of literature, it poses practical issues concerning the perception of creativity and the agent of authorship. Literary studies and the accompanying literary criticism, have always been viewed as a sphere of human activity that is positively biased, culture-oriented and politically inscribed. But the emergence of AI provokes another perspective because these devices today are able to produce interpretations themselves, making people wonder whether the AI is able to create literature at all.

It makes sense that a new recognition of AI devices raises questions concerning creativity and authorship. For example, as soon as AI starts composing poetry or takes thoughts to construct emerging ideas, the distinction between man and machine as the author of the text becomes vague, hence the very essence of being a creator becomes irrelevant! Such ethical issues and philosophical questions, concerning the changing role of AI in literary studies, are the ones which think about the technological effect in the classical notions of creativity or originality.

6.10 CHALLENGES AND LIMITATIONS OF AI IN LITERARY STUDIES

Although the use of AI in the study of literature is very beneficial, there are definitely some disadvantages. Perhaps the most worrisome is the threat of bias imbued in the very algorithms that power certain AI systems, to begin with. This is because these systems do not develop in a vacuum and are only as good as the data used to train them. For instance, if an AI model is trained on limited data, it will likely produce biased interpretations which may reinforce pre-existing biases within literature scholarship. Furthermore, while AI tends to look for trends and relations amongst the textual material, it is unable to understand the text deeply and appreciate complexities as a human would do in the act of literary criticism. This also indicates that AI should not be construed as the new standard for conducting an analysis of ceaseless and sophisticated literature but rather an additional aspect of current reading strategies (Jockers 78).

In addition, the heavy dependence of AI on quantitative information may pose a challenge in understanding the subjective and emotive aspects of literature which are often critical aspects in the interpretation of written works. Literature is not mere words organized together in book form; it captures the realities such as human thoughts, feelings and intricacies. While understanding AI can draw patterns and structures, comprehension of subtext within literary works remains a level out of reach. Therefore, the use of AI in literary studies should be approached with caution in order to achieve the desired objectives without losing the human aspect in line with progress.

6.11 LITERATURE REVIEW

In the field of literature, AI is increasingly utilized for computational text analysis, assisting in handling and comparing textual data to extract trends or patterns. As pointed out by Moretti (2005a,b), such change as, for example, the transition from close reading to distant reading that was enabled by the rise of computational techniques brought about a new perspective in literature studies. Distant reading is the use of AI to examine and analyse the blue trends that cut across the various literatures that have been produced in the past and predict the contemporary world. Nikki's writings lay down a foundation for the evolution of distant reading and for its dependence on numbers in the course of analysing literature on mass or common trends.

Other scholars have also delved into computational approaches in text analysis, building on Moretti's framework for distant reading. Jockers (2013a,b) introduced both sentiment and topic analysis in an era of literature studies never seen before, showing that sentiment and themes within a novel or several novels, and genres as a whole, can be analysed with the help of AI. Why is this important for his investigation? Jockers who conducted such an investigation emphasizes the fact that if sentiment analysis is thought of as a measure of emotional valence within texts, it can meaningfully advance one's understanding of characters, plot twists and tensions. For start, he noted that even entire genres of literature can be plot scenarios affected by trends in the particular genre, and therefore, such analysis is indicative of certain genres like sentiment analysis of gothics or sentimental fictions. These findings reinforce the promise of computational text analysis in addressing questions related to literary features that would otherwise be difficult to address.

Expert knowledge of linguistics and NLP has opened new and more complicated possibilities of textual studies. AI systems, in this regard, involve the computer's capability to encode human language and discern features such as the speaker's emotions, the context of communication and the meaning of the phrase. Underwood (2019) applied the techniques of Neural Poetry to great effect when analysing the transition from the 19th-century novel to the 21st-century one, noticing the interplay of language in certain evolutions of society. Underwood employed large amounts of text from various historically located documents and managed to establish patterns of selection and structure of discourse sequential to shifts in socio-ideological content. This research shows how the analysis of computational literary style change concerning social order is possible through computerized analyses of corpora.

For example, the rise of neural networks such as the various versions of GPT by OpenAI made it possible for machines to produce text closely approximating what one would call human language reasonably well. In a study done by Hammond (2019), the scope of AI producing creative writing was looked into, with results showing the capability of AI to formulate reasonable and coherent plots, poems and even speech so intelligent in expressing initiative as to be called human. It appears from Hammond's research that AI can actually produce magnificent works of literature, but simultaneously it brings ethical issues such as the author's rights and what is the place of man within the creative processes. AI producing customized content poses serious challenges towards the belief and hypothesis of authorship being one of the unique human attributes, if there is ever such a time when machines are able to produce literature, they will definitely change the landscape of authorship as we know it.

There are some recent studies that have looked into the impact of AI-generated literature on the process of creation itself. Davies (2020) sought to probe the issue of the collaborative authorship of humans and AI proposing that AI's role is as a creative augmenter rather than a creative author replacement. In these ways, AI solves the problem of writer's block, exploring the authors' ability to create in new times or unconventional plotline styles by generating narrative elements, such as ideas for plotlines or their style. Davies' research is about incorporating AI into literature, specifically how AI could be seen as an originating source rather than a replacement for authors.

The genre classification of AI has been developed exceptionally well by Algee-Hewitt et al. (2016) where they implemented machine learning in the stylistic analysis of 18th-century fiction. Their research proved that AI can classify novels by genre, pinpointing many subtle distinctions in style in all novels which belonged to sentiment, gothic or historical genres. The fact that it is possible to assign genres to works in a way that relies solely on textual characteristics makes it easier to address the development and the inter-relations of genres more satisfactorily. Algee-Hewitt's work demonstrates the potential of AI in genre studies by providing a genre classification system that is capable of improving the understanding of classical genre concepts.

Alongside genre categorization, AI has demonstrated its potential in thematic interpretation enabling researchers to look for themes in really large corpuses. For example, Stent et al. (2019) studied the topic of a discourse concerning British novels of the 19th century and found such subjects as gender, class and rural–urban migration. The researchers were able to discern such changes by using topic modelling, a method that aggregates words into relevant domains. Such modelling and burstiness factors allowed the researchers to identify thematic changes that are consistent with recent histories such as the representation of a social class – in this case the Industrial Revolution. The study done by Stent supports the idea about the multimedia dimension of AI and its applicability in understanding the messages which the author of the text was trying to portray.

As per Rockwell and Sinclair (2020), AI helps in the possible mass diffusion of literary analysis by addressing the entry barriers of being able to do thematic studies on a very large scale. Their work emphasizes what NLP platforms and machine learning algorithms have done, enabling even lay people devoid of advanced technical expertise to go out looking for nuts and bolts through computational analysis, which is an advancement that has done well to academia. This democratic wave has additional ramifications on the representation of voices within literary studies as it breaks boundaries within disciplines and broadens the reach of scholars and indeed practitioners within the domain.

The availability of these technologies has also made it possible to interlink disciplines such as computer science, linguistics and literary studies. Liu (2021), along with other researchers, is of the opinion that literary works cannot be fully understood in separation from one another since AI brings an interdisciplinary aspect into the field of literary criticism. The timeless dispute between quantitative and qualitative research can also be ended in this way: through the integration of computational methods with the traditional ones used in literary studies.

6.12 METHODOLOGY

The method of integrating AI language tools in bibliographic studies is a polemic which modifies both distant and close reading. With regard to close reading, AI helps in locating certain elements within a text such as styles, sequences and even patterns of languages which are likely to go unobserved. In distant reading, the application of AI in the study of large aggregates of texts makes it possible to trace changes in the basic themes, plots and typological features dominant in an extensive body of literary works. The integration of distant and close reading creates a new perspective in the study of the texts, where emphasis on the detailed study of a single aspect is integrated with the study of the work in its broad social context.

AI applications in close reading start with natural language processing (NLP), the third application technique which allows computers to read, understand and write human languages. NLP models are essential in disassembling and handling text since they assist in bridging the gaps for researchers in finding language patterns and even sentiment, syntax and tone in each piece. Close reading has witnessed the gradual incorporation of sentiment analysis, an NLP application, which works on identifying the mood accents in the text and how they change through the course of a story. For instance, it's possible to tell from the weight of the text when there was a shift from a more hopeful tone to a much darker one and this signals a change in how an author affects the audience's understanding. This makes it possible to provide an enumerative account of these elements which have been subject to qualitative analysis, thus enhancing our understanding of character development and the plot.

One of the NLP techniques applied in close reading is named entity recognition, or NER, which is a technology that detects and classifies people, organizations, locations and other relevant entities in a document. Classified in such a manner, NER entails focusing on the relationships between characters or the significance of certain locations. For example, if a particular location appears several times as the story progresses, NER allows them to be 'plotted' on a chart and analysed thematically. It can also determine the frequency of some characters in relation to others thus providing a web of relations that deepens understanding of the character. Particularly with complex works of literature with a large number of characters, this technique is valuable because it makes it possible to follow the relations and activities that the characters engage in with each other in an organized manner.

Outside the scope of NLP, many brokers use sophisticated algorithms and machine learning models to analyse the content of such elements as style. The classification of stylometry seeks to obtain the models and their stylolite levels by applying units of sentence length, punctuation marks, one or more adjectives and a selective preference for specific nouns. Based on these variables, it is possible to train a machine learning model that can tell the difference between writers or styles of writing. This sense of writing adds new information to studies of authorship but style classification also makes it possible to examine stylistic changes of each author throughout different pieces of literature or across different genres. For instance, it is possible to train a model to see the style of the work of a 19th-century author and see the content of the work in italics and evoke how the language was used to portray fear and mystery.

Topic modelling can be considered a further machine learning method that provides opportunities for both close and distant reading. It is a field within the unsupervised classification of models, where words that occur together frequently within a text are absorbed into a topic. In close reading, topic modelling can be of assistance in identifying themes or patterns that may be used to guide the reader in comprehending the aims of the author or that of the surrounding social environment within which the piece was written. For example, in a novel that engages with social class, topic modelling could indicate word groupings associated with wealth – which would imply that the story focuses on such classes or wealth hierarchies and labourers etc. Such a thematic focus has quite added to the scope of a theme not only making it an attribute of literary pieces but also as an attribute associated with the author's vision.

Even in this phase where even distant reading is left for interpretation by the readers, AI proves so useful in separating the aspects, which helps detect the message hidden in the bulk of documents. The core idea of clustering consists of assigning documents with similarities to a particular group which is used in distant reading. It is clear that if there is a thematic cluster, there should also be genre attributes typical for a certain time period. For instance, clustering can show how the idea of war or love is treated in different kinds of texts at different times. This strategy enables a large-scale understanding of literature that enables particular pieces of artwork to fit in with the flow of history and shows the patterns that shape a particular style of writing.

Network analysis is yet another robust approach as it allows one to see relationships between texts, authors or even themes connections in the form of graphs. It allows researchers to see how certain literary works are related, in terms of ideas, stylistics or orientations and so on. For example, network analysis can demonstrate the frequency with which certain authors or themes have appropriated and subsequently related to other authors, significant detail or dominance. This method is particularly beneficial for the exploration of intertextuality as the sites and relations of the transfer and transformation of ideas and style are analysed for different authors and epochs. In addition, network analysis helps to understand how literature influences each other like literary genre evolution by providing schemes of the evolution of literary modes through the chronotypes.

AI also makes it possible to classify texts into particular genres which is very important in distant reading for the reason that knowing structural and stylistic elements allows a researcher to place the text in a category. With specific genres in mind, one can also train models to classify texts and determine trends within a particular genre by employing machine learning techniques. For instance, genre classification might be able to determine that detective stories are likely to contain shorter sentences than other types of fiction in order to increase suspense, while a love story might not be as dull and bland because it relies on descriptive wording to bring out some passions. When such patterns are recognized, genre classification models help to achieve a more general level of order in the study of genre traditions revealing the ways writers adhere to or violate the principles of literature. This method not only advances classical genre theory but also opens up a field for the investigation of the trends in the development of genres and their transformations.

Stylometric analysis, which is a technique concerned with examining the writing style of an individual based on statistical implements, is another development of AI in order to conduct the research of the literature that is concerned with both distant and close reading. Stylometry is the study of measuring the features of the author's writing such as the structure and complexity of the sentences and paragraphs, the frequency of particular and general words and the use of punctuation marks and it is possible to indicate with a great accuracy the very authors of the texts. This case will be useful in such situations as promoting claims of authorship since there is ambiguity or anonymity. Moreover, stylometry can generalize the studies of writers in the sense of how their style changes regarding the sequence skirt affecting their writer's biography. The studies of the development of an author's style may also include the analysis of the author's earlier works and the latest ones and this will help highlight the complexity of the sentence structures that the author uses which demonstrates the growth of the author in terms of stylistic.

Analysing sentiment, first defined in the context of close reading, may also be employed in distant reading. When it is extended to large databases, be it a collection of texts or other sources, it assists in charting empires of emotions over genres, particular authors or even over epochs, and this points to the evolution of cultures. For instance, an evaluation of narratives produced in the 19th century would perhaps show that in the preceding decades, there was a growing tendency to portray social issues in a more optimistic light due to the effects of industrialization. This of course additional interpretation of sentiment analysis as goes beyond the standard definition adds a dimension to the interpretation of how cultures have evolved and responded to literature over a period of time. It also extends existing frameworks to comprehend the changes behind the literary phenomena as well as the patterns beneath them.

Natural language processing has also helped facilitate distant reading in the form of part of speech tagging and syntactic parsing. These techniques examine the textual features from the grammatical perspectives focusing on the functional patterns authors use to elicit tone, mood and style of the written language. For example, part of speech tagging classifies words into set grammatical categories in order to study their power of use in some context. It helps explain the dominance of certain linguistic constructions by specific genres or authors. For instance, the application of the tagging process may suggest that poets use a large number of adjectives in presenting photos and pictures while journalists use nouns and verbs to deliver their points. Such a process, as parsing, which finds the order of words in a sentence, gives us information about language complexity and how authors design and use different sentence structures to achieve certain purposes. This analysis has special applications when assessing changes in literary styles within cross-genre or cross-time periods from a more stylistic evolution point of view.

Embeddings, or word embeddings, are a subset of neural networks that transform words into vectors in n-dimensional space. They use the space surrounding words within the context to build up an understanding throughout the work. Specifically, when it comes to literary themes, they could tell which terms or phrases recurred in close context and over broader contexts. For example, in the case of word embeddings, terms may be culture, ideology, social class and literature. Or looking through a war-themed narrative, there might be high associated usage of the terms: valour and sacrifice. Apart from that – in analytics, word embeddings can actually sit well within sentiment analysis across times, demonstrating how societal meanings and associations towards certain terms evolve and/or shift. Through this process, experts and researchers come across how language, and in this case, literature, undergoes transformations throughout the ages, and most importantly transforming and correlating with cultural and historical contexts.

Distant reading has some advantages thanks to visualization tools that assist in the interpretation of bulky materials by displaying them in an easily understandable form. Visualizations like frequency charts, word clouds and heatmaps transform numerical data into a visual format. For example, a heat map depicting the frequency of certain words in distinct parts of a novel illustrates how the character of an author's theme is presented at the climax of the plot. Visualizations are also the most illustrative interfaces for understanding the other facets of the analysis conducted by AI and for correlating the figures with the text.

Through the use of AI technologies, the tasks of literature research can be solved in additional broad ways, each with its own distinctive analytical functions with respect to the text. Also, AI makes it possible for more detailed reading and contextual understanding at the same time leading to a more unified approach to the discipline of literary studies. Detailing through text, theme and genre is now enhanced through AI centres around natural language processing, machine learning and visual interpretation to amplify existing approaches. Our time has marked advancements in concepts and methodologies as both domains are blended to bring new angles of understanding literature without losing the nuances and intricacies of literary research.

6.13 THE ROLE OF AI IN CLOSE READING

For the past several years, scholars have increasingly turned to AI tools to assist in close reading, significantly altering its traditional methodologies, modes and methods. Close reading is a reading practice that engages with texts closely, and, in our days, it is augmented by means of AI featuring deep learning methodologies. Might be, advanced algorithms enable computers to detect more general patterns of language, hierarchies of themes and structural rhetorical devices which compose its deeper levels not restrained to the communication of meaning alone. Thus, with additional AI-contextualized commentary to human readings, it is not only possible, but also necessary, to address literature in such depth and detail that the manual readings of the texts could not achieve because of the additional amount of effort required. In addition, AI-based instruments provide an opportunity to focus on several texts at once, which enhances cross-examinations and, thus, improves the contextual interpretative quality. To summarize, AI completes the close reading stage, but also provides new alternatives in literature studies by integrating conventional and contemporary approaches to literary scholarship.

The world of digital technology is both parallel and interrelated with the world of language, and in this connection, it is noteworthy the role that AI tools have lately acquired within textual studies of English literature. These tools allow theorists to perform thorough searches of literary texts, which were tortuous, if not impossible before. As an example, distant reading methods afforded by AI facilitate such an investigation of considerable literary corpora, which fosters the possibility of discerning these patterns and trends on a macro scale among whole periods rather than individual works. This is particularly interesting in light of how literature is a vehicle of cultural processes, in this case of the animation about Paddington, mobility and cultural identity become one of the key issues (Buesa, 2024). In addition, the combination of sentiment interpretation and its visualization tools also adds to the pedanticism of analysis – the interpretation of emotions and themes embedded in the texts – angles which are usually distorted and rarely portrayed in isolation (Artem Suslov et al., 2023). Thus, it may be concluded that AI shifts the paradigm of literary studies, which has always been the conflict between close reading and distant reading.

6.14 ENHANCING READER ENGAGEMENT THROUGH AI

The progression of AI in the current landscape of English literature not only engages the reader in unique ways from what has been previously experienced but takes reader engagement to a whole new level. By making use of the functionalities offered by AI, it becomes possible for content and reading experiences to be personalized leading to stronger relationships between readers and texts. For example, recommendation systems relying on sophisticated algorithms assess the users' reading patterns and offer pieces of work that are likely to engage the users more and would be ideal pieces that the users' interest in them.

Moreover, it changes how consumers passively go through the text as the constructs the reader sees or interacts with matters. Additionally, AI-constructed interactive stories can be altered by the reader making the reading experience more engaging as the reader is now a part of the story rather than just consuming it. This trend shows how authors are now telling novel stories in new ways and thus literary emotions are felt uncommonly due to the fact that the literature world is opened up to the readers making their enjoyment of literature and emotional investment unprecedented. However, as AI integration into interactions between authors and readers increases, the advancements in potentiality will be enormous leaving the literary world highly engaged (Jones, 2013).

6.15 LIMITATIONS OF AI IN CLOSE READING

In principle AI has never backed off the close reading of the texts in practice due to the weaknesses which align the tools which would enable its wide use to the means available. There have been developments in pattern recognition, AI computers for example can readily scan the data provided, and pull out the relevant pieces of text or phrases but of the visual or essentialism properties of art. The cause in this case with AI is that it cannot engage with language on such a contextual level as it is needed, as literary text draws on culture, emotions, history and other aspects which requires much more than sheer statistics.

AI as of now cannot view them through a subjective lens as human readers do which is very important in closely reading the text in order to recognize the inherent motives and themes within symbols, the text itself and the narrator. This gap between the robotic nature of the code holder's disembodiment to poetry makes literature a less appealing substance for many as it brings a lot of complexities and ambiguity to the reader as well as unnecessary detail, which makes its purpose of enhancing the perception of the literature more or less ineffective (Underwood, 2019-02-14). So, while being wonderfully useful, AI is never going to step in for human readers and their critics as culturally and socially inflected variability is invaluable to readers.

6.16 THE IMPACT OF AI ON DISTANT READING

The field of distant reading is being enhanced by AI, thereby changing the methods of scholars with vast amounts of text in literary study. The application of algorithms and machine learning allows researchers to derive insights from a substantial literary corpus that no amount of close reading could uncover. This development not only saves time in terms of carrying out literary analysis but also broadens the reach of literary studies to people who belong to many different professions and do not have to devote themselves to manual labour for long periods of time. The Annual Reports of Education, Health and Sport, for example, argue the importance of AI over other means of accessing and interpreting literary texts demonstrating a global trend where AI is implemented in the field of literature. Others worry that the interpretation of literature in this way, the way literature is interpreted in a machine, will lead to an impoverished understanding of the texts. This is not the case, as AI only aids in providing a better understanding of the text, which is essential for further research (V. Romanchuk & Podgorna, 2013).

Although it's been around in Northern America as an academic term for too long, Franco Moretti proposes a new literary direction called distant reading which contrasts in strong opposition to the concept of 'close reading' which sequentially focuses on a singular text. Instead, distant reading advocates for the study of large bodies of literature, which may reveal the presence of various sociologies by using computational approaches and quantitative analytics to trace and analyse representatives across genres, time periods or geographic spaces. This method enables a better and deeper comprehension of the literary movement or the major theme or other cultural phenomena which would remain ungrazed whilst studying separate works in focus.

In this sense and through harmonizing the data from a multitude of texts, scientists are in a position to add how literature connects with social processes, stylistic changes or even with the language itself (Krouse & O'Callaghan, 2020). In this way, the distant sphere not only broadens the range of analysis desktop lie to people of literature but also quite controversially for some of them even jeopardizes spines to Edwardian perceptions of who is the author of the text and what its content and textual values in times of digital era.

6.17 AI TECHNIQUES FOR LARGE-SCALE TEXT ANALYSIS

The use of AI approaches in the analysis of a large body of texts focused on a single subject has completely altered the sphere of research in literature with the cut-off being supervised access to diverse texts. With the introduction of NLP, researchers have been able to use tools such as topic mapping and authors' identification, tasks that were previously tedious and time-consuming. This method increases the level of analysis but also frees literary scholars from the constraints of close reading and enables them to appreciate literature on a wider range. This is moments, it can include anything from mic – the ground and patterns to macro to units and relations focusing on structural integration within sociocultural settings. Furthermore, Anna Gallego Cuiñas et al. inhabit new prospects for the employment of computational techniques in the Humanities explaining that such integration between AI and literary studies adds depth to critical debates in the field and opens new areas for further development of neuroscientific approaches in the field.

6.18 CASE STUDIES OF AI IN DISTANT READING

AI, in its usage in distant reading, has provided a new understanding towards the study of literature, which has revolutionized textual studies. As shown in several case studies, AI has been proven to conduct reading with its ability to deal with large text collections, thereby cataloguing the dominant themes and the themes that are particular to cultural settings which would be lost in the process of reading. Most importantly, AI takes a broad view of language, incorporating the fact that languages and words are used in different ways and in different places.

In this regard, AI could be especially useful in analysing the relations between sociocultural contexts and literature by looking at patterns, such as the prevalence of certain linguistic styles. Based on what has been covered before, AI's only advantage in literary studies is that it can add a lot of material to existing literature and at the same time break traditional teaching methods. Therefore, as distant reading becomes more dependent on AI, it allows for a shift from traditional literary criticisms making a transition towards a more advanced and entertaining literary critique.

6.19 CHALLENGES AND ETHICAL CONSIDERATIONS IN DISTANT READING

The application of distant reading in the study of literature comes with its fair share of problems and moral dilemmas. Since AI easily makes distant reading possible, scholars have to at some point deal with algorithmic prejudice and the multilayered Ness of cultures' images and their histories. For example, AI may be able to analyse massive amounts of data, identifying various trends, but the viewpoints incorporated in the literary works may be reduced to reductive frameworks due to AI's limitation to certain broad narratives. This worry is similar to those in other parts of the academia that are concerned with the significance of technology in the study of humanities as well as the more practical issues of data rights and the risk of drowning individual voices in an ever-expanding sea of group narratives.

Besides, as Udo-Okon & Akpan (2024) observe, the question of how best to utilize AI without violating ethical norms in its use is not of literary significance only but extends to other domains that require caution with respect to privacy and responsibility. Several issues, however, raise cause for concern – and while distant reading presents immense opportunities for literary scholarship, these concerns have to be responded to in a flexible way to avoid compromising academic standards and inclusion.

6.20 COMPARATIVE ANALYSIS: CLOSE READING VS. DISTANT READING

The contrast between close reading and distant reading also opens a rich discourse on methods of interpretation in literary Criticism. Close reading entails, looking at the language, structure and minor details of a text; it, therefore, creates opportunities for the analysis of the view and style a text possesses, as demonstrated in the detailed critical assessment of D.H. Lawrence's short stories and paying attention to romance subtitle as well as gendered complexities (Kulakevych, 2021).

On the Other hand, Distant reading seeks to broaden the horizon. It assists in scanning and analysing huge volumes of texts making use of computers and bringing out findings which might be missed by reading one text. One of the contexts where this approach is effective is the modern debate of AI in the sphere of Privacy advocate's critique of the abuse of power by AI systems where history and literature offer perspective on the issues (Edmond et al., 2024). In the end, both approaches complement each other and make literature and the relationship between literature and new development more vivid.

6.21 METHODOLOGICAL DIFFERENCES

Considering the link whose convergence, i.e., collating AI with English literature and AI systems, one needs to bear in mind different approaches that lay the foundation of close reading and distant reading. Deep reading concisely means the gradual and slow looking into parts of the text, trying to seek reasons through the words, forms and structures. A method of its kind which gives particular weight to personal interpretation and always, or nearly always, integrates the context of the reader into the study.

On the other hand, distant reading uses machines to read hundreds of thousands, if not millions and allows historians to see trends that couldn't be noticed with the naked eye. This way of thinking advocates a more objective perspective. They employ statistics and a systematic approach to the study of literature to explain the emergence and evolution of the literary movement or theme in a certain period of time. Even though both methods can provide important information they complement each other which brings us to the question of the place of human interpretation and reading and the place of metrics in literary studies today (Jones, 2013).

6.22 THE ROLE OF HUMAN INTERPRETATION

As technology develops, so does its capability to analyse literature, but the role of human interpretation remains at the centre of its analysis. Due to this, readers thanks to their experience, emotion and worldview are able to engage with the text in a way that AI will find impossible. This immeasurable level of engagement is critical, especially in executing tasks such as analysing theme, tone and context which would often draw on factors outside the text that AI models may ignore.

The capacity to stay and make connections, as well as implicate meaning, elevates literature to a shared experience of language and thought. Thus, although AI may help to pinpoint patterns through distant reading, it is ultimately the readers' interpretations of literature that give it life and that make understanding and appreciation go beyond words and phrases found in the patterns of digital analysis (Jones, 2013). This relationship reinforces the argument that human involvement in literary studies is necessary.

6.23 FUTURE DIRECTIONS FOR READING PRACTICES

Although reading as an activity has not changed completely in human history, it is now transforming in unprecedented ways as we approach the age where AI will dominate entire societies. The inclusion of artificial or intelligent systems does not only increase the ease of accessibility but also increases the chance of engagement through targeted content serving, as evidenced by the astounding findings of AI application in the classroom (Suntharalingam, 2024). This phenomenon advances the logistics of reading and makes it possible for readers to transition from purely reading to embracing more sophisticated ways of interacting with technology. This, however, raises the need to look critically at what such shifts would entail and particularly how these respond to the ratio of literary text to non-literary text that informs the way audiences interpret a reader's expectations (Bajohr, 2023).

Handling these challenges will demand continuous engagement among the educators, the technological developers and the readers to improve the AI technologies in ways that embrace variety in literature. In this regard, the relationship between future reading practices and texts does not only aim at improving comprehension but also how we relate to texts.

6.24 ANALYSIS/DISCUSSION

The increasing use of AI in literary criticism is both in need of investigation as well as presents possibilities which are exciting. On the one hand, AI makes remote reading possible in which researchers scan millions or billions of words using computational means to discover trends and motifs that would be lost in conventional reading. This change not only expands the scope of text in relation to the analysis but also creates new challenges of what it means to interpret and what role humans' creativity has in understanding literature.

With the rapid adoption of AI technologies in research processes, more questions about authorship as well as the genuineness of literary analysis are raised with their growing capacity of producing perspectives from the data. More so, dependency on algorithms in some ways could homogenize interpretative frameworks to the extent that abundant voices and/ or perspectives are lost because of what trends AI analysis dictates. All in all, the intersection of AI and literary scholarship shall have to address these complex issues of change and continuity, technological advancement and human ideas and creativity.

As literary analysis is being taken to another level, AI offers new techniques that can help avoid the usual approaches to interpretation. Since the available databases are enormous, AI can discern themes and motifs that might escape even the most thorough human investigators.

This is because machine learning is capable of analysing linguistic and stylistic features of different texts in order to connect the dots between seemingly unrelated pieces and authors and thus broaden the contexts of textual comprehension. In thus addressing macro and micro in literary study, this development encourages a more advanced concept of literature, where literature distant reading may be automatically performed by AI systems while the actual reading of the text is more analytically intensive than before. Nevertheless, there are important issues that should be considered when discussing AI for interpretation, such as the presence of bias in algorithms or the potential erosion of qualitative aspects of literature (Jones, 2013). It is important in making interpretations of literature in this era that such progress is not allowed to override critical assumptions.

AI is incorporated into English literature as an ethical weapon with dual effects. On one hand, AI can revolutionize the interpretation of literature through approaches like distant reading whereby scholars identify patterns, themes and meaning of literature in its wider context which makes integration of AI into literature studies more beneficial. On the other hand, AI integration in other fields like healthcare (Chaurasia et al., 2024) or even into the speciality of ophthalmology paints an ethical picture which is likely to be similar in literature too. Concerns regarding AI generated data or artificial-intelligence-based algorithms and the possible scope of misinterpretation are some of the concerns which cast sigma over the authenticity of literary works. Besides, the over-dependence and erosion of intellectual critical dependence on AI may render unnecessary the role of man in the interpretation of literature. Ideally, great care must be taken in striking the balance between the benefits of the A.I and the moral ramifications of its use as only then can literary studies be able to forge ahead in the new digital age without any problems.

6.25 CONCLUSION

To conclude, the relationship between AI and English literature shows a paradigm shift in the approach towards literary studies in the future as it combines distant reading and close reading. This august domain enables academics to employ machine learning and data mining techniques which in turn reveal some trends and patterns that were previously hidden in large textual databases. Using AI in literary criticism allows researchers to address topics, examine stylistic features and trace changes in ideologies on a much larger scale and in precise detail. Furthermore, this connection helps to develop the understanding of storytelling and interaction with readers, thus making it more sophisticated than before. As we adopt these advancements in technology, we have to ensure a critical stance that the need for human interpretation remains intact. Therefore, as much as AI is an effective avenue in investigating literature, literature that is all-encompassing and informed by human emotion should be the compass (Bohr & Memarzadeh, 2020).

The use of AI in English Literature expands the horizons of both interpretation and composition in a considerable way. It showcases how AI can be an asset in the domain of distant reading, allowing historians to undertake voluminous texts in a matter of minutes and be able to unearth trends and patterns which would otherwise have been buried deep and unnoticed in close reading approaches. This development, however, not only shifts the techniques used in literary studies but also brings into question the conventional concepts of the creation of the Bard, for instance, as such can now be accomplished by algorithms.

Moreover, such arise also relating to the presence of AI in the literary world, particularly about the question of originality and the risk of creative work being reduced to a marketable product. Accordingly, this necessitates understanding what AI can offer but being conscious of what it cannot do, and an active engagement with that which AI continues to impact the literary space (Jones, 2013).

The last few decades have seen a huge proliferation of material new and old which is all available with just a mouse click. With all this data in mind, a literary study's scope has on the one hand greatly expanded but has also created a lack of focus on analysing the material. AI in literary studies has been proposed as a solution to both these problems and it seems to be working out so far. AI listens, helps enhance and definitely multiplies the manpower available to make an extensive reading and study of a work or even an entire corpus of work. Both close and distant readings can benefit from the new perspectives, even as they will still remain grounded and comprehensive.

But the problem AI solves in this case might be a whole different can of worms in other domains. AI poses a challenge of creativity and what it means to be an author, then consider the ethical dilemma AI raises, questions of data privacy and bias that need regulation at every level. As many questions remain unanswered for how AI will supplement traditional literary scholarship in widening the scope of readership it is clear that the reception of AI will be enhanced with collaboration across all disciplines while minimizing overload and expectations that come with critical and editorial work.

6.25.1 The Evolving Relationship between AI and Literature

With the continual development of AI, perhaps its most controversial perihelion is emerging in literature. Such integration into literary studies must be emphasized for it restructures the domain significantly. For instance, AI technologies can assist in perfecting distant reading techniques, whereby massive amounts of data and textual content can be scanned by scholars to unveil tenets that may elude human evaluation.

This dynamic shift is considerably validated by the existence of operatives who assist in carrying out various aspects of literary evaluation, hence facilitating the researchers to overcome the intricacies of data precision and integrity of AI systems. This progression of technology is an interesting one to pursue as it attempts to explain how these technologies can be used in altering the practices of evaluation in literary studies which can broaden understanding and lead to dilemmas of authenticity and authorship of artistic works (Bohni Nielsen et al., 2025). Still looking into the position of AI in literature in the future, it is requisite to contemplate its prospects as well as constraints so as to have an even-handed view.

6.25.2 Implications for Future Literary Studies

With the rapid growth of technology, literary studies are likely to undergo radical changes. Has the question of authorship been sufficiently dealt with? How are characters, plot development and narrative construction made? Such questions can be answered in social development of the society. Gō-chan is one of the first 'AI driven characters' designed to do two things, interact with users and grow and develop through interaction with users (Lamerichs, 2019). This makes us rethink the meaning of creativity and originality in writing.

More specifically, many countries have witnessed the deployment of hybrid environments in education and so the approach towards the pedagogical dimension of literary texts needs to change as well and educators will need to prepare students for a reality where a health-emotional dimension will be part of learning, especially in the post-pandemic world. Hence such a development makes changes in future literary studies equally adopting these changes and makes them relevant.

AI's impact can also be emphasized in analysing the numerous ways in which people relate to books. AI has the potential to approach reading in an innovative manner and incorporate new tools aimed at comprehensive narrative or theme interpretation. Of course, it is important to state that even when such technologies are successfully developed, AI will always lack the most important human features that are necessary for understanding literature, which are the emotional and empathetic elements.

As to the active capabilities of books, recipients can rely on AI-powered algorithms to consider thematic parallels, though the intricate layers that readers often respond to cannot be provided with only AI. Summing up, AI in literature shouldn't be seen as a threat but rather as a complimentary addition to the way books are written or read. All the aforementioned factors must be appreciated in understanding the historical circumstances that the new technological environment created for our reading relationships with literature.

BIBLIOGRAPHY

Journal Articles

Algee-Hewitt, M., & McGurl, M. (2016). Between canon and corpus: Six perspectives on 18th-century fiction. *Literary Lab Pamphlet, 13*, 1–20.

Bajohr, H. (2023). *Artificial and post-artificial texts: On machine learning and the reading expectations towards literary and non-literary writing (No. 007). Basel Media Culture and Cultural Techniques Working Papers*. University of Basel. https://doi.org/10.5451/bmcct.2023.007

Buesa, A. (2024). Facing apocalypse: Climate mobilities and the cinematic child. *Sociology Lens*, *37*(1), 105–115. https://doi.org/10.1111/johs.12438

Chaurasia, A., Curry, G., Zhao, Y., Dawoodbhoy, F. M., Green, J., Vaninetti, M., Shah, N., & Greer, O. (2024). Use of artificial intelligence in obstetric and gynaecological diagnostics: A protocol for a systematic review and meta-analysis. *BMJ Open, 14*(5), e082287. https://doi.org/10.1136/bmjopen-2023-082287

Davies, J. (2020). Collaborative creativity: Examining human-machine partnerships in creative writing. *Computers and Composition, 55*, 102–114.

Edmond, J., & Ketzan, E., & Yakupova, V. (2024). A "tyrannical societal something": Drawing lessons for 21st century "privacy-protecting" AI technology from long 19th-century literature. *Interdisciplinary Science Reviews*, 49(2), 189–207.

Ganascia, J. G. (2021). Artificial intelligence in literature: Creativity and authorship in the digital age. *AI & Society, 36*(3), 615–628.

Hammond, A. (2019). Generating narratives with neural networks: Literary creativity and AI. *Digital Scholarship in the Humanities, 34*(2), 233–248.

Jockers, M. L. (2013a). Macroanalysis and sentiment analysis in literary corpuses. *Journal of Digital Humanities, 5*(1), 1–19.

Kulakevych, L. M. (2021). Interpretation peculiarities of the Sleeping Beauty motif in D. H. Lawrence's short story "Princess". *Alfred Nobel University Journal of Philology, 2*(22). https://doi.org/10.32342/2523-4463-2021-2-22-3

Lamerichs, N. (2019). Characters of the future: Machine learning, data, and personality. *IMAGE*, 29, 100–119. doi:10.5167/uzh-182372

Liu, A. (2021). The AI-aided literary analysis: From theory to practice. *Digital Humanities Quarterly, 15*(4), 67–79.

Moretti, F. (2005a). Graphs, maps, trees: Abstract models for literary history. *New Left Review, 24*, 67–93.

Rockwell, G., & Sinclair, S. (2020). Democratizing access to textual analysis: AI in literary studies. *Journal of Digital Scholarship, 29*(3), Article 45, 445–462.

Romanchuk, A. P., & Podgorna, V. V. (2013). Pattern of spontaneous breathing and development of physical qualities of elementary school pupils. *Journal of Health Sciences*, 3(5), 7–14. https://doi.org/10.5281/zenodo.16142.

Stent, M., et al. (2019). Thematic analysis across British novels: A computational approach. *Journal of Cultural Analytics, 8*(4), 19–30.

Suntharalingam, H. (2024). Enhancing digital learning outcomes through the application of artificial intelligence: A comprehensive review. *International Journal of Innovative Science and Research Technology, 9*(4), 718–727. https://doi.org/10.38124/ijisrt/IJISRT24APR530

Suslov, A. V. (2023). Origins of Japanese archaic poetry: Japanese long song (nagauta/chōka). *Japanese Studies in Russia*, 3, 17–33. https://doi.org/10.55105/2500-2872-2023-3-17-33

Udo-Okon, T. N., & Akpan, E. E. (2024). The challenges of artificial intelligence in library management systems. *Intercontinental Academic Journal of Library and Information Science, 6*(1), 97.

Underwood, T. (2019). Machine learning and literary history: Predicting style and sentiment. *Modern Language Quarterly, 80*(1), 1–29.

Books

Bradley, A. (2016). *Critical tools for digital humanities*. Columbia University Press.

Eve, M. P. (2019). *Close reading with computers: Textual analysis for literary critics*. Stanford University Press.

Jockers, M. L. (2013b). *Macroanalysis: Digital methods and literary history*. University of Illinois Press.

Jones, Steven E.. (2013). *The Emergence of the Digital Humanities*. Routledge. DOI: 10.4324/9780203093085

Krouse, T., & O'Callaghan, T. F. (2020). *Introducing English Studies*. Bloomsbury Academic. https://doi.org/10.5040/9781350055438

Moretti, F. (2005b). *Graphs, maps, trees: Abstract models for a literary history*. Verso Books.

Renkema, M. (2024a). *The future of work for academics in the age of artificial intelligence*. Routledge.

Tuchman, G., & Fortin, N. E. (2012). *Edging women out*. Routledge.

Venturini, T., & Ricci, D. (2020). *Artificial intelligence in humanities: Machine learning and digital scholarship*. Cambridge University Press.

Edited Books

Adam, B. (Ed.). (2016). *Critical tools for AI in humanities*. Taylor & Francis.

Al-Khatib, A. W. (2024). Generative AI and the digital supply chain. In T. Venturini & D. Ricci (Eds.), *AI for industry and academia* (pp. xx–xx). Cambridge University Press.

Bohni Nielsen, S., Mazzeo Rinaldi, F., & Petersson, G. J. (2025). Evaluation in the era of artificial intelligence. In S. B. Nielsen, F. M. Rinaldi, G. J. Petersson (Eds.), *Artificial intelligence and evaluation: Emerging technologies and their implications for evaluation* (pp. 1–12). Routledge. https://doi.org/10.4324/9781003512493-1

Bohr, A., & Memarzadeh, K. (2020). The rise of artificial intelligence in healthcare applications. In A. Bohr & K. Memarzadeh (Eds.), *Artificial intelligence in healthcare* (pp. 25–60). Elsevier. https://doi.org/10.1016/B978-0-12-818438-7.00002-2

Konhäuser, K. (2024). AI for the analysis of textual data in the humanities. In M. Renkema (Ed.), *The AI revolution in humanities* (pp. xx–xx). Routledge.

Pleshko, E., & Podgorna, V. (Eds.). (2013). *Annual reports of education, health, and sport*. Global Publications.

Renkema, M. (2024b). AI's impact on literary scholarship. In A. Gendron (Ed.), *AI and academic research* (pp. xx–xx). Sage Publications.

Conference Proceedings and Symposiums

Chopra, R. (2024). Identifying bulls and bears: AI in stock market prediction. *Proceedings of the 2024 International Symposium on Data Science*, Hong Kong, China, 22–37.

Fu, M. (2024). Speech recognition in interactive AI systems. *Proceedings of the 2024 IEEE International Conference on Education Technology*, Shanghai, China, 19–30.

Patent Reference

Gendron, Y. (2024). *System and method for machine learning in textual data analysis* (Patent No. US8798659B2).

7 Transforming Classrooms
AI-Driven Innovations in School Education

Siddhi Sood and Poonam Angurwala

7.1 INTRODUCTION

In contemporary times, there is a discernible trend in society toward a pervasive and extensive process of technological integration across several domains, involving but not limited to politics, economics, education, and social interactions. The phenomenon of adjusting to novel technological interaction communities has given rise to a diverse range of technologies known as "virtual assistants." These virtual assistants employ computer algorithms to simulate human intelligence, thereby creating an illusion of interpersonal interaction for users (Doleck et al. 2017). This notion is referred to as "artificial intelligence" (Ocaña et al., 2019). In educational backgrounds, there has been a notable focus on the integration of artificial intelligence (AI), particularly on the potential for enhanced communication between students and teachers using virtual information assistants (Martinez et al., 2023; Petko, Prasse, & Cantieni, 2018) (Figure 7.1 showing an interaction among AI, education, and society). These assistants simulate human-like responses, thereby facilitating a more natural conversational experience. Furthermore, as these tools are utilized, they possess the ability to learn and intuitively recognize user interactions.

However, within the present global framework of the technological revolution, certain human attributes, such as creativity, the capacity to generate novel concepts, and the ability to improvise and adapt continuously, remain beyond the capabilities of AI. The current direction of AI and computational science is progressively emphasizing human-centered AI, which considers the characteristics and circumstances of persons to address any biases that may exist in the implementation and execution of supporting algorithms (Yang et al., 2021).

The connection between AI and education encompasses three key domains, as outlined by UNESCO (2019). These domains include the utilization of AI tools in the classroom to facilitate learning, the acquisition of knowledge about AI technologies and approaches, and the provision of opportunities for all individuals to comprehend the prospective ramifications of AI on human existence. Furthermore, it is posited that AI possesses the capacity to tackle prevailing educational obstacles. Specifically, it can foster inventive pedagogical approaches that adhere to the core tenets of inclusivity and equity, hence expediting advancements in education (Martínez et al., 2023).

DOI: 10.1201/9781003560463-8

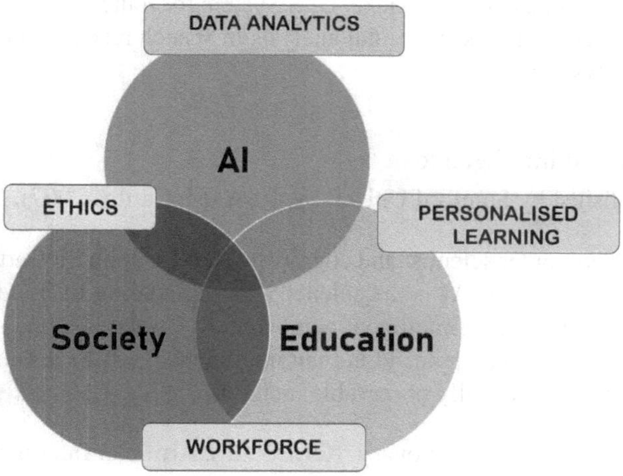

FIGURE 7.1 Conceptual framework diagram representing the interconnections between AI, education, and society.

7.1.1 Theoretical Background

In the modern, dynamic, and ever-transforming world, changes in technology are taking place rapidly while focusing on "the future is today and not tomorrow" (Pavlova, 2021). In this context, there is a pressing need for AI technology in education that goes beyond the instruction of fundamental reading (Srinivasan & Murthy, 2021) and numeracy skills to instill moral, ethical, and human views.

Aiming that education with technology is one of the strong foundations that must help build 21st-century skills, the Union Cabinet of India put forth the first education policy of the 21st century on July 29, 2020, entitled National Education Policy 2020. In this advanced and modern era, education is not only focusing on what to learn but rather on how to learn. Children must not become rote learners, but analytical people (Ministry of Human Resource Development, Government of India, 2020). Thus, education must help build critical thinking, higher-order cognitive abilities, creativity, and ethics (Liang & Lin, 2015; Aggarwal, 2021).

AI is a term that came into existence in the 1950s. There is no evidence of this field in the age of philosophers since Aristotle. The first system called "intelligence" was a Turing test: a simulated game proposed by the British mathematician Alan Turing. He suggested creating a simulation to test a listener's capacity to distinguish conversations with machines from those with humans; if the system is unable to do so, AI would be acknowledged (Russel & Norvig, 2010). This term was first used by one of the founding fathers of AI, John McCarthy, after discussions at the seminal event named Dartmouth College, USA in 1956.

This conference was a pivotal event in the history of AI which set out the vision for AI that included the formation of intelligent devices that could think, learn, and interface like human beings (Haenlein & Kaplan, 2019). The emerging technology: AI systems interact with us directly or indirectly in our environments. Think of Google

Assistant, Google Translate, Siri, Alexa, GPS System for route directions, image classification, face recognition, sentimental analysis in speech recognition, etc. are some examples of AI systems.

7.1.1.1 Artificial Intelligence

In a report published by Stanford University, USA (McCarthy, 2007),

> an American computer scientist and cognitive scientist, John McCarthy, defined Artificial Intelligence as AI is the science and engineering of building intelligent machines, especially intelligent computers. It is related to the similar task of using computers to understand human intelligence, but AI does not have to confine itself to biologically observable methods.

In the Oxford Learner's Dictionary (OLD, n.d.), Artificial Intelligence is defined as "the study and development of computer systems that can copy intelligent human behavior." To put it simply, the design and advancement of computer systems proficient of accomplishing tasks often rely on human cognitive abilities, including visual perception, recognition of speech, constructing decisions, and translating language. IBM (Stryker & Kavlakoglu, 2024) defined Artificial intelligence as powers enabling computers and machines to simulate the problem-solving and decision-making potentials of the human mind. Additionally, it includes the subjects of machine learning (ML) and deep learning (DL), which are frequently cited in the research works related to the AI.

As outlined by Niti Aayog in the National Strategy for AI, "AI represents the ability of machines to execute cognitive processes like thinking, perceiving, learning, problem-solving, and decision-making. Although originally developed to mimic human intelligence, AI has evolved far beyond its initial purpose". Advances in data acquisition, processing capabilities, and computational power have enabled intelligent systems to perform a wide range of functions, enhance connectivity, and drive productivity improvements (Lei et al., 2020).

As per the EuropeanAI leadership, "the path for an integrated vision AI is not a well-defined technology and no universally agreed definition exists. It is rather a cover term for techniques associated with data analysis and pattern recognition. AI is not a new technology, having existed since the 1950s. While some markets, sectors and individual businesses are more advanced than others, AI is still at a relatively early stage of development, so that the range of potential applications, and the quality of most existing applications, have ample margins left for further development and improvement."

In a nutshell, AI represents a form of intelligence, a branch of technology, and a field of study. It involves the design and expansion of computer systems—both hardware and software—capable of executing tasks that typically need human intelligence. Encompassing a wide array of domains and applications, AI is anticipated to influence virtually every field in the future. At its essence, AI emphases on creating machines and algorithms that can execute computational tasks requiring functions similar to those of the human brain (Logg, Minson, & Moore, 2019).

7.1.1.2 Categorization of AI

Different authors introduce different categories of AI on various basis. They all are different in terms of their philosophies, purpose, goals, and method of operations. AI can be categorized in a number of varied ways, but it can be thought-provoking to do so accurately and without specific criteria. Kyunghwa Chung et al. (2022) categorize AI in three types based on the degree of technological advancement: Weak AI, Strong AI and Superintelligence.

Hintze (2016) defines AI by distinguishing the different types of AI based on their application as reactive machines, limited memory, theory of mind, and self-awareness. Reactive machines are the most fundamental sort of AI, and self-awareness is the most crucial type and acts close and even better to the human mind.

Rouse (2020) divides AI based on its applications as Weak AI and Strong AI. Weak AI can perform only a specific form of tasks and Strong AI can perform much more complex tasks with high decision-making ability. Shopping bot and Self Car Driving are the examples, respectively.

In an article in the magazine *Forbes* AI was classified into two categories based on the AI-enabled machines that show the likeness of human and another based on the technology used (Joshi, 2019). He claimed seven types of AI combining the four types given by Hintze in 2016 and three types given by Kyunghwa Chung in 2022, under one roof.

AI can be categorized on the basis of techniques used, such as rule-based AI that follows predefined rules to make decisions; ML that learns from data and updates itself; DL as the subset of ML that uses Artificial Neural Networks; and reinforcement learning that learns by interacting with their environment through feedback (Khan, 2021). Some (Stewart et al., 2020) differentiate ML and DL as a component of AI or subdisciplines of AI, and some (Madakam et al., 2022) even define ML and DL as sister technological footprints (Figure 7.2).

In general, we can conclude that some researchers differentiate AI on what they can do and how they can do it. So, we categorize AI into two parts based on the capabilities and techniques used as shown in Figure 7.1.

7.1.1.3 Domains of AI

AI trains itself using previously fed data. Data can be in numbers, text, speech, or visual. So, we can also categorize AI based on the data fed into it. Each classified part is called the domain of AI. Intel (2019) defines the domains of AI into three parts: Statistical Data Domain, Natural Language Data Domain, and Computer vision Data Domain. A detailed concept of domains of artificial intelligence is tabulated in Table 7.1.

7.1.1.4 Artificial Intelligence in Educational Context

The educational environment is always growing and adapting to effectively address the educational needs of future generations (Halili, 2019). Given that all of these changes coincide with technological advancements, people may claim that the rate of change in education may be the quickest ever (Harrison, 2020). Even if these innovations come in a variety of shapes, AI is a common element in all of them. AI

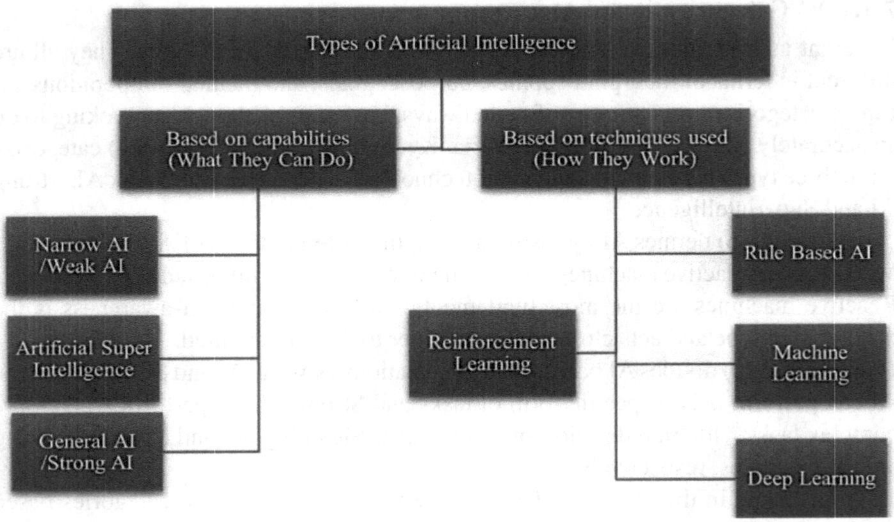

FIGURE 7.2 Types of artificial intelligence.

TABLE 7.1
Domains of Artificial Intelligence

Type of Data	Domain	Usage	Examples
Numeric or alphanumeric	Statistical Data Domain	Used for predictions based on previously found patterns	Weather predictions, students' score predictions
Text or speech	Natural Language Processing Domain	Used for conversational chatbots	AI Tutors, SIRI, ALEXA
Images and videos	Computer Vision Domain	Used for image and videos by pixel or resolution theory	Artify, photo identification in gallery, Snapchat, thumbprint

has long been a strange field (Allen, 1998), where researchers have extremely varied perspectives on fundamental issues.

The idea of AI, as mentioned above, must be identified to incorporate any tool or computer that performs human activity. These devices are made by humans to automate daily activities so that they may do more work in a shorter amount of time. It is defined by Popenici and Kerr (2017) as "computer systems that can do operations like learning, adapting, synthesizing, self-correcting, and utilizing data for intricate processing tasks." Educational artificial intelligence (EAI) is more precisely used in education to enable tailored and computerized responses and assistance in the learning sphere (Song & Wang, 2020). A booming new field of study that combines AI with education has emerged in response to the recent increase in demand for education, and this has led to an extension of the body of knowledge on EAI.

7.1.1.5 AI in Education and AI as intervention***

The AI-based teaching–learning model offers a promising approach to enhance the teaching and learning process in programming education wherein such model promotes collaborative and active learning and helps students identify areas for improvement to facilitate the understanding and evaluation of programming concepts among students (Huddar et al., 2023). However, personalized learning, improved student engagement, increased efficiency in administrative tasks, and the challenges that comprise the need for proper training and support for teachers, apprehensions about data-privacy and security, and the probability for spreading the digital-divide. AI has the potential to transform education and contribute to the advancement of human development, but it requires careful planning and consideration of ethical and social implications (Alam et al., 2022). So, in this world of cutting-edge technologies where teachers are yet to comprehend completely the potential or utility of AI in education, they need support to be well-organized and actively effective in their work exercise for which it can be predicted that this support could have been provided by the AI (Chounta et al., 2022) (Figure 7.3).

Increased usage of AI and robotics in Indian classrooms can help instructors focus on more complicated tasks and make better use of their expertise, although the role of teachers is irreplaceable (Kumar et al., 2022). Therefore, it is significant to address the advantages and difficulties connected with the usage of AI and emerging technologies in many sociocultural contexts, while also considering their impact on marginalized student populations (Salas-Pilco et al., 2022). By looking into the application of AI in the area of schooling, improved theoretical foundations can create a promising path for future research partnerships between educators and AI developers (Zhai et al., 2021). The concepts of AI even when applied in a game-based environment to clarify the math concepts can pave a way to enhance student engagement and understanding (Lee & Lee, 2017; Greenwald, 2021). Therefore, it can be concluded that AI has the potential to revolutionize education by introducing changes to teaching methods (Ashraf, 2020).

There is a need to prepare the generation for future workforce and introducing new courses at school, college, university, and even MOOCs will help them in becoming AI ready and make them able to develop AI applications for future use (Florea & Radu, 2019). According to studies on boosting academic performance, the utilization of the online life-crafting intervention has shown positive outcomes in terms of

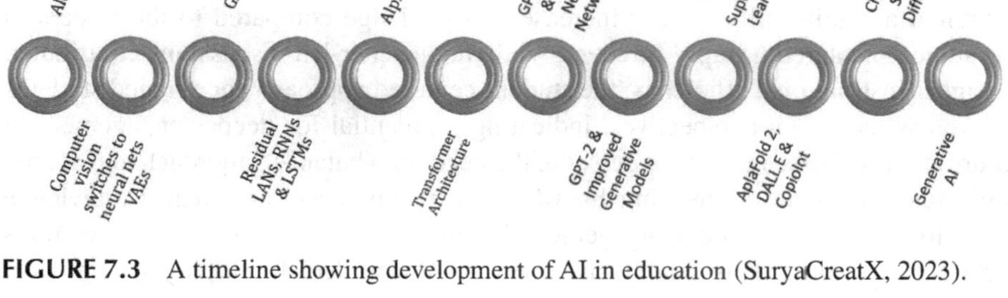

FIGURE 7.3 A timeline showing development of AI in education (SuryaCreatX, 2023).

student retention rates and academic performance (Dekker et al., 2018). Moreover, the relevance of actively supporting students who are engaged in online learning processes has increased along with the growth of online education (Zacharis, 2016).

In the present article, AI in education discusses to the integration of AI technologies into education systems to augment experiences in the learning, teaching methods, and organizational tasks, while AI as intervention refers to the strategic use of AI technologies to address the needs of 21st-century classrooms and students who are digital natives for teaching any subject rather than teaching AI as a subject. There is certain statistical evidence which reveals the extensive use of AI in education and its integration in classrooms as interventions. According to a recent survey by the Digital Education Council, around 86 percent of students admitted that they use AI platforms in their studies (Kelly, 2024). Furthermore, the survey also revealed that on average students reported using 2.1 AI tools for their coursework. Among these, ChatGPT emerged as the most frequently used, with 66 percent of respondents mentioning it, while Grammarly and Microsoft Copilot were each utilized by 25 percent. In addition to this, it was also found that AI is being integrated in education frequently for searching information (69 percent), checking grammar (42 percent), summarizing documents (33 percent), paraphrasing (28 percent), and creating the first draft of any document (24 percent) (Digital Education Council, 2024).

AI has the potential to transform education, by providing students with more personalized and effective instruction (Grover & Pea, 2015). Technology for the right use or the positive use can be the right direction to convey the deeper potential of technology to education (Joseph & Thomas, 2020). To achieve the goals of India's AI strategy: #AIforALL, the Govt. of India has given guidelines to the education sector in various reports like NEP2020, which aimed at the transformation of the teaching–learning process (Report, 2019). Central Board of Secondary Education (CBSE) has recently announced the integration of AI with other school subjects with an aim to make school students "AI Aware" or "AI Ready."

7.2 TRANSFORMING CLASSROOMS WITH AI IN SCHOOL EDUCATION

Goldschmidt et al. (2015) draws on research conducted with Bavarian tenth-grade students (N = 583) on plant genetic engineering within an out-of-school laboratory setting. The study employed three distinct instructional methods: teacher-guided discussions, student-centered investigations, and traditional text-based learning, aiming to explore their effects on cognitive achievement, cognitive load, and instructional efficiency. The findings revealed that students in the teacher-guided group exhibited a significantly greater increase in knowledge compared to their peers in the student-centered group. Moreover, while the teacher-led discussions contributed to higher instructional efficiency, the student-centered approach encouraged students to express their own perspectives, indicating a potential for deeper engagement in future studies. These insights underscore the need for a balanced approach to teaching socio-scientific issues, suggesting that while teacher guidance can enhance knowledge acquisition, fostering student engagement through student-centered methods warrants further exploration. This dual focus opens pathways for developing pedagogical practices that both promote cognitive achievement and empower students as active contributors to societal discussions.

Geethalakshmi (2021) studied how AI is transforming classroom environments, examining its potential to redefine the roles of educators and students while fostering a more innovative and technology-driven approach to learning. The COVID-19 pandemic has underscored the urgent need for tech-based educational solutions. The widespread closure of schools and universities revealed the vulnerabilities of conventional systems and accelerated the adoption of digital tools. These developments highlight the growing importance of AI in shaping the future of education. As AI continues to evolve, it is poised to bring transformative changes to education, reshaping traditional methods of teaching and learning. Despite these advancements, teachers will remain central to education. However, their roles will adapt to complement AI-driven tools and systems. The classroom of the future will be vastly different, embracing technology to create dynamic, personalized, and efficient learning experiences. In this redefined educational paradigm, both students and teachers will take on new roles, leveraging AI to enhance collaboration and outcomes (Čaplinskas, 1998). Some of the anticipated changes include the integration of AI-driven tools such as voice assistants and speech recognition software to support teaching. Facial recognition technology could streamline attendance processes, while AI-enhanced systems may revolutionize traditional evaluation methods, ensuring accuracy and reducing errors. Even examination supervision could be modernized with AI-enabled surveillance tools, offering real-time monitoring to uphold integrity.

Alam et al. (2022) examined how incorporating AI into classrooms can enhance teaching and learning by assisting both educators and students through the use of robotic technologies and sensors in teacher-led and student-focused settings. AI technologies promote inclusive, equitable, and high-quality education while ensuring global access to lifelong learning opportunities. Advanced AI systems are capable of detecting students' gestures, moods, and comfort levels during lessons, even analyzing facial expressions and postures to identify challenges they face. These systems can suggest adjustments to the lesson accordingly. Although students find AI tools to be valuable additions to traditional classroom learning, they continue to appreciate teacher–student interactions and the mentorship provided by educators. The integration of AI in education has transformed teaching approaches, redefining the teacher's role as a facilitator and significantly improving the overall quality of education worldwide.

Chiu et al. (2023) investigated the impact of teacher support on student motivation in learning through AI technologies within a K–12 educational context. The study involved 123 Grade 10 students, and the experimental nature of the study focused on AI-based chatbots. The research contributes to understanding how educational technologies can be better integrated into instructional practices to enhance learning experiences. Utilizing self-determination theory, the study demonstrates that both teacher support and student expertise significantly influence intrinsic motivation and competence when engaging with AI tools. The findings underscored the potential of AI integration to transform classroom practices but often lack integration with pedagogical theories in K–12 education. Teacher support plays a crucial role in enhancing student motivation and competence in learning with AI tools. Results indicated that intrinsic motivation to learn with chatbots was affected by both teacher support and student self-regulation skills. Teacher support effectively satisfied the need for relatedness among students but was less successful in promoting their autonomy.

Makwana and Band (2024) studied the impact of digitalization and AI in CBSE and state board schools in Nagpur City through mixed-method approach, which revealed several significant findings regarding the exploration of AI strategies. It highlighted that AI technologies enhance adapted learning experiences by personalizing/individualizing the educational content to individual student needs through customizable learning platforms, which analyze learning patterns and deliver real-time feedback, thereby improving student engagement and learning efficiency. Educators reported improved teaching effectiveness as AI tools automated administrative tasks, letting them to focus more on differentiated instruction and skill development. Additionally, the integration of AI and digital tools increased student engagement through interactive and immersive technologies, while also broadening access to educational resources, particularly in remote areas where qualified teachers may be scarce. The study emphasized that AI could help mitigate educational inequities by providing personalized and remote learning opportunities, especially for students in underserved areas. However, it also noted variability in the implementation and effectiveness of AI strategies between CBSE and state board schools, suggesting the need for tailored approaches to maximize benefits. Furthermore, the findings underscored the importance of ongoing professional enhancement for educators who have the potential of utilizing AI tools effectively, ensuring that they can fully leverage the capabilities of AI in enhancing instructional practices. Overall, the exploration of AI strategies demonstrated their transformative potential in improving educational experiences and outcomes while promoting equity in access to quality education.

Kotsis (2024) discussed the potential applications of AI in primary education science teaching, highlighting its ability to enhance personalized learning experiences and improve student engagement and academic performance. It employs a comprehensive analysis of existing literature on AI applications, revealing that AI can adapt to individual student needs, facilitate hands-on learning, and streamline administrative tasks for teachers. The findings indicate improved student engagement and performance, while also noting concerns about AI's lack of personalization and emotional intelligence. Ultimately, the chapter explores various AI tools, including virtual laboratories and simulations, which provide hands-on learning experiences in science education. Additionally, it discusses AI tools for lesson planning and delivery that analyze student data to personalize lesson plans and suggest appropriate interventions. These tools aim to enhance student engagement and support individualized learning needs. The authors suggest that integrating AI fosters a more student-centered environment, allowing students to take an active role in their education, while teachers can focus on facilitating and supporting learning.

7.3 AI-DRIVEN INNOVATIONS

A conceptual framework for computational thinking as a pedagogical device stated that by combining techniques like data analysis, which generates new information from a thorough study of enormous volumes of data and insightful analysis, AI technology offers valuable tools for treatments (Wing, 2006; Chhabra et al.,

2023). Similarly, the application of robots in the field of mathematics can significantly enhance the learning process by fostering problem-solving skills (ML et al., 2019). Furthermore, perception of teachers and students toward integration of AI in the school curriculum revealed that AI can be an interesting subject and it should be taught as a compulsory subject (Soesanto, Dirgantoro, & Priyanti, 2022; Mandal & Mete, 2023, Digital Education Council, 2024). The act or fact of interposing the AI tools in the teaching–learning process is known as AI intervention in education. AI intervention can majorly be taken into two perspectives: one is a teacher-controlled AI intervention and the other is a student-focused AI intervention.

7.3.1 Teacher-Controlled AI Intervention

In this paradigm, AI serves as a tool to support and enhance the teacher's role in the educational process. The focus is on teacher-directed learning, where AI systems are employed to design, implement, and manage instructional strategies (Salas-Pilco, 2019). Educators maintain control over how AI is integrated into the classroom, using it to automate administrative tasks, create adaptive learning pathways, and monitor student progress. This can also be named as the AI-directed paradigm (Ouyang & Jiao, 2021). The following are the key features of this paradigm:

- *AI as a Knowledge Modeler:* AI is used to represent and structure knowledge, enabling the delivery of personalized content that aligns with curriculum standards.
- *Cognitive Learning Support:* AI-driven systems such as intelligent tutoring platforms and adaptive testing modules provide personalized/individualized feedback, thus supporting teachers to recognize areas where students require supportive add-ons.
- *Teacher Decision-Making:* Educators use insights from AI analytics to make informed decisions about instructional practices and interventions.

In this model, the learner's role is largely guided by the teacher, with AI acting as an intermediary to streamline educational delivery and reinforce learning objectives. This paradigm ensures consistency and standardization while maximizing the teacher's ability to cater to diverse learning needs.

7.3.2 Student-Focused AI Intervention

The student-focused paradigm shifts the emphasis from teacher-led instruction to learner autonomy and empowerment. Here, AI acts as a collaborator or enabler, placing students at the center of the learning experience. This approach fosters active engagement, critical thinking, and personalized exploration, allowing learners to take greater control of their educational journey. This paradigm can also be named as AI-supported paradigm (Ouyang & Jiao, 2021). The major features of this paradigm can be listed as follows:

- *AI as a Learning Partner:* AI systems work alongside students, supporting problem-solving, experimentation, and discovery. Examples include interactive virtual labs, gamified learning platforms, and AI-driven chatbots for instant feedback.

- *Empowered Learner Agency:* Students take an active role in setting learning goals, exploring content, and reflecting on their progress. AI adapts dynamically to individual preferences, offering customized learning experiences.
- *Personalized and Adaptive Learning:* Advanced AI algorithms analyze student data to create unique learning pathways, enabling self-paced and interest-driven exploration.

In this paradigm, educators take on the role of facilitators and mentors, guiding students in leveraging AI effectively while encouraging independent thinking and creativity. The focus is on developing lifelong learning skills, with AI serving as a tool to amplify student engagement and achievement.

7.3.3 AI + X or X + AI Paradigm for Education

Under the guidance of Ministry of Human Resource Development (MHRD) of India, CBSE has taken a "twin initiative" in this regard. CBSE introduced AI as an optional subject from ninth std. from session 2019 to 2020 as mentioned in *Artificial Education Curriculum- Class 9 Facilitator Handbook, 2019-20* [CBSE and Intel (n.d.)]. In 2021, CBSE has introduced two different teaching perspectives at the middle school level (sixth class to eighth class) for four fundamental subjects i.e. Math, Language, Science, and Social Science, which will be taught through AI interventions. The CBSE has introduced AI as a multidisciplinary, integrated pedagogical approach under the "X+AI paradigm," aligning with the recommendations of the National Education Policy 2020 (NEP 2020). The following steps outline how AI can be incorporated as a teaching tool or integrated into various subjects:

Identify a topic within the subject where the teacher already employs a specific teaching methodology, referred to as "X."
Conduct research to discover AI concepts that share a conceptual connection with the subject and topic.
Integrate the identified AI concepts "AI" with "X" during lesson planning.

Moreover, a document released by CBSE, 2022 titled AI integration with other subjects also raised a point that all CBSE instructors shall be acquainted with the existing AI expertise and apply it to make studying their topics more successful and student-centered. As the process of education mainly involves teachers and students, it has also been revealed that transforming the teaching and learning process always has two perceptions – one is on teachers and the other focuses on students.

7.4 CHALLENGES IN IMPLEMENTING AI IN INDIAN EDUCATION SYSTEM

In order to propagate AI in education, proper assistance from educational institutions in terms of money, time, and recognition, as well as the creation of AI solutions that are compatible with instructors' current work styles is highly required (Gupta & Bhaskar, 2020). Incorporating AI into classrooms can enhance the effectiveness of teaching and learning by assisting both teachers and students through the use of robotic technology and sensors (Sadiku et al., 2021), which suggests that AI intervention in the education system can be a disruptive technological intervention that makes education holistic and accessible to everyone (Roy, 2020). Teachers express interest in exploring peer-teaching and approaches based on games in the classroom to introduce AI. The challenges reported by teachers include gaps in policy dissemination, educational infrastructure, teaching methods, content delivery, and cultural influences within the framework of Indian schools (Anu et al., 2019). However, there are several challenges must be overcome before AI can be broadly adopted in education, including the necessity for further research to develop and implement AI systems effectively and the importance of addressing associated ethical issues (Grover & Pea, 2015). Additionally, despite the many advantages offered, integrating these technologies into teaching procedures poses a serious ethical and pedagogical test for instructors due to the complexity of their design and implementation, which calls for additional investigation from the field of education research (Martínez et al., 2023). Hence, it is important to reserve and use AI resources, enhance instructors' and students' digital literacy, employ AI to support teaching and learning, and enhance students' creativity and academic achievement (Wang et al., 2023). Nevertheless, it is imperative to establish a curriculum that incorporates sophisticated robotic kits, given the increasing prevalence of AI (Weng, 2022).

The use of AI in education creates both obstacles and possibilities that change the landscape of contemporary learning environments. Many instructors lack the expertise required to properly incorporate AI technologies into their teaching practices, emphasizing the importance of continual professional development. Ethical issues emerge as a key difficulty, underlining the significance of guaranteeing Fairness, Accountability, Transparency, and Ethical (FATE) behaviors in AI applications in educational contexts (Chounta et al., 2022). Building confidence in these technologies requires balancing the potential benefits of AI with ethical issues. Therefore, integrating AI into education offers significant benefits but also raises ethical concerns. While using AI in its ethical ways, it definitely includes a personalized learning environment thus supporting inclusive classrooms, extended support system for administrative tasks and early intervention (Yates & Tawil, 2019). Research on ethical AI in education highlights the importance of using AI responsibly to support student success (Akgun & Greenhow, 2022). On the other hand, there are also some unethical practices which have been witnessed by the educationists which include privacy breach, academic dishonesty, and plagiarism. A survey discovered that 78 percent of parents believe that the practice of AI generative tools in school assignments promotes cheating (Adamson, 2023). This signifies the ethical challenges surrounding AI in education.

Another barrier to successful AI intervention adoption is a lack of infrastructure and resources as successful adoption of AI treatments frequently necessitates solid technological support and consistent internet connectivity (Chaudhri et al., 2014). Schools in resource-constrained areas may have difficulties in adopting and sustaining AI-driven instructional innovations. This possibility offers the ability to cater to a wide range of learning preferences and talents, resulting in more effective educational results. Another prospect is increased teaching efficiency, as AI interventions can aid instructors in automating repetitive chores, giving up time for more individualized interactions with students (Alam et al., 2022). This efficiency has the potential to improve the overall quality of the teaching and learning experience. AI-driven activities provide students with opportunity to learn important abilities such as computational thinking, problem-solving, and teamwork (Atmatzidou & Demetriadis, 2016; Doleck et al. 2017; Luo, Liu, & Luo, 2019; Yilmaz & Koc, 2021). These abilities are critical for success in today's fast-changing digital world. Furthermore, the incorporation of AI opens the door to new teaching approaches, such as intelligent tutoring systems and game-based learning, which make education more interesting and participatory (Chaudhri, 2014; Leonard, 2016). AI-enabled global learning networks connect students with peers all around the world, facilitating cross-cultural understanding and collaborative learning experiences. Overall, navigating hurdles and exploring possibilities is critical for realizing the full potential of AI in education.

Despite these obstacles, using AI in education presents significant prospects. Personalized learning experiences stand out as a big benefit because they allow AI to personalize instructional content to particular student requirements and learning styles (Huddar et al., 2023). This possibility offers the ability to cater to a wide range of learning preferences and talents, resulting in more effective educational results.

7.5 CONCLUSION

Schools play a vital and responsible role in creating meaningful learning experiences for students, guiding them toward success. It is essential to direct education toward fostering innovative skills and practical applications that contribute to a resilient and sustainable future with AI. In order to flourish education, AI has contributed through several means. Platforms like DreamBox and Carnegie Learning use AI to provide real-time feedback, adaptive learning modules, curriculum, and professional learning solutions. ChatGPT or platforms like Squirrel AI in China are acting as virtually personal tutors, while there is Gradescope which pulls AI to modernize grading processes. Moreover, when it comes to AI as intervention or integrated technology, several examples can be listed out. Like Microsoft's Seeing AI app supports visually impaired students by describing their surroundings in real time, AI-powered text simplification tools like Lexile and Affectiva's emotion AI analyzes facial expressions to gauge classroom engagement. Additionally, students must be encouraged to critically assess the benefits and challenges of AI and develop the ability to discern what is ethical and acceptable in its application. AI is becoming increasingly prominent in various aspects of our personal and professional lives. It is essential to focus on equipping ourselves and our students to navigate and thrive in the future (Artificial Intelligence Curriculum, 2019–20). In the existing system, AI in education enhances personalization, automates tasks, and supports accessibility through tools like adaptive learning and virtual tutors. Future AI promises the powered AR/VR, holistic student analytics, equity in education, AI-driven pedagogical models, and lifelong learning platforms. The shift from progressive improvements to transformative changes requires advancements in ethical AI, global reach, and innovative pedagogies.

REFERENCES

Adamson, D. (2023, Apr 16). *Survey results highlight parents' thoughts on AI writing, like ChatGPT, in education*. Turnitin. Retrieved on 15 October, 2024 from www.turnitin.com/blog/survey-results-highlight-parents-thoughts-on-ai-writing-in-education

Aggarwal, R. (2021, November 17). NEP 2020 and NSDP 2015 can cultivate 21st century skills including critical thin (R. Roy, Editor). Retrieved from https://government.economictimes.indiatimes.com/news/digital-india/nep-2020-and-nsdp-2015-can-cultivate-21st-century-skills-including-critical-thinking-and-digital-literacy-msde-secretary/87752126

Akgun, S., & Greenhow, C. (2022). Artificial intelligence in education: Addressing ethical challenges in K-12 settings. *AI and Ethics*, 2, 431–440. https://doi.org/10.1007/s43681-021-00096-7

Alam, A., Hasan, M., & Raza, M. (2022). Impact of artificial intelligence (AI) on education: Changing paradigms and approaches. *Towards Excellence*, 14(1). https://doi.org/10.13140/RG.2.2.15581.28646/1

Allen, J. F. (1998). AI growing up: The changes and opportunities. *AI Magazine*, 19(4), 13–23. https://doi.org/10.1609/aimag.v19i4.1422

Ashraf, C. (2020). Artificial intelligence and the rights to assembly and association. *Journal of Cyber Policy*, 5(2), 163–179. https://doi.org/10.1080/23738871.2020.1778760

Atmatzidou, S., & Demetriadis, S. (2016). Advancing students' computational thinking skills through educational robotics: A study on age and gender relevant differences. *Robotics and Autonomous Systems*, 75, 661–670. https://doi.org/10.1016/j.robot.2015.10.008

Čaplinskas, A. (1998). AI paradigms. *Journal of Intelligent Manufacturing*, 9, 493–502. Available at: https://doi.org/10.1023/A:1008880017722

CBSE and Intel (n.d.) *Artificial Education Curriculum - Class 9 Facilitator Handbook for 2019-20*. Central Board of Secondary Education, New Delhi. Retrieved on 3 September, 2024 from https://cbseacademic.nic.in/web_material/Curriculum20/AI_Curriculum_Handbook.pdf

Chaudhri, V., Heymans, S., Overholtzer, A., Spaulding, A., & Wessel, M. (2014). Large-Scale Analogical Reasoning. *Proceedings of the AAAI Conference on Artificial Intelligence*, 28(1). https://doi.org/10.1609/aaai.v28i1.8752

Chhabra, G., Kumar, S., Gupta, S., & Nagpal, P. (2023). Artificial Intelligence for Understanding Human Behavior and Psychology. In *Artificial Intelligence to Analyze Psychophysical and Human Lifestyle,* pp. 15–28. Springer, Singapore. https://doi.org/10.1007/978-981-99-3039-5_2

Chiu, T. K. F., Moorhouse, B. L., Chai, C. S., & Ismailov, M. (2023). Teacher support and student motivation to learn with Artificial Intelligence (AI) based chatbot. *Interactive Learning Environments*, 32(7), 3240–3256. https://doi.org/10.1080/10494820.2023.2172044

Chounta, I. A., Bardone, E., Raudsep, A., & Pedaste, M. (2022). Exploring teachers' perceptions of artificial intelligence as a tool to support their practice in Estonian K-12 education. *International Journal of Artificial Intelligence in Education*, 32(3), 725–755. https://doi.org/10.1007/s40593-021-00243-5

Chung, K., Le, J.D., Thaichon, P., & Quach, S. (2022). Types of Artificial Intelligence (AI) in Marketing Management. In Park Thaichon and Sara Quach (Eds.) *Artificial Intelligence for Marketing Management* (pp. 28–39). Routledge. https://doi.org/10.4324/9781003280392

Dekker, I., De Jong, E. M., Schippers, M. C., De Bruijn-Smolders, M., Alexiou, A., & Giesbers, B. (2018). Optimizing students' mental health and academic performance: AI-enhanced life crafting. *Frontiers in Psychology*, 11, 1063. https://doi.org/10.3389/fpsyg.2020.01063

Digital Education Council. (2024). *The Digital Education Council global AI student survey*. Retrieved from www.digitaleducationcouncil.com/post/digital-education-council-global-ai-student-survey-2024

Doleck, T., Bazelais, P., Lemay, D. J., Saxena, A., & Basnet, R. B. (2017). Algorithmic thinking, cooperativity, creativity, critical thinking, and problem solving: Exploring the relationship between computational thinking skills and academic performance. *Journal of Computers in Education*, 4(4), 355–369. https://doi.org/10.1007/s40692-017-0090-9

Florea, A. M., & Radu, S. (2019, May). Artificial Intelligence and Education. In *2019 22nd International Conference on Control Systems and Computer Science (CSCS)* (pp. 381–382). IEEE. https://doi.org/10.1109/CSCS.2019.00069

Geethalakshmi, R. (2021). Artificial intelligence as a tool to education in changing the classroom environment. *International Journal of Creative Research Thoughts (IJCRT)*, 9(2), 3046–3050. Retrieved on November 30, 2024 from https://ijcrt.org/papers/IJCRT2102367.pdf

Goldschmidt, M., Scharfenberg, F. J., & Bogner, F. X. (2015). Instructional efficiency of different discussion approaches in an outreach laboratory: Teacher-guided versus student-centered. *The Journal of Educational Research*, 109(1), 27–36. https://doi.org/10.1080/00220671.2014.917601

Government of India, Ministry of Human Resource Development. (2020). *National Education Policy 2020,* pp. 3. Retrieved on 20 Aug, 2024 from https://ncert.nic.in/pdf/nep//NEP_2020.pdf

Greenwald, E., Leitner, M., & Wang, N. (2021). Learning Artificial Intelligence: Insights into How Youth Encounter and Build Understanding of AI Concepts. *Proceedings of the AAAI Conference on Artificial Intelligence*, 35(17), 15526–15533. https://doi.org/10.1609/aaai.v35i17.17828

Grover, S., & Pea, R. (2015). Using artificial intelligence to learn and teach: A review of the field. *Educational Psychologist*, 50(4), 247–266.

Gupta, K. P., & Bhaskar, P. (2020). Inhibiting and motivating factors influencing teachers' adoption of AI-based teaching and learning solutions: Prioritization using analytic hierarchy process. *Journal of Information Technology Education: Research*, 19, 693–723. https://doi.org/10.28945/4640

Haenlein, M., & Kaplan, A. (2019). A brief history of artificial intelligence: On the past, present, and future of artificial intelligence, 61(4), 5–14. https://doi.org/10.1177/0008125619864925

Halili, S. H. (2019). Technological advancements in education 4.0. *The Online Journal of Distance Education and e-Learning*, 7(1), 63–69.

Harrison, N. (2020). Patterns of participation in higher education for care-experienced students in England: Why has there not been more progress? *Studies in Higher Education*, 45(9), 1986–2000. https://doi.org/10.1080/03075079.2019.1582014

Hintze, A. (2016, November 14). Understanding the Four Types of AI, from Reactive Robots to Self-Aware Beings. *The Conversation.* https://theconversation.com/understanding-the-four-types-of-ai-from-reactive-robots-to-self-aware-beings-67616

Huddar, M. R., Kharade, G. K., & Kamat, K. R. (2023). Designing of AI-based teaching-learning model for revitalizing education. In International Conference on the Future Global Business and Technology, Kolhapur (91–98). Retrieved from www.researchgate.net/publication/370561052_Designing_of_AI-Based_Teaching-Learning_Model_for_Revitalizing_Education

Joseph, G. V. & Thomas, K. A. (2020). Moderating effect of social media usage on technology barriers to technology adoption by teachers. *International Journal of Advanced Science and Technology*, 29(3), 5504–5512.

Joshi, N. (2019, June 19). 7 Types of Artificial Intelligence. *Forbes*. Retrieved on September 15, 2024 from www.forbes.com/sites/cognitiveworld/2019/06/19/7-types-of-artificial-intelligence/

Kelly, R. (2024, 28 August). *Survey: 86% of Students Already Use AI in Their Studies.* Digital Education Council. Campus Technology News, 1105 Media, Woodland Hills, CA. Retrieved on 10 October, 2024 from https://campustechnology.com/articles/2024/08/28/survey-86-of-students-already-use-ai-in-their-studies.aspx

Khan, H. (2021, September). Types of AI: Different Types of Artificial Intelligence Systems. *Fossguru.com.* Retrieved from www.researchgate.net/publication/355021812_Types_of_AI_Different_Types_of_Artificial_Intelligence_Systems_fossgurucomtypes-of-ai-different-types-of-artificial-intelligence-systems

Kotsis, K. T. (2024). Integrating ChatGPT into the inquiry-based science curriculum for primary education. *European Journal of Education and Pedagogy*, 5(6), 28–34. https://doi.org/10.24018/ejedu.2024.5.6.891

Kumar, S., Prabu, M., Selvan, A., Kalayiarasan, G., Ramnath, R., & Kumar, S. N. (2022). Exploring artificial intelligence and robotics in Indian education: Potential for teacher substitution? *International Journal of Emerging Knowledge Studies*, 1(1), 14–19.

Lee, I., & Lee, J. (2017). The effects of artificial intelligence-based intelligent tutoring systems on students' learning outcomes in mathematics: A meta-analysis. *Educational Technology Research and Development*, 65(4), 1425–1447.

Lei, H., Chiu, M. M., Li, F., Wang, X., & Geng, Y. J. (2020). Computational thinking and academic achievement: A meta-analysis among students. *Children and Youth Services Review*, 118, 105439. https://doi.org/10.1016/j.childyouth.2020.105439

Leonard, J., Buss, A., Gamboa, R., Mitchell, M., Fashola, O. S., Hubert, T., & Almughyirah, S. (2016). Using robotics and game design to enhance children's self-efficacy, STEM attitudes, and computational thinking skills. Journal of Science Education and Technology, 25, 860–876. https://doi.org/10.1007/s10956-016-9628-2

Liang, C., & Lin, W.-S. (2015). The interplay of creativity, imagination, personality traits, and academic performance. *Imagination Cognition and Personality*, 34(3), 270–290. https://doi.org/10.1177/0276236614568638

Logg, J. M., Minson, J. A., & Moore, D. A. (2019). Algorithm appreciation: People prefer algorithmic to human judgment. *Organizational Behavior and Human Decision Processes*, 151, 90–103. https://doi. org/10.1016/j.obhdp.2018.12.005

Luo, H., Liu, J., & Luo, Y. (2019). The necessary mental literacy in the era of artificial intelligence: Computational thinking. *Modern Educational Technology*, 6, 26–33.

Madakam, S., Uchiya, T., Mark, S., & Lurie, Y. (2022). Artificial intelligence, machine learning and deep learning (Literature: Review and metrics). *Asia-Pacific Journal of Management Research and Innovation*, 18(1–2), 7–23. https://doi.org/10.1177/2319510X221136682

Makwana, P., & Band, G. (2024). Enhancing learning through digitalization and AI: A study of CBSE and state board schools in Nagpur City. *Nanotechnology Perceptions*, 20(S5), 642–647. https://doi.org/10.62441/nano-ntp.v20iS5.60

Mandal, R., & Mete, J. (2023). Teachers' and students' perception towards integration of artificial intelligence in school curriculum: A Survey. *International Journal of Multidisciplinary Educational Research*, 12(7(5)), 2277–7881. https://s3-ap-southeast-1.amazonaws.com/ijmer/pdf/volume12/volume12-issue7(5)/9.pdf

Martínez, I. G., Batanero, J. M. F., Cerero, J. F., & León, S. P. (2023). Analysing the impact of artificial intelligence and computational sciences on student performance: Systematic review and meta-analysis. *NAER: Journal of New Approaches in Educational Research*, 12(1), 171–197.

McCarthy, J. (2007). What Is Artificial Intelligence?. Retrieved from www-formal.stanford.edu/jmc/whatisai.pdf

ML, S. I., Andrade, W. L., & MR, S. L. (2019, October). Analyzing the effect of computational thinking on mathematics through educational robotics. In 2019 IEEE Frontiers in Education Conference (FIE) (pp. 1–7). IEEE. https://doi.org/10.1109/FIE43999.2019.9028384

Ocaña-Fernández, Y., Valenzuela-Fernández, L. A., & Garro-Aburto, L. L. (2019). Artificial intelligence and its implications in higher education. *Journal of Educational Psychology - Propositosy Representaciones*, 7(2), 553–568. Retrieved from https://dx.doi.org/10.20511/pyr2019.v7n2.274

Ouyang, F., & Jiao, P. (2021). Artificial intelligence in education: The three paradigms. *Computers and Education: Artificial Intelligence*, 2, 100020. Available at: https://doi.org/10.1016/j.caeai.2021.100020

Oxford Learner's Dictionary (OLD). (n.d.). Artificial Intelligence. Retrieved from www.oxfordlearnersdictionaries.com/definition/english/artificial-intelligence

Pavlova, A. (2021). Education in 21st century – Challenges and transformations. *KNOWLEDGE – International Journal*, 47(2), 287–292. Retrieved from https://ojs.ikm.mk/index.php/kij/article/view/4694

Petko, D., Prasse, D., & Cantieni, A. (2018). The interplay of school readiness and teacher readiness for educational technology integration: A structural equation model. *Computers in the Schools*, 35(1), 1–18. https://doi.org/10.1080/07380569.2018.1428007

Popenici, S. A., & Kerr, S. (2017). Exploring the impact of artificial intelligence on teaching and learning in higher education. *Research and Practice in Technology Enhanced Learning*, 12(1), 1–13. https://doi.org/10.1186/s41039-017-0062-8

Roberts-Yates, C., & Silvera-Tawil, D. (2019). Better education opportunities for students with autism and intellectual disabilities through digital technology. *International Journal of Special Education*, 34(1), 197–210.

Rouse, M. (2020 April). Artificial intelligence. TechTarget.com. Retrieved May 20, 2020, from https://searchenterpriseai.techtarget.com/definition/AI-Artificial-Intelligence

Roy, M. (2020). AI Intervention in education systems of India: An analysis. *Solid State Technology*, 62(2s). Retrieved from www.researchgate.net/publication/344739714

Russel, S., & Norvig, P. (2010). *Artificial intelligence: A modern approach.* Pearson, London.

Sadiku, M. N., Musa, S. M., & Chukwu, U. C. (2021). *Artificial intelligence in education.* iUniverse. Retrieved from https://doi.org/10.51542/ijscia.v2i1.2

Salas-Pilco, S. Z. (2019). Introducing technology into learning designs for indigenous contexts. *Interaction Design and Architecture(s) Journal – IxD&A*, 41, 62–77.

Salas-Pilco, S. Z., Xiao, K., & Oshima, J. (2022). Artificial intelligence and new technologies in inclusive education for minority students: A systematic review. *Sustainability*, 14(20), 13572. https://doi.org/10.3390/su142013572

Soesanto, R. H., Dirgantoro, K. P. S., & Priyanti, N. (2022). Indonesian students' perceptions towards AI-based learning in mathematics. *Journal on Mathematics Education*, 13(3), 531–548. https://files.eric.ed.gov/fulltext/EJ1379744.pdf

Song, P., & Wang, X. (2020). A bibliometric analysis of worldwide educational artificial intelligence research development in recent twenty years. *Asia Pacific Education Review*, 21, 473–486. https://doi.org/10.1007/s12564-020-09640-2

Srinivasan, V., & Murthy, H. (2021). Improving reading and comprehension in K-12: Evidence from a large-scale AI technology intervention in India. *Computers and Education: Artificial Intelligence,* 2, 2–22). https://doi.org/10.1016/j.caeai.2021.100019

Stewart, J. C., David, G. A., & Igoche, D. A. (2020). AI, IoT, and AIoT: Definitions and impacts on the artificial intelligence curriculum. *Issues in Information Systems,* 21(4), 135–142. https://iacis.org/iis/2020/4_iis_2020_135-142.pdf

Stryker, C. & Kavlakoglu, E. (2024, August 9). What Is Artificial Intelligence (AI)? IBM. Retrieved from https://www.ibm.com/think/topics/artificial-intelligence

SuryaCreatX. (2023, July 30). *AI through the ages: Unveiling the unforgettable milestones in history*. Medium. Retrieved from https://suryacreatx.medium.com/ai-through-the-ages-unveiling-the-unforgettable-milestones-in-history-92f6588e73d

UNESCO. (2019). Beijing Consensus on artificial intelligence and education. *Outcome document of the International Conference on Artificial Intelligence and Education* (Sustainable development goals), 16–19 May, 2019. Retrieved from https://unesdoc.unesco.org/ark:/48223/pf0000368303

Vazhayil, A., Shetty, R., Bhavani, R. R., & Akshay, N. (2019). Focusing on teacher education to introduce AI in schools: Perspectives and illustrative findings. *IEEE Tenth International Conference on Technology for Education (T4E)*, Goa, India, *IEEE Xplore*, pp. 71–77. https://doi.org/10.1109/T4E.2019.00021

Wang, S., Sun, Z., & Chen, Y. (2023). Effects of higher education institutes' artificial intelligence capability on students' self-efficacy, creativity and learning performance. *Education and Information Technologies*, 28(5), 4919–4939. https://doi.org/10.1007/s10639-022-11338-4

Weng, C., Matere, I. M., Hsia, C. H., Wang, M. Y., & Weng, A. (2022). Effects of LEGO robotics on freshmen students' computational thinking and programming learning attitudes in Taiwan. *Library Hi Tech,* 40(4), 947–962. https://doi.org/10.1108/LHT-01-2021-0027

Wing, J. M. (2006). Computational thinking. *Communications of the ACM*, 49(3), 33–35. https://doi.org/10.1145/1118178.1118215

Yang, S. J. H., Ogata, H., Matsui, T., & Chen, N. S. (2021). Human-centered artificial intelligence in education: Seeing the invisible through the visible. *Computers and Education: Artificial Intelligence*, 2, 1–5. https://doi.org/10.1016/j.caeai.2021.100008

Yilmaz Ince, E., & Koc, M. (2021). The consequences of robotics programming education on computational thinking skills: An intervention of the Young Engineer's Workshop (YEW). *Computer Applications in Engineering Education*, 29(1), 191–208. https://doi.org/10.1002/cae.22321

Zacharis, N. Z. (2016). Predicting student academic performance in blended learning using artificial neural networks. *International Journal of Artificial Intelligence and Applications*, 7(5), 17–29.

Zhai, X., Chu, X., Chai, C. S., Jong, M. S. Y., Istenic, A., Spector, M., ... & Li, Y. (2021). A review of artificial intelligence (AI) in education from 2010 to 2020. *Complexity*, 2021, 1–18.

8 Digital Renaissance

AI's Role in Academic Transformation and Career Innovation

Aashmi V., Pushpitha Reddy Patur, and Janardhana Kumar Reddy P.

8.1 INTRODUCTION

AI is a tool and a transformative force for reshaping education, academia, and career landscapes. Digital skills are needed for tomorrow's workforce to revolutionize student learning. The digital renaissance, characterized by the rapid development of advanced digital technology, has profoundly changed the academic landscape and career paths. Artificial intelligence (AI) is the crux of bringing about the change that is redefining how knowledge is disseminated, learned, and applied in various fields. During this modern age, AI will be an instrument as well as an agent of change for innovation, creativity, and enhancement of human capabilities. This chapter includes content about AI shaping innovative career pathways and changing present academic practices, thus redefining and enhancing career pathways. AI helps in strengthening education's digital renaissance by improving traditional methods rather than replacing them completely. Educational experiences like classroom management, student learning, and teacher assistance are transformed using modern AI tools. It is also expected that AI will have a positive effect on career and employment options improving the existing ones. The improvement in technology due to AI enhances the skills and knowledge an individual will need to succeed in a specific job. AI helps educators create new trends and patterns in curriculum development which will assist students in achieving future career goals. To accomplish these large volumes of data must be analyzed to enhance teaching and learning practices. Understanding AI's role in these domains can be better to prepare for a future where learning and profession intertwine with AI technologies. Therefore, this chapter will draw a clear understanding of how AI advancements are setting the platform for the future of education and careers.

8.2 DIGITAL RENAISSANCE

A period where sectors like healthcare, business, and education are markedly transforming with advancements in AI is called the "Digital Renaissance." It is associated with the Renaissance period of history, which saw significant advancements in knowledge and culture.

 DOI: 10.1201/9781003560463-9

8.2.1 Defining the Digital Renaissance

The term digital renaissance refers to both technological growth and comprehensive change in how we approach innovation, education, and business structures.

- ***Technological Integration***: The modern digital renaissance integration of AI with regular protocols is one of the qualities in the workplace. This includes generative algorithms that can encourage effective content in all professions.
- ***Cultural Shift***: Like the historical Renaissance which emphasized discovery and compassion similarly digital transformation strengthens creativity and innovative solutions. Advancements in digital transformation help organizations to use strategy as a necessity rather than an add-on.
- ***Educational Transformation***: AI technologies will not replace conventional teaching approaches in education rather, they are augmenting them. This change facilitates individualized learning experiences and it helps teachers to foster students' critical thinking abilities.
- ***Business Evolution***: AI motivates customer satisfaction in business firms to facilitate the process and encourage innovation. The future leaders need to grasp the new technologies to adapt to the rapidly changing business patterns, which is a critical need in today's developing digital atmosphere.

8.3 ARTIFICIAL INTELLIGENCE INTEGRATION IN NEP 2020

A remarkable transformation has occurred in Indian education. The National Education Policy (NEP) 2020 insists on the integration of both technological tools and automated intelligence. It also mentions improving education by technology-enhancing education. The NEP outlines the need to transform the old "one size fits all" approach to education by integrating internet, e-learning, and digital technologies into the new model. Moreover, the first education policy of the 21st century recognized the huge role of AI. AI influences many sectors. So it's no surprise that in the field of NEP 2020 the role of AI in education is emphasized. It discusses a number of projects that should be implemented for the better use of AI in education. The NEP plans to utilize AI and technology to improve the educational process. NEP includes the use of technology and online platforms for education. The policy suggests the inclusion of AI and computer programming in the school syllabus starting from the range of Class VI so that students can be useful members of a global workforce in the digital age.

Therefore, AI is a powerful tool for customizing the learning system as per the requirements of individual learner. AI is an effective tool to create personalized learning experiences for the requirements and skills of every individual. Using games, mobile apps, adaptive learning systems, and social media, AI can assess what students are mastering and areas of strength and weakness and provide the delivered content accordingly. These methods align with the NEP's focus on providing each learner with an education tailored to learning needs. This policy favorably views the use of AI in education to develop tools that will reduce the administrative burden both on educators and learners by providing customized education to each individual.

Multidisciplinary education is a key professional development component in National Education Policy (NEP). AI is essential in multiple disciplines because it allows one to get complete knowledge with unique ideas and additional information for future growth in education.

8.3.1 Personalized Learning Vision Using Artificial Intelligence for NEP 2020

The NEP strongly highlights the potential of technology to enhance both the availability and quality of education. It is NEP 2020's vision of education to achieve each individual learner in all their uniqueness and possibilities, the reason why National Educational Policy is being focused on personalized education. AI is positioned as a critical factor in achieving this purpose through the following means:

- Intelligent Content Creation Available for All Learners: These systems employ AI algorithms that evaluate the student's performance instantly thus enabling the provision of the instruction based on the student's strengths and weaknesses.
- Equitable Access to Quality Learning: Due to AI and its ability to gamify interactive tools, learning can be advanced which enhances interest, collaboration, and participation in the class by the learners.
- Skill Building: Providing personalized learning based on AI helps in developing crucial skills such as enhancement of digital literacy, logical reasoning abilities, and problem-solving skills within the context of the NEP.

In the framework of the Indian Education System, NEP 2020 seeks to be more adaptable, inclusive, and student-centered, with AI acting as a transformative force. This policy emphasizes personalization in learning adjusted to the needs of every learner with AI. The best LMS (Learning Management System) and ERP (Enterprise Resource Planning) software can be integrated with advanced AI applications to deliver a more personalized and effective learning environment in educational institutions. The NEP 2020 delineates targeted, personalized learning experiences for every learner. This transformation is a movement from the conventional, one-size-fits-all teaching approach. Method and instead fostering the learner's own pace and time. NEP 2020 is designed to improve student collaboration and interaction through personalized learning and outcomes by tailoring digital content and support. With the help of such advanced tools, educational institutions may build more interactive, encouraging, and effective learning environments that cater to individual student's needs. As per the National Education Policy (NEP), the introduction of coding and AI should begin from the primary. The curriculum imparts basic skills in learners to survive in this digital-oriented world. Understanding AI and coding will help the learners make more significant contributions to the digital economy as well as prepare for tomorrow's landscape.

8.3.2 National Education Policy Incorporate Artificial Intelligence into the School Curriculum

According to NEP 2020 suggested that future, innovative subjects like AI can be incorporated into the regular curriculum. As an outcome, the NCERT (National Council of Educational Research and Training) is currently creating a new National Curriculum Framework for School Education that is in line with NEP 2020's objectives. The NEP emphasized the necessity of teaching the required skills at all educational levels to adequately prepare learners for a developing AI economy and align India's curriculum accordingly.

National Education Policy NEP 2020 suggests that AI can be molded into the education system in all levels of learning such as the following:

- Modern disciplines like AI and design thinking will be taught to young learners at an early age, introducing them to abilities like digital literacy, programming, and logical reasoning.
- The NEP is designed to implement AI-powered solutions to establish multilingual as well as holistic education. Among school students, multilingualism promotes an endeavor to promote natural language processing capabilities for the diverse languages in the country.
- AI, big data analysis, machine learning, and 3-D machine learning will all be included in undergraduate courses to prepare students for the workforce.
- In addition, Master's and doctorate programs in fundamental AI fields like machine learning will be offered by all institutions.

8.4 Artificial Intelligence is Integrated into the Global Economy through the Digital Transformation

AI is growing day by day and is being integrated into the global economy through digital transformation. Through digital transformation, education develops by personalizing the learning experience for learners and enabling the teaching methods among educators. It is already changing the world in a variety of ways and is becoming increasingly significant in our daily lives and economy. "AI is rapidly reshaping the global education landscape," said Saadia Zahidi, Managing Director of the World Economic Forum. A report published by "The World Economic Forum" has released a paper that explores the possible effects of AI on educational institutions and mentions several ways that AI might improve the educational process for both teachers and students. It highlights the wide range of AI's potential, including the ability to customize learning experiences, streamline administrative duties, incorporate AI into the curriculum, and more. It implies that a new age of global education may be ushered in by the judicious application of these technologies. The incorporation of AI into curricula also gives students the information and skills they will need in the future, and individualized learning experiences and materials offer specialized educational routes to accommodate a range of student demands. Generative AI promotes creativity and innovation among students through digital transformation. Tailored educational development enables the technologies to analyze the big data and approach to streamline administrative tasks and remote learning and it will provide innovative and quality learning environments that can raise the economy.

8.5 REDEFINING STUDENT ENGAGEMENT AND INTERACTION WITH ARTIFICIAL INTELLIGENCE

Imagine a world where students absorb knowledge and interact, engage, and grow within a tailored educational ecosystem. A world where AI seamlessly integrates into every facet of learning, from personalized education paths to nuanced feedback mechanisms. AI-powered social networks can facilitate deeper connections by analyzing student interactions and suggesting groups, study partners, or mentors that align with their academic and personal interests. It makes smart adaptive learning and meaningful connections and collaborations. Understanding student engagement has always been a challenge for educators. Traditional metrics often fail to capture the full picture. AI changes this providing deep analytics on how students interact with content, participate in discussions, and perform assessments. These insights allow educators to identify at-risk students, refine their methods, and enhance overall interaction. Chabot's and virtual assistance can provide instant answers to students' queries, facilitate interaction between learners and educators to focus on teaching, and allow students to get the help they need without delays. The use of interactive technology helps prepare students for the future. It plays a main role in many students' future careers, whether they are using them. According to the World Economic Forum, employers will be looking for both familiarities with such tech and soft skills that automation can't replicate, like problem-solving, self-management, active learning, and flexibility.

8.5.1 Educator's Role in Preparing for Future Careers

Half of the world's jobs will disappear by 2030 due to automation, according to an Education Commission's report. Employers will be looking for workers with the kinds of soft skills that will help them become highly relevant and sought after. A global initiative that's all about quality education and continuous learning for everyone is reported by the Education Commission. This generation of learners possesses abilities, but they are extremely challenging to acquire and can be elusive and challenging to evaluate. So educators should be focused on helping learners build a bedrock of soft skills to help them prepare for evolving future careers with AI. All assessments should be delivered through learning management.

8.5.2 Artificial Intelligence and the Evolution of Learning Models

Since its early days, AI has undergone a remarkable transformation. One of the main forces behind industrial progress, it also plays a crucial role in encouraging the integration of new technologies, including the Internet of Things (IoT), graphic processing units, cloud computing, and the blockchain. The rapid pace of technological innovation requires that a career should be upskilled and be able to adapt to changing practices. Personalized learning needs more training. Raising deep learning based on algorithms needs development in training initially in the field of AI training has continually evolved, through technological advancement and the growing amount of data. As we look to the future, it is important to understand the current trends in AI training to grasp new trends and predictions that will mold the next level of AI.

8.6 HISTORICAL CONTEXT

The historical context of technological evolution is essential for understanding the rapid and ongoing transformation within education and career development. The history of digital learning can be traced back to the late 20th century when educational institutions first attempted to integrate computers into their programs. This pivotal development was the beginning of a broader shift in how educational content was delivered and accessed. In the beginning, computers were developed to facilitate multiple modes of education by providing basic educational applications and word processors. However, it was with the explosion of the internet in the mid-1990s that the full potential of digital learning was anticipated. Quite rapidly, the internet increased the range of resources exponentially, helping to bring together a wide variety of materials, research, and all other assets to interactive learning on a global scale. This was an important step in the globalization of knowledge because now learners did not have problems of limited geography or access to printed resources. As the century turned, broadband access as well as mobile technologies have facilitated a surge in digital transformation, enabling a wider audience to benefit from e-learning and e-classrooms or e-collaborative tools. The single most important progress in the last few decades, however, has been witnessed by the birth and growth of AI which is now beginning to lead the change in education and career direction. AI encompasses a variety of tools that take analyses to another level, including automated scoring of learning outcomes, dynamic adaptation of tasks, and providing reliable predictions about students' performance and workforce trends. This capability allows for the creation of tailored learning paths that optimize individual growth, fostering deeper engagement and more effective outcomes. These technological advances have also transformed career development, with AI-supported personalized career counseling and job placement tools. To fully appreciate the current state of digital transformation, it is crucial to understand these historical milestones. By contextualizing the evolution of educational technologies, one can better grasp the magnitude of current innovations and anticipate future trends in digital learning and workforce development.

8.6.1 Pre-Internet Era (1950–1989)

The development of semiconductors and microchips allowed the shift from manual to digital technologies, laying the groundwork for digital transformation. This era marked the beginning of a shift toward digitization, as organizations began to recognize the potential benefits of converting outdated processes into digital formats. Businesses concentrated on changing their processes to leverage digital data, laying the groundwork for upcoming developments in corporate procedures and technology.

- 1950: The concept of Computer-Assisted Instruction (CAI) emerges with early experiments using mainframe computers to provide individualized learning experiences.

- 1960: Educational psychologists advocate for the use of machines in teaching, leading to the development of programmed instruction and early educational software.
- 1970: The introduction of microcomputers began to make computers more accessible in schools, although their impact on learning remains limited.
- 1980: The personal computer revolution leads to increased investment in educational technology, with initiatives like one-to-one laptop programs beginning to take shape.

8.6.2 Post-Internet Era (1990–2006)

The development of the internet in the early 1990s completely changed how people communicated and shared information. This period saw a significant increase in personal computer ownership in households accessing PCs by 2000. The launch of major platforms like Google (1998) and social networks such as Facebook (2004) further accelerated digital connectivity. Providing organizations with scalable resources that would become essential for digital transformation efforts across various sectors.

- 1991: The World Wide Web is introduced, dramatically changing access to information and resources for students and educators.
- 1995: The first online courses are offered by universities, marking the beginning of formal online education, although many programs struggle to gain widespread acceptance.
- 2002: The term "E-Learning" gains popularity as more institutions adopt online learning platforms, leading to the establishment of dedicated e-learning companies.

8.6.3 Mobile Era (2007–2018)

The launch of the iPhone in 2007 created a new era dominated by mobile technology. This changed how consumers interacted with digital content but also promoted businesses to adapt their strategies to accommodate mobile users. The term "Digital Transformation" was coined around when software began to rule every industry by encapsulating the ongoing changes organizations faced as they adapted to new technological realities.

- 2008: Massive Open Online Courses (MOOCs) emerge, providing free access to high-quality educational resources and courses from top universities.
- 2010: The proliferation of smartphones and tablets to the rise of mobile learning applications, allowing students to learn anytime and anywhere.
- 2015: Adaptive learning technologies gain traction, using data analytics to personalize learning experiences based on individual student performance.

8.6.4 Post-Pandemic Era (2019–Present)

The COVID-19 pandemic accelerated digital change among intuitions. In a socially distant environment, organizations were obliged to reconsider their interactions with educators and students. This period has seen an explosion of digital innovation, with spending on digital transformation reaching new heights. The pandemic highlighted the significance of agility and adaptability, promoting educational institutions to invest extensively in technology that enables remote work and online interaction.

- 2020: The COVID-19 pandemic forces educational institutions worldwide to transition to remote learning almost overnight, highlighting both the potential and challenges of digital education.
- 2021–Present: Increased investment in AI-driven educational technologies as schools seek to enhance personalized learning, improve assessment methods, and address equity issues in access to technology.

8.7 TRANSFORMATIVE OF DIGITIZATION IN EDUCATIONAL INSTITUTIONS

AI is bringing about a digital revolution in education, fundamentally transforming teacher preparation, school education, and higher education. The potential of AI to improve administrative efficiency, personalize learning, and prepare students for the workforce of the future is highlighted in this chapter, which examines the technology's varied effects on academic settings and professional innovation.

8.7.1 The Impact of Artificial Intelligence (AI) on Higher Education

AI will improve educational institutions by providing appropriate educational encounters to each student's needs. In specific, through personalized learning, students' datasets such as their learning preferences or learning strengths and weaknesses are used to inform the materials provided by an adaptive learning system. This allows the student to go through the work on their own time and be able to get help where necessary. AI-powered rapid feedback systems enhance effectiveness and efficiency, on average learning duration taken by learners in the educational context. Virtual and computer-aided AI systems view real-time and past student responses. It allows students to improve through doing even in a not-so-supportive learning environment. Opening new AI opportunities brings together innovation and training. By automating activities such as grading and attendance tracking, educators may be able to devote more time to guiding student's involvement and help. This modification allows for a more engaging and encouraging learning environment, which is vital for fostering students' creativity and critical thinking.

8.7.2 Transformations in Teacher Education

AI is similarly changing teacher education. Teachers are increasingly utilizing AI-powered tools to enhance their pedagogical approaches. Intelligent tutoring systems are virtual assistants offering teachers and students individualized training. This material can recognize when learners are having trouble understanding a particular concept and deliver suitable solutions. AI helps to build the curriculum by examining the student performance data from trends in education. Teachers may modify the curriculum to keep current sources relevant to educational standards by using a data-driven approach. Therefore, as a result, educators may train aspiring educators in future by using technology in the classroom.

8.7.3 Innovations in School Education

AI has a significant impact on education as it can be used to design individualized learning environments. By using algorithms to choose instructional materials according to each student's profile, adaptive learning systems increase student engagement and enhance academic performance. Addressing a range of learning needs in the classroom requires this personalization.

Furthermore, AI technologies support the implementation of early intervention strategies by identifying these students early on; teachers can give targeted encouragement programs that help them overcome gaps in education before they grow. This proactive approach enhances classroom performance overall in addition to raising individual student progress.

8.8 ROLE OF ARTIFICIAL INTELLIGENCE IN ACADEMIC TRANSFORMATION

The most important applications of AI are undoubtedly changing education as regular jobs are simple to automate. These days, AI handles all administrative tasks, including course enrollment in courses and attendance tracking. This allows learners to save time and concentrate on important activities, such as student interaction. With AI in education, customized learning is enhanced by algorithms that offer tailored suggestions. Students are more motivated and engaged while using the tailored method, which improves their academic achievement. Technologies like machine learning and AI in education help students gain better practical knowledge and techniques to reduce the workload. AI has become the most integrated subject in school education and virtual classrooms. In education, AI solutions can enhance personalized teaching methods, design engaging tests, assess homework, oversee research initiatives, monitor progress reports, generate presentations and digital notes, and handle various administrative tasks efficiently.

8.8.1 Personalized Learning Experiences

AI plays a crucial role in the academic transformation to adapt content to individualized needs for enhanced outcomes with personalized learning. Each student reacts to knowledge uniquely. Some individuals learn skills quickly, while others take their own time. In the traditional curriculum, the concept of individualized teaching for every student disappeared. AI ensures that instructional material is tailored to every student, which may be helpful in the field of education used in online learning. In addition, the framework helps students understand different classes and adjust to that process to lessen the strain by utilizing other technologies, such as machine learning, in education. To empower individuals by offering personalized suggestions for training, skill improvement and professional growth, we will comprehensively assess every individual's achievement. Furthermore, individualized instruction driven by AI has effects that go beyond academic achievement. Based on the study conducted by McKinsey & Company analysis, individualized instruction, student motivation, and interest in learning by up to 72% increased. Additionally, a study conducted by Oxford University Press indicates that personalized learning may greatly reduce the rate of dropout and boost student satisfaction. Educational institutions are transforming how students acquire knowledge and succeed in a world that is becoming more competitive and dynamic by utilizing AI to develop personalized learning experiences.

8.8.2 Enhanced Teaching Methods

AI integration in teaching improves instructional strategies by providing insights into students' performance trends and enabling targeted support measures. AI for education simplifies a variety of activities and makes learning fun and interesting. Content creation is indeed one of the most powerful AI applications in education. Modern technology aids educators in producing intelligent information. The only visual aids available under traditional teaching approaches are lap trials. Visualized web-based study settings were enhanced by AI-powered intelligent content generation. With the aid of technology, students may see knowledge in two or three dimensions and develop new perspectives. Digital Curriculum is aided by AI in education. With AI in virtual classrooms, education, and schools, the aid and technologies take up most value-added tasks. It also plays a vital role in teaching visually to hearing-impaired students. AI-powered converter tools like Presentation Translator provide real-time subtitles for virtual lectures. Teachers frequently use AI-powered educational games, adaptive learning platforms, and automated grading and feedback systems in the classroom. Gamified learning platforms, powered by AI, incorporate rewards, challenges, and leaderboards, promoting active participation. Predictive analytics can identify at-risk students, allowing for personalized learning methods. AI in education creates interactive learning environments, boosting student engagement and motivation.

8.8.3 Administrative Efficiency

By automating routine tasks such as grading and scheduling, teachers focus on interactive teaching techniques and train to invest valuable time in AI tools. This change makes it possible to focus on encouraging educators to concentrate on improving critical thinking and creativity among students through an interactive classroom environment.

8.8.3.1 Skill Gap

AI in academics is a crucial approach for academics to bridge the digital divide to enhance students' skills. AI and machine learning provide access and cost-effective upskilling future possibilities through application development and software for future learners. This approach is beneficial for both teachers and students. This collaborative learning environment encourages learners at all levels to contribute unique ideas and adapt modern technologies to problem-solving initiatives. In the end, this effort promotes institutions and strengthens learning growth, which places us in a position to quickly adapt to promote future success.

8.8.3.2 Raising Academic Standards and Educational Quality

Improve the overall quality of education in career innovation. The integration of AI has the potential to raise the academic standards in the field of education. It can develop academic careers by providing high-quality resources and it can help ensure consistency to support educators in delivering effective instruction. By leveraging AI technologies, educational intuitions can offer a deeper analysis and broad curriculum that meets students' evolving needs to build academic and career innovations.

8.8.3.3 Curriculum Development

AI plays a very important role in curriculum planning and development by analyzing the educational patterns and performance of student data. Educators may use these skills to adapt the curriculum to assign the resource. By continuously assessing learning outcomes, educational content remains relevant and aligned with current standards to ensure with the help of AI. These techniques not only improve the standard of instruction but also fill the gaps of students' practical knowledge.

8.8.3.4 Crafting the Future Workforce

The workforce of the future needs abilities that are both flexible and adaptable. Based on industry expectations and individual requirements, AI improved the personalized learning paths and developed the professional training. It is essential to our modern world to promote a lifelong and flexible learning environment. In our contemporary world, as well as in any specialized field, fostering a flexible and lasting learning environment is crucial.

8.8.3.5 Knowledge Retention

The real indicator of schooling measures true knowledge retention. AI brings innovative techniques for improving retention through individualized learning experiences. AI guarantees that knowledge is truly learnt but also kept by spotting knowledge gaps and offering focused corrections. These techniques strengthen the institutions' collective knowledge base in addition to improving individual learning outcomes.

8.8.3.6 A New Age of Learning

Digital transformation aims to improve traditional education rather than replace it. In this progression, AI is essential to enhance the educational process through AI platforms and tools. The aims of the digital transformation in education include anything from virtual reality (VR) in the classroom to AI-driven assessments to create inclusive, accessible, and effective learning environments.

8.8.3.7 Beyond the Grade

The traditional examinations do not capture intricate understanding of content deeply. An assessment of knowledge dose not measure understanding, especially when it relies of factual recall and is restricted to rigid criteria. On the other hand, AI provides a different dataset to create a global picture of learning outcomes. This encompasses everything from evaluating the use of Problem-solving methods of the metrics of interaction, which enables the shift from using grades to understanding real level of learners understanding.

8.9 INNOVATION IN CAREER DEVELOPMENT

Rapid changes in AI are impacting educational intuitions. AI integration into education is changing career pathways routes after graduates in a positive way. To develop digital literacy and problem-solving skills, education institutions are modifying their curriculum. Additionally, through group projects and peer assessments, AI encourages students to use these platforms for collaborative learning. Apart from the presence of various options, the knowledge of AI tools for career guidance can lead career professionals to effectively tailor their interests to learners. Here are some AI application platforms that offer interest and skill assessments to analyze the current trends and career guidance.

8.9.1 Career Copilot

It launched in September 2023, Career Copilot offers individuals personalized career guidance and support with resumes, cover letters, and interview preparation. It is also providing organizations with features to support the professional development of their workforce.

8.9.2 Career Explorer

Users can take a test to find their top career matches based on their interests, goals, history, workplace preferences, and personality. Career Explorer uses machine learning models and career satisfaction data to provide personalized career matches. Jobseekers at any stage of their career can use this tool.

8.9.3 Existential

This is another AI-powered tool that analyzes an individual's interests, skills, values, and experience to suggest results and determine whether those align with their goals. Students at any stage in their career can use tools.

8.10 JOB MATCHING ALGORITHMS

AI is transforming the recruitment landscape through sophisticated job-matching algorithms that connect candidates with suitable roles. These are mentioned in the following sections.

8.10.1 LinkedIn AI Career Coach

AI Career Coach, which allows users to interact with a Chabot about open job opportunities, evaluate whether a posted role is a good fit based on their LinkedIn profile, and identify the skills they need to become better candidates for their desired role. The tools are available to premium members through the LinkedIn Learning platform.

8.10.2 Talentprise

People can take an assessment that will show their strengths and areas to work on for their job search journey. They can join a talent pool to connect with employers and learn about industry trends to help improve their job search.

8.10.3 Jobiri

This e-career guide collects job advertisements and applies machine learning to match job requirements with a candidate's qualifications. Additionally, it provides CV comments and classes to fix skill gaps.

8.10.4 Personalized Career Coaching

The coach of Career Village is an AI-based career assistant that provides personalized advice to students and job seekers to help them to become good job searchers and navigate their routes throughout their careers.

8.10.5 Board Infinity

The platform employs AI algorithms to recommend the ideal career paths based on users' profiles and interests. It guides users through a structured process from finding the paths to acquiring the necessary skills.

8.10.6 Duolingo

This AI is utilized by the educational platform in order to customize and modify the learning pathways of the users. It does so by conducting an initial assessment of the learner's skill set and setting the level of exercise appropriate to what the user managed to do while also being kept interesting as it is leaderboard/game point-based.

8.10.7 StepWise

This AI platform developed by Querium has its core objectives around STEM (Science, Technology, Engineering, and Mathematics) education. It applies analytics in order to monitor each student's progress and give them specific instructions geared toward their individual learning objectives in order for them to learn difficult ideas appropriately.

8.11 ETHICAL CONSIDERATIONS AND DIFFICULTIES

The deployment of AI in education is accompanied by a multitude of ethical concerns. These ethical concerns stem from how AI systems are trained which often involves using large sets of data making issues such as who owns this data, if consent was given to use this data, and how transparent these operations are, for example, this could result in student's private data being gathered even when such students never consented to this action which could also result in the misuse of their sensitive information. In addition, AI algorithms can be biased and this bias can lead to disparities in education level across different regions. If such AI systems are educated using these skewed datasets, they could end up reinforcing the existing stereotypes and biases or would deny the minority groups the chance to learn. It is therefore important that educational intuitions create appropriate policies that address such ethical issues in the use of AI's that guarantee the private information of the students is respected and the possibility of discrimination in the algorithms is reduced. Integration of AI into education raises important ethical considerations that must be addressed to ensure equitable, fair, and effective use of the technology. This section outlines the key ethical aspects associated with AI in educational settings, focusing on the principles of fairness, privacy, transparency, and responsibility.

8.11.1 Fairness and Bias

Some algorithms may contain biased content that does not satisfy the ethical concerns one has to keep in mind. AI has developed over a while and depends on historical data to predict and come to conclusions, which can affect the result as it has existing biased content. For example, if the AI algorithm was programmed to reflect on certain social issues, it may consider the same biased program to come out with educational outcomes that may affect certain underprivileged groups. The European Commission's Digital Education Action Plan was implemented for training data to reduce bias. AI systems should undergo a regular check to make sure equity is maintained and to avoid discrepancy in the education system.

8.11.2 Data Privacy

Data privacy is an important ethical consideration when adopting AI in education. The very first consideration that comes to one's mind is the collection of personal data and the functioning of educational institutions to avoid cyber-attacks. Such centers are required to keep sensitive data safe, such as the information of students which is submitted to them. This ultimately leads the students and teachers to be impacted gravely since there lies a risk of data breaches. Furthermore, the shortcomings of adequate data protection legislations in various regions add to despair. Institutions are expected to employ effective data management strategies to protect students' data while abiding by the law. The problem is not just storing the data, it's also the fact that how would young impressionable students be able to grasp the understanding of such concepts related to their data. AI systems frequently require access to sensitive student information, creating questions about how it is acquired, stored, and used. To avoid breaches and exploitation, schools and technology businesses must prioritize student data security. However, education should teach students about data privacy and help them to understand how their information is used and how to protect them. Transparency in data handling processes is important to build confidence between students, parents, and instructors.

8.11.3 Digital Divide

Challenges associated with the digital divide that constitutes one of the trends where AI's employed in education. There is no doubt that technologies related to AI can reshape the way students learn, but also tackle the digital inequalities that exist among students. AI-powered tools for education can be used by students who have access to the required hardware, software, or a stable network, which can be a challenge for many students situated in rural areas. This can lower their grades or academic achievements and later on their professional career, making it a vicious circle.

8.11.4 Transparency and Accountability

Transparency in AI systems is important for developing trust among users. Educators and learners must understand how AI tools operate and make judgments. Initiatives like the AI4T project aims to ensure that educational tools are explainable and accessible, allowing users to grasp the underlying process of AI applications. Accountability mechanisms are also crucial; stakeholders must establish clear guidelines on who is responsible for decisions made by AI systems. This addresses issues related to errors or biases in AI-generated outcomes and ensures that there are avenues for redress when necessary.

8.11.5 Human Interaction vs Technology

AI tools never replace the importance of human interaction in the teaching and learning process. While they contribute to social and emotional development through personalized instruction, AI tools provide effective learning experiences. In the education field technology balances the major challenge with human interaction to make the classroom environment meaningful. AI technologies are crucial in developing well-rounded learners who can successfully navigate their complex social situations and resolve issues through the specified solutions to encourage learners to create creative and critical thinking.

8.11.6 Ethical Use of Generative AI

In the modern era, the raising of AI tools in the education field provides major moral issues. These generative AI tools provide misleading content on the training data. Educators must ensure the accuracy and relevance of output from generated AI technologies to evaluate the critical limitations while teaching to the students.

8.12 CASE STUDIES

Some of the studies that demonstrate the transformative role of AI in academic settings, focusing on its impact on educational outcomes and career innovation:

- **Zouhaier Slimi (2023)** conducted study on The Impact of AI on Higher Education: An Empirical Study. This study examines the effects of AI on teaching and learning, assessment, ethics, required skills, and future careers in higher education. The researcher used a qualitative approach. Based on the survey, the finding highlights that AI positively impacts the learning experience by facilitating knowledge acquisitions and skill development and future careers. It also emphasizes the importance of considering the ethical implications of AI. This study reveals that higher education intuitions need to integrate AI more extensively in their programs to prepare graduates for the future workforce.

- **Evangelos Katsamakas et al. (2024)** investigated AI and the Transformation of Higher Education Institutions. This study explores how AI technology advances, particularly generative AI tools like ChatGPT, present opportunities and challenges for higher education intuitions. It develops a causal loop diagram to map the feedback mechanism of AI transformation in HEIs, emphasizing the need for academic integrity measures and the adaption of curricula to prepare students for AI influenced job markets.
- **Suvrat Jain and Dr. Roshita Jain (2019)** investigated about Role of AI in Higher Education – An Empirical Investigation. This study explores how AI can be applied in teaching and learning within higher education. This study examines the learning implications of evolving technologies on educational methods. This study suggests that implementing AI significantly enhances students' learning capacities and holds promising future prospects for higher education.
- **Iffath Unnisa Begum (2024)** investigates about Role of AI in Higher Education – An Empirical Investigation. This study compares traditional teaching methods with those augmented by AI, examining their efficacy in higher education settings. Through these qualitative and quantitative analyses, the researcher found that the students exposed to AI-assisted instruction showed improved understanding and retention of material compared to those taught through conventional methods. The investigation emphasizes the transformative potential of AI in enhancing educational practices and preparing students for future careers.

8.13 FUTURE TRENDS AND PREDICTIONS

As we look toward the future of education, especially in the context of AI's influence, several key trends and predictions are emerging that will shape the academic landscape over the next decade. These trends highlight the increasing integration of technology into educational environments, the shift toward personalized learning, and the need for continued skill development in a rapidly changing job market.

8.13.1 Immersive Virtual Learning Environments

By 2035, VR and augmented reality (AR) technologies are predicted to alert the experiences of education. Students will participate in interactive learning environments where they will be able to conduct experiments while virtually exploring historical locations. This change will not only make education more interesting, but it will also be effective in communicating complicated topics in a simpler way.

8.13.2 AI-Driven Adaptive Learning

AI will play a significant role in personalized education. Individual learning styles and performance data will be analyzed by advanced algorithms in real time to modify instructional content accordingly. These adaptive learning techniques ensure that each student receives an education adapted to their own needs, potentially increasing academic achievements. Furthermore, AI tutors will provide both emotional guidance as well as academic support, resulting in a more complete holistic educational experience.

8.13.3 Continuous Learning and Upskilling

Individuals completing their studies and entering a single career path is becoming common. As technology advances and industries develop, continuous learning will become more essential. Educational intuitions will increasingly offer modular courses and micro-learning opportunities that allow professionals to improve their skills as needed. This shift toward lifelong learning will enable individuals to enhance better careers throughout their lives.

8.13.4 Expansion of Open Educational Resources (OER)

The rising cost of traditional educational materials is speeding up toward Open Educational Resources (OER). These free and customizable resources help to access high-quality academic resources for all students, reducing the financial barriers to traditional textbooks.

8.13.5 Flexible Learning Environments

As technology advances, flexibility in learning will gain traction as technology enables education to access time and place more feasible. Students will be able to learn through various formats like online courses, in-person classes, or hybrid models according to their preferences.

8.14 CONCLUSION

The integration of AI into education and career development signifies a transformation era that is reshaping learning, teaching, and interaction with technology as known as the digital renaissance. The incorporation of AI within the framework of NEP 2020 is the first step to modernizing education in India. By focusing on customized learning approaches, improving teacher education, adding relevant topics to the curricula, NEP 2020 seeks to prepare learners for the challenges of the rapidly evolving technological world. This overall strategy not only equips students for prospective employment but also promotes the education system to be more flexible and responsive to the diverse needs of different learners. This chapter explores the various roles of AI in academic transformation and career innovation and also highlights the potential of personalized learning experiences, encourages continuous professional development, and streamlines administrative processes. AI's integration into education can revolutionize learning and equip students well for their careers. However, it is necessary to evaluate the challenges that come along with it including ethical issues, privacy of data, and the technological gap educational stakeholders can optimize the benefits of AI and minimize the risk of its application by developing sound frameworks that foster ethical standards, protect individuals privacy, and ensure fair distribution of technologists and communities so that AI is a means of effective and inclusive education for all learners.

As we have seen various case studies exploring AI technologies are not merely theoretical concepts, worldwide educational institutions are actively implementing AI technology to yield tangible benefits. Every individual learner needs AI-powered Chabot that should provide instant support for the adaptive learning platform that is personalized learning. These unique demonstrations improve student interactive learning outcomes through technology. An immersive learning environment will be expressed in education for the future, by focusing on upskilling and soft skills development to be continued. By these trends, enabling personalized learning experiences, the evolution of AI will drive the unique needs for each individual to prepare them to increase the complex job market. In conclusion, we cherish the opportunities presented by AI in education and career development, because it is important to be responsible and thoughtful in the transformation of this approach by giving the most importance to cultivating collaboration among educators, technologies, and policymakers from ethical considerations. In an ever-changing world we can create a more inclusive and effective educational landscape to build every individual. The AI technological advancements in the digital renaissance will enhance human potential and foster lifelong success by reimagining learning for the future.

BIBLIOGRAPHY

Awofiranye, M. (2024, September 13). The challenges of using AI in education. *After School Africa*. www.afterschoolafrica.com/78994/the-challenges-of-using-ai-in-education/

Begum, N. I. U. (2024). Role of artificial intelligence in higher education- an empirical investigation. *Deleted Journal*, *2*(03), 49–53. https://doi.org/10.47392/irjaem.2024.0009

Birchwood. (2024, March 27). *What are the opportunities and challenges for AI in education?* Birchwood University. www.birchwoodu.org/what-are-the-opportunities-and-challenges-for-ai-in-education/

Chirag. (2024, October 24). 12 Ways AI in education is transforming the industry. *Appinventiv*. https://appinventiv.com/blog/artificial-intelligence-in-education/

Eden, N. C. A., Chisom, N. O. N., & Adeniyi, N. I. S. (2024). Integrating AI in education: Opportunities, challenges, and ethical considerations. *Magna Scientia Advanced Research and Reviews*, *10*(2), 006–013. https://doi.org/10.30574/msarr.2024.10.2.0039

EIC Challenge. (n.d.). *About Jobiri: The 1 AI-based digital career advisor*. EIC Challenge. https://eic.eismea.eu/challenges/solution/jobiri-1degai-based-digital-career-advisor/about

eSchool News. (2024, January 31). *Impact of artificial intelligence in education*. www.eschoolnews.com/digital-learning/2024/02/05/impact-of-artificial-intelligence-in-education/

European Commission. (2024, April 25). *Ethical considerations in educational AI | European School Education Platform*. https://school-education.ec.europa.eu/en/discover/news/ethical-considerations-educational-ai

Goyal, A., Goyal, A., & Goyal, A. (2024, October 24). *AI's role in Personalised Learning under NEP 2020*. Extramarks Blogs: Weaving Stories for Schools, Students, and Parents. www.extramarks.com/blogs/ai-personalized-learning-nep2020/

Gupta, D. S. (n.d.). National education policy 2020: Technology use and integration. *National Education Policy 2020: Technology Use and Integration*, 131. www.sieallahabad.org/hrt-admin/book/book_file/86384aa21a6e87317f4b16982e3bd594.pdf

Haileybury Pangea. (n.d.). *Automation is changing the skills that matter*. Haileybury Pangea. www.haileyburypangea.com/news/how-automation-is-changing-the-skills-that-matter-for-the-next-generation

ibl.ai. (n.d.). *Redefining student interaction: The rise of AI-Driven Social Networking and education*. https://ibl.ai/blog/redefining-student-interaction-the-rise-of-ai-driven-social-networking-and-education

ibl.ai. (n.d.). *Unveiling the digital renaissance: AI's transformative role in educational innovation*. https://ibl.ai/blog/unveiling-the-digital-renaissance-ai-s-transformative-role-in-educational-innovation

Innosight. (2022, August 24). *Three steps to jumpstarting digital transformation | Innosigh*. Innosight. www.innosight.com/insight/three-steps-to-jumpstarting-digital-transformation/

Jain, S., Jain, R., Neerja Modi School, & Director Finaxis Consultancy Services. (2019). Role of artificial intelligence in higher education- an empirical investigation. *IJRAR-International Journal of Research and Analytical Reviews*, *6*(2), 144–145. http://ijrar.com/upload_issue/ijrar_issue_20544069.pdf

Katsamakas, E., Pavlov, O. V., & Saklad, R. (2024, February 13). Artificial intelligence and the transformation of higher education institutions. *Sustainability*, *16*(14), 6118. https://doi.org/10.3390/su16146118

Khan, S. (2024, September 18). Why generative AI? Reasons to embrace the AI renaissance. *Delta News*. https://news.delta.ncsu.edu/2024/09/18/why-generative-ai-reasons-to-embrace-the-ai-renaissance/

Lu, Y. (2019). Artificial intelligence: A survey on evolution, models, applications and future trends. *Journal of Management Analytics*, *6*(1), 1–29. https://doi.org/10.1080/23270012.2019.1570365

Maity, S. (2019). Identifying opportunities for artificial intelligence in the evolution of training and development practices. *Journal of Management Development*, *38*(8), 651–663. https://doi.org/10.1108/jmd-03-2019-0069

Marr, B. (2024a, September 11). *The future of education: 2035 trends* [Video]. YouTube. www.youtube.com/watch?v=UrvFotxTxF4

Marr, B. (2024b, August 27). The biggest education trends of the next 10 years. *Forbes*. www.forbes.com/sites/bernardmarr/2024/07/09/the-biggest-education-trends-of-the-next-10-years/

Modern Diplomacy. (2024, April 29). *How artificial intelligence is reshaping global education*. https://moderndiplomacy.eu/2024/04/30/how-artificial-intelligence-is-reshaping-global-education/

Paige, M., & Paige, M. (2024, May 20). *The evolution of digital transformation history: From pre-internet to generative AI*. HatchWorks. https://hatchworks.com/blog/product-design/history-digital-transformation/

Pioneer, D. (2024, February 7). Integration of AI in NAT'l edu policy 2020. *The Pioneer*. www.dailypioneer.com/2024/state-editions/integration-of-ai-in-nat-l-edu-policy-2020.html

Rathi, P. (2020). Artificial intelligence based learning: NEP 2020 enhancing Indian initiatives for multidisciplinary education. In *St. Vincent Pallotti College* [Journal-article]. www.theprogressjournals.com/admin_panel/Gijournals/journalsfile/3551406202406200715.pdf

Roy Chowdhury Ghosh, S. (2023). Integration of technology in education in NEP-2020. *The Impression, X*, 34–36. [Journal-article]. https://sreechaitanyamahavidyalaya.in/ejournal/journal/journalbkend/paper_list/Sonali_roy_ch_ghosh.pdf

Rozal, K. (2023, November 9). *5 AI tools for career exploration – CareerWise*. CareerWise. https://careerwise.ceric.ca/2023/11/28/5-ai-tools-for-career-exploration/

Schiller International University. (2023, August 24). *The impact of artificial intelligence on higher education*. Schiller International University. www.schiller.edu/blog/the-impact-of-artificial-intelligence-on-higher-education-how-it-is-transforming-learning/

Slimi, Z. (2023). The impact of artificial intelligence on higher education: An empirical study. *European Journal of Educational Sciences, 10*(1). https://doi.org/10.19044/ejes.v10no1a17

Solutions, P., Solutions, P., & Solutions, P. (2024, May 15). *The evolution of AI training: from basic algorithms to deep learning and beyond*. Problem Solutions – Problem Solutions Meets Our Clients at the Intersection of the Physical and Digital. Specializing in Artificial Intelligence (AI) and Immersive Technologies That Drive Business Results. https://problemsolutions.net/2024/05/15/the-evolution-of-ai-training-from-basic-algorithms-to-deep-learning-and-beyond/#:~:text=Artificial%20Intelligence%20has%20undergone%20a,ever%2Dincreasing%20amount%20of%20data

S-Square. (2023, January 23). *Digital transformation in higher education: Key trends & examples*. S-Square. www.s-squaresystems.com/digital-transformation-in-higher-education-key-trends-and-examples/

Stella, M. (2024, April 29). Revolutionizing education: The role of AI in transforming learning. *Karpagam Academy of Higher Education*. https://kahedu.edu.in/revolutionizing-education-the-role-of-ai-in-transforming-learning/

Team DigitalDefynd. (2024, September 6). Use of AI in schools [25 case studies] [2024] – DigitalDefynd. DigitalDefynd. https://digitaldefynd.com/IQ/ai-in-schools-case-studies/

Team, N. I. C. A. (2024, June 21). Artificial intelligence in education – current affairs. *Current Affairs – NEXT IAS*. www.nextias.com/ca/editorial-analysis/21-06-2024/artificial-intelligence-in-education

Tiwari, M. (2024, September 6). *Using AI for personalized learning in NEP 2020*. LearnQoch. www.learnqoch.com/using-ai-for-personalized-learning-in-nep-2020/

Toi-Online. (2022, September 15). How did the NEP incorporate AI into the regular study curriculum? *The Times of India*. https://timesofindia.indiatimes.com/education/online-schooling/how-did-the-nep-incorporate-ai-into-the-regular-study-curriculum/articleshow/94221017.cms

University Canada West (UCW). (2024, July 19). *Advantages and disadvantages of AI in education*. University Canada West (UCW). www.ucanwest.ca/blog/education-careers-tips/advantages-and-disadvantages-of-ai-in-education/

Van Rijmenam, M. (2023, July 14). *Entering the digital Renaissance: Embracing the future of business*. Datafloq. https://datafloq.com/read/entering-digital-renaissance/

World Economic Forum. (2024a, September 10). *From virtual tutors to accessible textbooks: 5 ways AI is transforming education*. World Economic Forum. www.weforum.org/agenda/2024/05/ways-ai-can-benefit-education/

World Economic Forum. (2024b, September 12). *The future of learning: AI is revolutionizing education 4.0*. World Economic Forum. www.weforum.org/agenda/2024/04/future-learning-ai-revolutionizing-education-4-0/

World Economic Forum. (2024c, September 10). *From virtual tutors to accessible textbooks: 5 ways AI is transforming education*. World Economic Forum. www.weforum.org/stories/2024/05/ways-ai-can-benefit-education

www.researchgate.net/publication/378575426_Digital_Renaissance_in_Education_Unveiling_the_Transformative_Potential_of_Digitization_in_Educational_Institutions. https://qentelli.com/thought-leadership/insights/digital-transformation-strategy-key-components

Part II

AI in Healthcare and Social Services

9 AI-Enhanced Health Communication

A Netnographic Analysis of Indian Pediatrician Influencers on Instagram Promoting Child Health Literacy

Sylvia Raha

9.1 INTRODUCTION

Health literacy has evolved beyond conventional boundaries in the information age with the advancement of artificial intelligence (AI) by refocusing on social media and online communities. In India, Instagram has become a powerful tool for pediatric doctors to bridge the gap by delivering accessible health information to a broader audience that acts as consumers (Das & Faxvaag, 2014). AI technologies have improved health communication in social media and promoted doctor influencers. It has led to a rise of pediatric doctors as influencers, improving digital health literacy by empowering parents to make better health decisions and fostering a proactive approach to managing their children's health (Norman & Skinner, 2006). Health literacy refers to a person's ability to access, understand, and apply healthcare information to make appropriate health decisions. High levels of health literacy led to better health outcomes, reduced healthcare costs, and enhanced quality of life (Nutbeam, 2000; Norman & Skinner, 2006; Berkman et al., 2011). It also improves chronic disease management, medication adherence, and the ability to follow medical advice (Kutner et al., 2006). Child health literacy is closely linked to the health literacy of parents and caregivers because it lays the foundation for the child's lifelong health behaviors and attitudes (Kickbusch, 2001). New parents often seek immediate and reliable information and turn to online resources for guidance on child health issues. In this context, the integration of AI in social media has significantly transformed healthcare communication, particularly in addressing misinformation concerning child health. By offering concise summaries of relevant health information, AI helps to alleviate the challenges associated with disseminating accurate data on various healths. Child health literacy refers to the capacity of parents to locate, comprehend, and utilize health information to make well-informed choices regarding their child's health. This evolving focus is largely influenced by the immediate availability

DOI: 10.1201/9781003560463-11

of extensive data on internet, allowing parents to easily access information whenever needed (Korda & Itani, 2013). Pediatric influencers utilize their social media Instagram platforms to share professional knowledge on important topics like vaccination, disease prevention, and medical conditions. They aim to present this information in an engaging and easily accessible manner, making it more relatable and understandable for their consumers (audience) (Househ et al., 2014).

This study aims to analyze how the top five Indian pediatrician influencers on Instagram use content strategies to effectively simplify and communicate complex medical information, ultimately promoting child health literacy among parents. This research investigates the specific techniques used by pediatric influencers, including relatable stories, visual aids, and interactive content, to make health information more accessible and engaging. Additionally, the study examines the growing role of AI in assisting these influencers to tailor content based on user preferences, trending health topics, and engagement metrics. By analyzing parent engagement through likes and comments, the research seeks to understand the trust and credibility these influencers have built within their online communities. Further, the study applied sentiment analysis to uncover the emotional tone of parent discussions, exploring how these online platforms function as support networks for sharing experiences, asking questions, and seeking advice related to child health.

9.1.1 Growing Role of Artificial Intelligence (AI) in the Health Sector and Instagram

AI is transforming the healthcare system by making it accessible to the public through clinical decision-making and diagnostics (Mesko et al., 2017). This swift progress of AI technologies has substantially impacted various industries, particularly healthcare and online media, by using augmented human clinic decision-making and continuously learning from the interactions with humans and the environment (Väänänen et al., 2021). Within this transformation, Instagram pediatric doctor influencers are at the forefront of this transition, redefining the approach to child's health and wellness. Their unique ability to connect with parents and share valuable insights is making a significant impact in the community. They are using the power of AI to broaden their reach to ensure that their health educational information resonates deeply on social media. This thoughtful approach allows them to use social media platforms to address the concerns of new parents—who turn to platforms like Instagram for guidance. This also creates a publicity network where new parents end up meeting the influencer doctor in person at their clinic. This technology benefits pediatric doctor influencers by addressing new parents (as Instagram users) queries related to children's health outcomes.

Natural language processing (NLP) is a key subfield of AI that holds significant importance in healthcare applications. NLP algorithms specialize in processing, interpreting, and analyzing human language by enabling applications such as predictive analytics (Agarwal, 2022). In the healthcare industry, NLP can extract significant insights from spoken language and electronic health records to anticipate possible ailments (Chowdhury, 2003). These capabilities enable more personalized and prompt treatments that improve patient care quality. AI additionally serves a critical part in reshaping online media, particularly social media, where it is used for content personalization and increased user engagement. Personalized recommendations and content moderation are examples of AI applications in social media, all of which promote improving the user experience and marketing campaigns (Salma et al., 2024). These influencers use NLP-powered techniques for personalized content creation and patient engagement, which strengthens their influence and credibility. As these influencers continue to use AI in their practices, they not only improve their online presence, but also make significant contributions to digital health communication to promote child health literacy to new parents (Hussain et al., 2024).

In brief, social media has transformed healthcare communication where doctors increasingly utilize platforms like Instagram to reach a wider audience to share their medical knowledge and experiences (Vaca et al., 2017; Hopkins et al., 2020).

The analysis explores two main themes—doctors as influencers on social media and netnographic methodological approaches for studying this phenomenon. Recent studies highlight the growing influence of Instagram on the healthcare sector where the influencers use visually appealing content, such as explanatory short videos, interactive posts, and infographics, which simplifies complex medical information (Butdisuwan et al., 2024). Borchers (2019) discussed that visually oriented content is more engaging and likely to be shared on social media platforms extending its reach (Korda & Itani, 2013). In addition to visual content, many pediatricians use a storytelling approach to connect with their audience on a subjective level to make it more relatable and impactful. This technique educates and empowers parents and caregivers by providing real-world examples of managing child health issues. The concept of cultural capital, originally introduced by Pierre Bourdieu, offers a valuable lens to understand these non-financial social assets, such as education, intellect, style of speech, and appearance that promote interpersonal networks (Erickson, 1996) with social media users. This approach effectively addresses new parents' inquiries and preferences by offering solutions that align with their cultural framework (Chen, 2015, p. 185). In today's media-saturated society, media plays a crucial role in shaping interactions between individuals and society (Chen, 2015; Livingstone, 2009; Thompson, 2013). In doing so, these influencers influence their cultural capital to communicate in ways that are both authoritative and accessible, ensuring that their advice resonates and is followed. Their role in board-style questions is to increase pediatricians' engagement on social media (Little-Wienert et al., 2021).

9.2 METHODS

This study employs a netnographic approach to explore the role of pediatric doctor influencers on Instagram. Netnography, an adaptation of ethnographic methods to study online communities, allows for the in-depth examination of social media interactions, focusing on how health information is communicated and engaged with by parents. This study employed the "netnography" method coined by Robert V. Kozinets for market research. This online ethnographic research approach was developed to explore the practices, cultures, and communities that arise from computer-mediated communication (Kozinets, 2000, p. 7). Unlike other forms of ethnography conducted on the internet, it is "primarily based upon observation of textual discourse" (Kozinets, 2000, p. 9) and provides specific guidelines for adapting observational netnography, participant-observation netnography and autoethnography techniques to the unique dynamics of online communities and cultures that succeed through digital interactions (Kozinets, 2010). This method seeks to uncover the intricate human behaviors, opinions, concerns, and motivations that progress within online communities. Using language, images, and engagement strategies is an analyzing element for netnography. Wittgenstein's (1953) work further supports this approach by framing the posting of online content as a social action where every aspect—such as the act, type, and content of the posting, as well as the medium—is valuable observational data. Kozinets (2015) describes netnography as particularly effective in capturing the "embedded, contextualized, and emergent" nature of online interactions, making it an ideal approach for examining how health narratives are constructed and disseminated in social media contexts. This perspective reinforces that netnography provides insights into the digital interactions that shape social understanding and influence social behavior (Kozinets, 2000, pp. 9, 10) in terms of child health. Thus, the decision to use a netnographic approach for this study on influential pediatrician influencers on Instagram is based on several reasons. This method enables a comprehensive understanding of the multifaceted impact that pediatrician influencers have on their followers (Kozinets, 2010, p. 55).

9.2.1 Data Collection

This study employs a multi-stage sampling method to select and analyze top pediatrician influencers on Instagram. The process involves three stages: identifying the top influencers, selecting specific posts, and conducting a qualitative analysis of posts.

Stage 1: Identification of Top Influencers: The first stage involves web scraping to identify the top five pediatrician influencers from a broader list of the top 50 medical influencer doctors on social media in India, as reported by SPAG FINN Partners in their "Health Change Makers: Top 50 Influential Doctors on Social Media—India 2024" report, published on July 1, 2024. An observational method was used to select the top five influencers based on the engagement rate of Instagram followers. The influencers chosen for this study are based on the number of followers (see Figure 9.1).

Stage 2: Selection of Instagram Posts: The observational netnographic method is used to select all Instagram posts made by pediatric influencers during the specified ten days (1 August 2024 to 10 August 2024). This time frame ensures the analysis is based on recent and relevant content. For each influencer, all posts within this period are collected and documented.

Stage 3: Qualitative Analysis of Posts: I have collected all 22 video posts from the five doctors from the total 24 posts which included 1 infographic and 1 image. This ensures that every video post within the last ten days is included, providing a complete dataset for qualitative analysis. The selection of the likes and comments on posts offers insight into the content that resonates most with followers, indicating higher engagement. I will be qualitatively examining these 22 videos based on the average

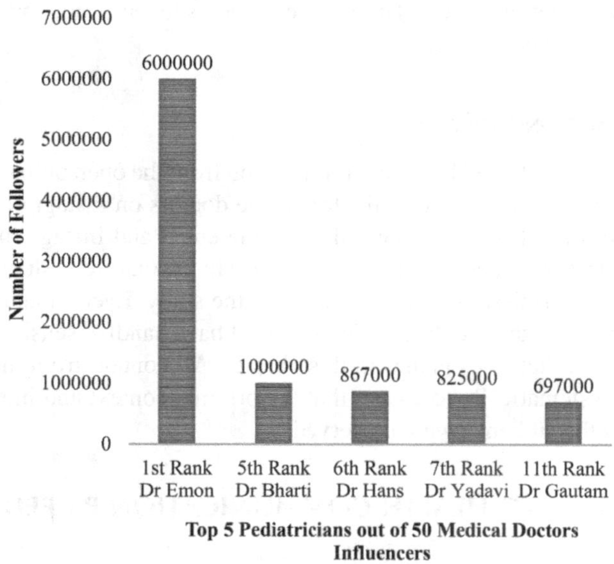

FIGURE 9.1 Top five pediatric influencers.

likes and comments. To guarantee accuracy and relevance, a thorough data cleaning and data pre-processing step was conducted after data collection. This required fixing typographical errors, standardizing content for consistency, and eliminating redundant entries like emojis and non-relevant comments.

In accordance with the study's objectives, these above stages ensured that the sentiment analysis and content classification were based on significant and relevant data.

By employing a multi-stage sampling methodology, this study ensures a systematic and comprehensive analysis of the top pediatrician doctor-influencers on Instagram (Garcia et al., 2009, p. 60; Boyd & Ellison, 2007, p .213; Kozinets, 2015, pp. 133, 135, 140; Muniz & O'Guinn, 2001, p. 413).

9.2.2 Interpretative Analysis

It was employed to explore the deeper meanings and cultural significance behind the interactions between pediatric doctor influencers and their followers on Instagram. This method involved identifying recurring themes in the content, such as prevalent health concerns and frequently asked questions, and interpreting these within the broader social and cultural contexts. This process is used to understand how parents engaged with the posts and the nature of their questions revealed insights into their health literacy and trust in these influencers.

9.2.3 Sentimental Analysis

Sentiment analysis was conducted using Boardflare's Sentiment Analysis for Excel Add-in, which is directly integrated within Microsoft Excel. This tool is used as automated categorization of text data from comments taken from, sorting them into positive, negative, or neutral sentiments.

9.2.4 Ethical Consideration

This research was conducted by taking information from the open online forum where pediatrics influencers are ranked as the top active doctors on Instagram. As a part of ethical practice, I reached out to these doctors via email and Instagram messages to personally request their permission to use their names. Only one doctor responded and gave explicit permission to use his name in the study. Even though there are no privacy restrictions in these public online spaces, I have handled sensitive issues with care and have used fictitious names in this chapter. All content from the discussions was managed systematically to ensure that the original context and intentions of the influencers and the audience were preserved.

9.3 AI-ENHANCED HEALTH COMMUNICATION BY PEDIATRIC INFLUENCERS

AI-enhanced health communication is important for influencers who have built a loyal following by creating an authentic online identity (Tafesse & Wood, 2020). By holding a significant following these top five pediatricians have followers ranging from 728k to 6.8 million (Inflact, 2024) on Instagram. These numbers reflect their extensive reach and substantial influence on public health perceptions and behaviors (Chen et al., 2021). Their profiles have become a significant space for engaging parents by providing educational content, child health tips, and teleconsultation services which solidify their role as key influencers in the digital health environment.

9.3.1 Presentation of Self by Pediatric Influencers

Goffman's (1959) theory of self-presentation suggests that people behave in ways that correspond to how they want to be viewed by others, much like actors performing on a stage, consciously regulating their appearances to correspond with social circumstances. This theory has proven highly adaptable to AI-enhanced health communication where social media platforms like Instagram provide new arenas for identity construction and impression management. These pediatric influencers create their profiles by selecting photos, taking up health-related issues for the newborn child and new moms, and creating captions, and narratives that align with desired public, shaping not only how they are perceived but also increasing their connections and credibility (Bardhan, 2022; Hongs et al., 2020). One way to measure their influence is by the number of followers they have, which demonstrates their reach and potential impact. For instance, Dr Emon has an impressive 6.8 million followers, indicating a substantial and potentially diverse audience. However, reach does not equate to influence. The engagement rate calculated based on the ratio of likes and comments to followers (Inflact, 2024) provides a better understanding of the active consumers interacting with the content. Despite having 1.2 million followers, Dr Bharti has the highest engagement rate of the five doctors analyzed, at 70.54%. This demonstrates a

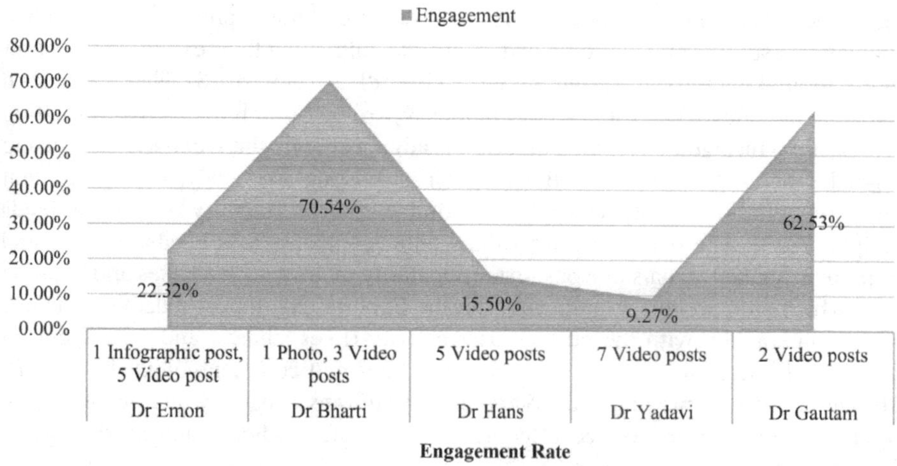

FIGURE 9.2 AI-estimated engagement rate.

strong connection with her audience, who are actively engaging with her content. So, the type of content used significantly enables impression management and influences engagement. These doctors mainly utilize videos, photos, and infographics, each yielding different levels of success. Again, it is found that Dr Hans has the highest number of posts at 938, indicating a consistent presence. However, his engagement rate is relatively low at 15.50%. This suggests that posting frequently does not guarantee high interaction; quality and relevance of content are also crucial. On the other hand, Dr Gautam, who has fewer posts, enjoys a high engagement rate of 62.53%. This indicates that his content resonates well with his audience, possibly due to its visual appeal or relevance. Additionally, his high average likes of 37.6k support this, showing that his audience finds his posts valuable and engaging (see Figure 9.2).

9.3.2 Content for Communication

Content driven by AI has modified health communication by allowing influencers to actively alter the content in response to patients as consumer preferences and basic medical topics related to newborn child's health and toddlers. In the case of pediatrician influencers on Instagram, applying AI promotes messaging through personalized suggestions that align with browsing history or preferences in previous interactions. Content serves as a mirror of the real-life questions and challenges that parents face every day. It engages directly with culturally ingrained practices. It addresses cultural norms where parents access and engage with this expert knowledge and are likely to enhance their cultural capital (Bourdieu, 1986, p. 17). The significance of contents of ten days of Instagram posts by these five pediatric influencers are categorized as "very important", "important", and "not important" based on the significance of the topic and the nature of discussion in the comment section, as reflected in the question asked. The comparison of the contents of the five pediatricians offers valuable insights into effective child healthcare communication on Instagram. Additionally, these pediatrician influencers on child health gain cultural capital in the embodied state by investing significant time and effort into cultivating their expertise and communication skills through Instagram posts (Bourdieu, 1986, p. 18). New parents on social media often have distinct expectations regarding how child health information is presented. Instagram's ranking algorithms align well with these expectations, using "signals" derived from user activity—such as posting frequency, timing, content type preferences, and interactions—to optimize content visibility on various feeds (Gupta, 2023). The platform's algorithms tailor content across Feeds, Stories, and Explore using key signals like post data (popularity metrics such as likes and location tags, which are particularly influential on the Explore page), poster data (frequency of user interaction with the poster), user activity (likes, shares, and time spent on posts), and interaction history (such as prior likes and comments with the poster). Through AI-driven predictive analytics, Instagram ranks posts by aggregating these signals, ensuring users see recent content from friends in Feeds and Stories, while Explore displays posts likely to engage them based on broader signals (Gupta, 2023). In this environment, influencers are motivated to maintain high engagement, often receiving cash, products, or experiences from companies as incentives for their promotional efforts. Instagram's algorithms thus serve as a critical tool for both users and influencers, with each section of the platform—Feed, Explore, Stories, and Reels—operating with unique algorithms refined to present content that aligns with user interests (Campbell & Farrell, 2020).

All these contents in Table 9.1 are the result from previous interactions in the comment sections. AI tools modify those comments and suggest influencers to predict which topics will perform best for future posts. Moreover, influencers sponsor posts by providing Hashtags which are driven by personalized ads, targeting specific communities for increasing relevance and engagement. The organization of content into these categories reveals a thoughtful strategy by pediatrician influencers. By assigning topics to levels of importance, they direct parents' attention toward the most critical areas of child health and development. The "Very Important" and "Important" categories focus on empowering parents with essential knowledge on preventive health, developmental milestones, and holistic family dynamics, while the single "Not Important" topic suggests a selective approach to content prioritization.

9.3.3 Promoting Child Health Literacy as Educator by Simplifying Complex Content

The pediatricians featured in this study are essential educators who excel at simplifying medical concepts into easily understandable content. This approach is particularly evident in posts that provide explanations of medical conditions like *Cradle Cap* or *Functional Constipation*. Dr Hans's post on *Cradle Cap*, for example, breaks down the condition of illness for parents into simple terms. This reassures parents by clarifying that it is a common, non-contagious skin issue seen in babies. The use of *Question-and-answer* formats makes the information more relatable and easier to comprehend. Similarly, Dr Emon's clarification of *Functional Constipation* goes beyond medical terminology; his practical advice in the video demystifies the illness.

TABLE 9.1
Type of Content

Type of Video	The Topic of the Post
Very important	Make parenting easy for newborn workshop
	Club foot
	A brief guide to all breastfeeding mums
	Steps to ensure successful breastfeeding
	Functional constipation
	Cradle cap
	Understanding hand, foot, and mouth disease
	Fungal infection on the ear leans behind the ear
	Mindful parenting: Be it screen, biscuits, scream or sugary things (chocolates), we keep doing in front of our kids
	Celebrating Breastfeeding Week with an amazing session on breastfeeding and lactation guidance by Dr Santosh, Dr Parth, and Dr Bina! Supporting our moms and moms-to-be every step of the way
	Baby need any supplements?
	Cold, cough, and curd
	Do babies rub their ears when teething?
	Emergency medicines to take with your child
Important	Childhood obesity
	Beating postpartum blues with awesome weather and papa's new song
	Always pay more or equal attention to the elder one! Once you are dealing with two
	Should we apply kajal to the eyes?
	Impact of phone on interpersonal relationships
Not important	Putting on a vaccine video

Dr Yadavi's topic *Do babies rub their ears when teething? and Mindful parenting: Be it screens, biscuits, screaming or sugary things (chocolates) we keep doing in front of our kids* is important as it addresses a common concern and actions among parents. This educational content guided them to distinguish between normal teething behavior and potential signs of ear infections as well as improved their ability to interpret their baby's signals more accurately. Understanding these types of child behavior can ease parental anxiety, lead to better-informed health decisions, and ensure proper care for the child.

By using simple language and visual aids, these pediatricians address common parental concerns to ensure that their followers are actively involved in understanding child health, rather than just being passive recipients of information. Their ability to refine complex medical concepts into understandable and culturally sensitive content is a direct result of this personal investment, which in turn enhances their credibility and influence in the digital health community. As a result, they are building a well-informed community of new parents who gain valuable health literacy, thus increasing their cultural capital (Bourdieu, 1986, p. 18). As a result, the use of AI in healthcare communication changes the way patients interact with health services, making them more efficient and effective (Sun & Zhou, 2023). As these influencers continue to use AI in their practices, they not only boost their online presence, but also contribute significantly to digital health communication and patient education. The combination of AI and social media platforms such as Instagram is ushering in a new era in healthcare, with medical influencers playing an important role in disseminating precise and relevant health information, hence enhancing digital health communication.

9.3.4 Sentimental Analysis of the Comments and Queries

As part of the data extraction and cleaning process, comments and queries were gathered from the top pediatric influencers Instagram posts. Parental responses, questions, and input regarding child health issues are included in these comments. Boardflare's Sentiment Analysis for Excel Add-in was one of the tools used to evaluate and categorize comments according to sentiment markers.

Positive Sentiments: Convey thanks, admiration, or agreement.
Negative Sentiments: Express worries, apprehensions, or doubts.
Neutral Sentiments: This table does not include queries devoid of emotional context.

The resulting analysis of Table 9.2 provided insight into parents' engagement by allowing it to measure the audience's emotional responses to pediatric health content.

The sentiment analysis shows that parents frequently express appreciation for the guidance and concerns shown by pediatric influencers about their children's health. Positive comments stereotypically express appreciation or gratitude, while negative comments reflect concerns about child health and child development. For instance, positive comments are associated with myth-busting content: comments such as "Thank you so much for giving all facts and figures" show appreciation for guidance that challenges traditional beliefs, while negative comments indicate a need for empathetic responses to common parenting anxieties, comments related to children's behavior (e.g., developmental milestones) and health conditions (e.g., dietary allergies or skin issues) are frequently marked with negative sentiment. This shows that followers are not only seeking reassurance but also want effective advice for dealing with specific issues. They also value the guidance that challenges or questions traditional beliefs related to child health.

9.3.4.1 Insights from Sentiment Analysis

High Proportion of Negative Sentiments: A considerable number of negative comments were found using sentiment analysis, often expressing parental worries about aggressive behavior, food allergies, or developmental abnormalities in children. These remarks highlight the worries and doubts parents have when deciding how best to care for their kids' health and welfare. **Implication:** These observations draw attention to areas in which parents actively look for advice, highlighting the significance of sympathetic and understandable answers from doctor influencers. Influencers

TABLE 9.2
Sentiment Analysis of User Comments on Pediatrician Instagram Posts

Comment[a] and Query Test	Sentiment Label
You are the only one who breaks the old stereotype and old school thinking. Thank you so much for giving all the facts and figures, hope slowly Indian mothers through the process will change. All the best with a new journey	POSITIVE
Does wearing a belt after delivery reduce mom's pouch, or is exercising more important? Belts have no role in reducing weight or mom's pouch. Exercise and the right nutrition are the keys	NEGATIVE
My nephew is about 6 years old, but he is so aggressive and stubborn. He does not respond like 6-year-old children …. He shouts so loudly and out of control even biting his own hands …. I can't express how we're worried about him. Can you suggest it or put a video on your wall about that?	NEGATIVE
Hi doctor, my 4-month-old screams and occasionally coos. Is screaming normal?	NEGATIVE
Sir, this is happening behind my baby's ears, it is a little cut, how can I fix it?	NEGATIVE
We understand but what about those who are grandmothers in our house, they say that we have imposed nothing will happen, we have raised more children than you	POSITIVE
Doctor, my baby just turned 8. I fed her Greek yogurt yesterday and she started developing rashes. Does that mean she is allergic to milk products or it was not advisable to introduce yogurt this early?	NEGATIVE
Most of the positioning of the child in the uterus and less space lead to club foot	NEGATIVE
A baby's health condition depends on the mother's food consumption. "*doodh khaneshe breast milk hota hain? The answer is no*"	NEGATIVE
Skin to skin contact with the child will increase the mother's hormone which will start the cycle of breastfeeding	POSITIVE
Addressing functional constipation in children requires a multifaceted approach that includes dietary adjustments, psychological support, and proper hydration. By following these guidelines, parents can help their children maintain healthy bowel habits and overall digestive health	POSITIVE
Cradle cap is a common, non-contagious skin condition seen in babies, usually within the first few months of life	NEGATIVE
Hand, Foot and Mouth Disease, despite its alarming symptoms, is generally a mild condition that resolves without significant intervention. Ensuring the child stays hydrated and comfortable is key to a smooth recovery	POSITIVE
Reason visualization baby stores information (memory) through visualization, *Monkey See Monkey Do*	NEGATIVE
Vitamin D is necessary	POSITIVE
Kajal in eyes is not recommended please avoid	NEGATIVE
Thank you so much for giving all facts and figures	POSITIVE

Note:
[a] Public comments copied as it is from Instagram.

may build a closer relationship with their audience and offer much-needed support by sensitively addressing these issues.

Positive Sentiments Drive Trust and Engagement: Myth-busting post or evidence-based advice from influencers were often met with positive comments. For example, parents frequently expressed gratitude when old assumptions were explained in an approachable way. These remarks demonstrate how an audience can be moved by the clear and concise presentation of reliable, scientifically supported information. Implication: In addition to fostering trust, this kind of content strengthens the influencers standing as trustworthy providers of health-related information. Maintaining relationships and increasing follower retention can be achieved by consistently interacting with followers through uplifting, instructive postings.

Recurring Themes in Negative Sentiments: "*Does wearing a belt after delivery lessen the mom pouch?*" and "*Does breast milk quality rely on what the mother eats?*" are two examples of queries that revealed common misconceptions among parents revealed by the analysis. These widespread worries highlight the extent of false information on the health of mothers and children. **Implication:** Influencers have an opportunity to produce focused instructional content that dispels common myths and misconceptions in response to these frequently asked questions. Personalized content aimed at dispelling these stereotypes might be a useful tool for parents looking for correct information, strengthening the influencers function as teachers.

Constructive Criticism from Mixed Sentiments: Mixed sentiments often contained constructive criticism, pointing to cultural conflicts in parenting practices. Comments like, "*We understand, but what about those who are grandmothers in our house? They say that we have imposed; nothing will happen*", reflecting intergenerational conflicts and the challenge of balancing modern parenting advice with traditional family beliefs. **Implication:** These findings suggest the necessity for a culturally sensitive way of communication. Pediatrician influencers consider addressing cultural nuances and generational differences in their content, offering respectful and inclusive guidance to bridge the gap between traditional beliefs and contemporary evidence-based practices.

By integrating these insights, pediatrician influencers on Instagram are crafting content that resonates with their audience and addresses parental concerns in a culturally aware and empathetic manner, ultimately fostering trust and enhancing child health literacy. Thus, the study findings demonstrate that pediatric influencers on Instagram play a crucial role as a support network for new parents. It fosters new connections among parents or caregivers facing similar challenges. It serves as an essential resource for accessing reliable child health information. This dual role of pediatric influencers enhances the well-being of new parents by providing both emotional and practical guidance about child health. By understanding what resonates positively or negatively, influencers can tailor their content to address concerns, provide clarity, and foster trust.

9.3.5 Clarifying Child Health: Myth-Busting with Expert Answers

Debunking and misguided practices play an important role in the field of medical education (Martimianakis et al., 2020) in developing scientific awareness for the betterment of health. Myth is an "idea and belief that we inherit as a part of our shared culture" (Loughlin et al., 2013) even though we find them unreasonable (Durkheim, 1964). They influence the underpinning behaviors of our actions by repeating practices over the generations. There are some hereditary practices, some are beneficial to the child's health like oil massage, and some may have detrimental effects on the child, like shaving the hair of the newborn. In the general scenario, these are discussed during the counseling of the mothers. India has a robust network of ASHA and Anganwadi workers under the National flagship program for Reproductive, Maternal, Newborn, Child, Adolescent Health and Nutrition. They take up the primary responsibility of imparting health education to the mothers and break the harmful practices that continued over generations. These programs are more successful in the rural areas and the suburban areas as the government healthcare delivery system reaches out to each expecting mother. The expected mothers are enrolled and followed up throughout pregnancy, childbirth, and thereafter till the child has attained 5 years of age. But as there is fast urbanization, a rise in the number of urban middle class, private sector caring for 60%–70% of the healthcare needs according to NFHS 5 data. there is a sharp contrast in the services received by the mothers in the urban areas. As a matter of fact, many women are not even aware of the government programs to support the mother and child.

This gap is a reality and the need for guidance is evident, along with the digital evolution, the online information is the best solution. The only catalyst to this is to bridge the gap between the information available online and a reliable source to reach out to those who need it. The social media along with AI-curated suggestions to connect the influencer to those mothers searching for guidance.

Overcoming all the hurdles these influencer doctors address such common practices and provide evidence-based suggestions to the new mother. This not only gives the mother confidence in following or avoiding the common practices but also helps them grow their knowledge when it comes from the experts. The medium of learning has extended beyond the traditional counseling ways where some village elders or the Anganwadi workers will teach the new mother dos and don'ts for the child. The information on these social media platforms is short, targeted to specific needs, reliable, convincing, relatable, and easy to access.

Therefore, in our fast-paced information society, it is crucial to debunk myths and misinformation. Pediatricians like Dr Bharti and Dr Yadavi have taken on the responsibility of debunking common myths related to child health, particularly those around breastfeeding and diet. For example, Dr Bharadwaj's post addressing the myth that a mother's diet directly influences breast milk quality and provides a clear, evidence-based counterargument. By stating, "*Doodh khaneshe breast milk hota hain or Does breast milk quality depend on what the mother eats*? *Answer is no*", she challenges a pervasive misconception, offering clarity in a space where misinformation often thrives. This direct approach not only educates parents but also positions Dr Bharti as a trusted source of information. This form of information strengthened its influence within the online community. Her posts often incorporate examples of misinformation, followed by clear explanations and scientific evidence that refute these false claims. Another example is, Dr Yadavi also engages in myth-busting, addresses practices such as the use of kajal to babies' eyes and the association of curd with cold and cough, medical advice on a culturally rooted practice with potential health risks, carefully addressing the associated sensitivities. This approach allows him to challenge traditional beliefs (*Kajal does harm babies' eyes and curd is not associated with cold and cough*) while respecting cultural nuances. Balancing respect for cultural practices with accurate medical information highlights the need for a nuanced approach to effectively educate and influence parental behavior. This method of myth-busting is crucial in guiding parents toward making informed decisions in an environment saturated with conflicting advice. Moreover, the engagement these posts generate reflects the importance of their role in shaping health narratives within the online community.

9.4 QUALITATIVE NARRATION AND COMMUNITY IMPACT

Qualitative analysis of parent–pediatrician interactions on Instagram provides valuable insights into their impact on health literacy and community support among parents. By observing posts and comments on pediatricians Instagram profiles, a significant pattern in audience engagement with the content is uncovered through netnographic research. Dr Emon takes on a more passive role when it comes to engaging with parents. Although his content is informative, it lacks interactive depth. For example, one post mentioned that "Most of the positioning of the child in the uterus and less space led to clubfoot", but despite this information, there were no follow-up questions or further interactions that would reflect an engagement. This indicates that parents may consume the content but are less likely to engage in deeper discussions or seek further advice from Dr Emon's posts.

In contrast, Dr Bharti encourages active participation from her audience by engaging parents and addressing common concerns related to baby health and breastfeeding. For instance, she directly answers questions such as, "*Does wearing a belt after delivery reduce mom's pouch, or is exercising more important?*" by explaining that "*Belt has no role in reducing weight or mom's pouch. Exercise and right nutrition are the key*". Her posts cover topics like the impact of a mother's diet on breastfeeding and the importance of skin-to-skin contact, encouraging parents to engage actively. The direct answering of questions in the comments further enhances the perception of her approachability and expertise, making her a trusted figure in this online community.

Dr Hans takes a comprehensive approach in his posts, addressing complex health issues such as functional constipation. He explains that "*Addressing functional constipation in children requires a multifaceted approach that includes dietary adjustments, psychological support, and proper hydration*". His thoroughness in his posts encourages parents to seek personalized advice, as reflected in comments like, "*Can you suggest it or put any video on your wall about that?*". Here, the detailed content creates a sense of trust and reliance on his expertise, prompting parents to reach out to him with personal queries, seeking further guidance or requesting specific content.

Dr Yadavi effectively uses relatable analogies, such as "*Monkey see, Monkey do*", to explain child behavior and health, which resonates well with his audience. His posts not only receive direct engagement from parents but also stimulate peer discussions. One parent expressed, "*We understand, but what about those who are grandmothers in our house? They say that we had imposed... nothing will happen, we have raised more children than you*", shedding light on the cultural context and familial influence on child-rearing practices. This indicates that his content resonates on a personal level, encouraging parents to contribute their experiences and thereby enriching the communal knowledge. Dr Yadavi's platform serves as a space for dispensing medical advice but also for navigating cultural beliefs and parenting practices.

Lastly, Dr Gautam's posts strike a balance between professional advice and community interaction. His content often results in both parent-to-parent engagement and direct queries to the doctor, indicating that his approach effectively bridges the gap between expert advice and communal support.

Therefore, when pediatricians actively engage with parents by addressing concerns and providing detailed advice, they significantly enhance parents' cultural capital. This means they empower parents with the knowledge and skills needed for effective parenting. They offer clear, actionable guidance on topics like breastfeeding and child health aligns with Bourdieu's concept of cultural capital as accumulated knowledge that confers social advantage (Bourdieu, 1986, p. 47). Moreover, those who incorporate cultural nuances and use relatable analogies not only provide medical advice but also foster a space for peer-to-peer interaction. This further enriches the communities' collective cultural capital where parents share and reinforce each other's learning through discussions in the comments section (Bourdieu, 1986, p. 49). Conversely, pediatricians who take a more passive approach such as merely disseminating information without encouraging interaction contribute less community impact. Thus, engaged pediatricians play a pivotal role in shaping community health outcomes. Their commitment to enhancing cultural capital, health literacy, and community support networks creates a healthier, more informed society—one where parents thrive and children flourish.

9.5 DISCUSSION AND CONCLUSION

The research findings underscore the utility of netnography as a method for studying the intersection of Instagram and healthcare communication. This method is particularly suitable for understanding the way health information is communicated, consumed, and acted upon in a virtual environment. The strength of netnography lies in its ability to capture the real-time dynamics of online interactions, including the engagement between content creators (pediatricians) and consumers (parents) (Kozinets, 2010, p. 57). The study's validity was guaranteed by methodically choosing the leading pediatric influencers based on follower counts and engagement rates. The authenticity of the content was preserved by emphasizing realistic parental concerns identified through chats and feedback. The study was made more credible by ethical considerations like anonymizing the names of the doctor influencers. The educational background of the five MBBS doctors highlights that three hold MD degrees in pediatrics, and a few have additional specializations in neonatology, allergy management, and child nutrition. These advanced qualifications and specializations not only enhance their credibility but also grant them significant professional power in influencing health discussions, particularly in child health, on platforms like Instagram. This expertise likely strengthens their ability to engage and educate followers, boosting their authority in promoting child health literacy.

The analysis highlights that true influence is not merely reflected by the number of followers but in a way the parents as consumers actively and positively interact regarding the content. These influencers curate their posts, engage with their targeted community, and use their expertise to contribute significantly to child health literacy, building cultural capital by enhancing the knowledge and health practices of their audience (Bourdieu, 1986, p. 243). Thus, the role of Instagram in healthcare is continuously evolving by connecting the power of visual health communication through actively engaging and making a substantial change in child health outcomes (Kozinets, 2015, p. 85). For example, the recommendation for parents to add captions to videos for new mothers using their phones in silent mode demonstrates the real-world application of netnographic findings. This type of feedback is crucial for understanding the effectiveness of communication strategies by adapting to the consumer (parents) needs, aligning with Bourdieu's concept that cultural capital is accumulated through engagement with culturally relevant content (Bourdieu, 1986, p. 49). Moreover, sentimental analysis of Instagram comments helps to understand parents' engagement with pediatric health content. It found that parents often appreciate the guidance and concerns provided by pediatric influencers about their child's health. Positive comments show appreciation for myth-busting content, while negative comments reflect concerns about child health and development. Therefore, pediatric influencers act as a support network for new parents, fostering connections and providing reliable information. This dual role enhances the well-being of new parents by providing emotional and practical guidance about child health.

For these reasons, Instagram has provided valuable insights for healthcare professionals seeking to enhance online influence (Kozinets, 2010, p. 57; Berger, 2008; Zarei & Rakhshani, 2015) to expand their reach in disseminating child health literacy and their personal growth as popularity. The study reveals two major aspects:

First aspect—Shift in Parental Reliance (trust) on Online Health Information: The rise in the use of Instagram has significantly altered parents' accessibility in seeking health information and relying on online health information. Traditionally, parents depended on pediatricians during office visits or consulted books and pamphlets for information. However, the widespread availability of health information on social media platforms (Ventola, 2014; Smailhodzic et al., 2016) like Instagram has revolutionized the accessibility of timely updates, educational content, and direct engagement with pediatricians. This empowerment allows parents to actively seek out information and fact-check child health advice by reducing reliance on traditional sources and enabling them to make more informed decisions regarding their children's health. Interpretative analysis shows that parents have increasingly turned to Instagram to access health tips and educational content from a variety of doctors. This allows them to cross-reference the information with other pediatric influencers online and consult their family pediatrician for verification. This has become a common tendency to seek and trust online resources for health-related advice. The accessibility and immediacy of online information have made it a go-to resource for many, especially in times of urgent need or when traditional consultation is not feasible. Therefore, building trust is pivotal for pediatric influencers to establish credibility and maintain their power as reliable health communicators.

Second aspect—Implications for Healthcare Providers: The broader application of Instagram in child health literacy is substantial. It provides a unique opportunity for healthcare professionals to extend their reach beyond traditional settings. Pediatricians can use these platforms not only for disseminating information but also for community engagement and education. For instance, Dr Gautam's use of visually appealing infographics demonstrates how effective social media can be in conveying complex medical information in an understandable format. The potential for engaging with a global audience and addressing widespread health concerns makes it a powerful tool for public health initiatives (Ventola, 2014). The responsibility to provide accurate and reliable content is paramount, as misinformation can have serious consequences for child health. These pediatric influencers possess objectified and embodied forms of cultural capital by producing content (such as videos, infographics, and posts) that serve as the objectified form of cultural capital. The true value of their content—the embodied cultural capital—lies in the ability of the consumers as followers (parents) to understand, interpret, and apply the information to their child's health. So, their responsibility increases to empower parents with the embodied cultural capital to make informed health decisions for their children. In this way, pediatric influencers play a crucial role in enhancing the health literacy queries of their followers (audience), contributing to the collective cultural capital of the community (Bourdieu, 1986).

Moreover, pediatrician influencers are efficiently utilizing digital spaces not merely to disseminate information but to create interactive, community-driven dialogues around health. For instance, when they address myths about vaccinations or provide parenting tips through visually engaging content, they directly influence the perceptions of their followers. Next, the use of visual content is a powerful tool for educating parents and overcoming barriers to understanding that may exist with text alone. The interactive aspect of social media, where parents can ask questions and engage in discussions, adds another layer to this dynamic. These interactions represent active participation in a community where health knowledge is co-constructed, which is critical in netnographic studies aiming to understand how online interactions translate into real-world behaviors. Bourdieu emphasizes that cultural capital is not static; it is dynamically constructed through social interaction and engagement with cultural content (Bourdieu, 1984, p. 170).

Through this lens, pediatricians' influence extends beyond the screen—they shape health practices by engaging with cultural norms, dispelling myths, and fostering community dialogue. Social media has become not just a channel for information distribution but a transformative tool for public health education, capable of altering how individuals and communities perceive and act upon health information. This aligns with Bourdieu's assertion that cultural capital includes the knowledge and skills necessary to navigate social spaces, which, in this case, is enhanced through the effective use of digital media in healthcare communication (Bourdieu, 1986).

In conclusion, this study validates the important role Instagram plays in shaping contemporary health communication, particularly in the case of child health. The high levels of engagement between parents and pediatrician influencers underscore a significant shift toward digital platforms as trusted sources of health information. Through likes, comments, and shares, parents not only consume content but also actively participate in building a community of collective learning and support. This interactive dynamic reinforces the credibility of the influencers and amplifies the impact of the health information shared. As more parents turn to social media for guidance, the potential for these platforms to enhance child health literacy and improve health outcomes becomes increasingly evident.

This circular diagram, as shown in Figure 9.3 illustrates that AI plays a significant role in creating a dynamic content cycle that enhances engagement of the consumers as audience members by helping the influencers who act as a producer of new posts on Instagram. AI makes it easier for pediatric influencers to create content based on the discussions and queries made in comments by their audience's (followers) needs and interests. It guides them to prioritize the most relevant topics so that the content resonates with their followers. It also helps them to review which posts receive the most engagement, such as likes and comments, and helps them to adjust their content strategy accordingly. Therefore, after the content is shared, AI uses algorithms to maximize the visibility of pediatric influencers by reaching those audiences who seek similar forms of specific information. In this way, a loop is created that is continuous

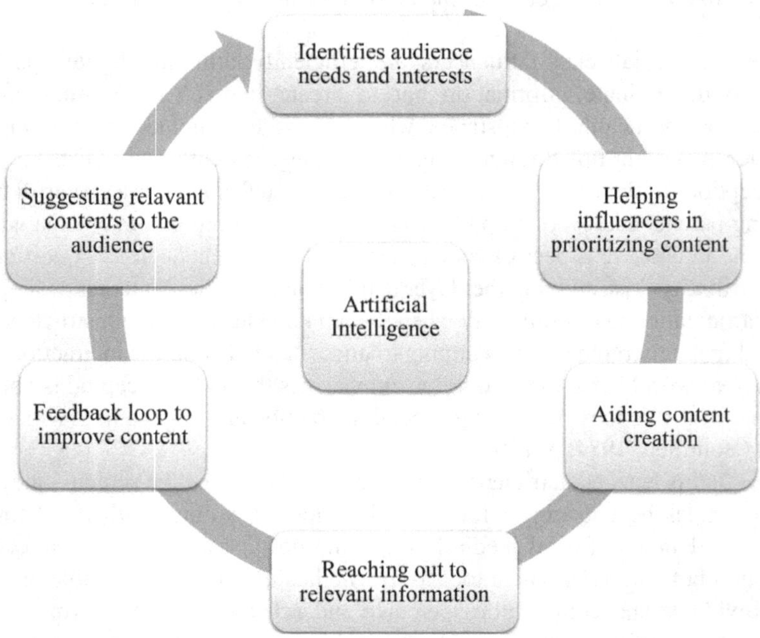

FIGURE 9.3 Continuous content cycle.

to capture audience reactions and allow AI to refine its suggestions to guide the pediatric influencers toward more relevant topics related to child health.

Berger explores in his book *Ways of Seeing* that the images contain some meanings that are shaped by cultural and social aspects. This act of seeing is influenced by societal power systems and inherited values. By using his concept, the role of AI in content creation is a new approach toward health communication. It is shaping and clarifying the consumer's (followers) ways of seeing and perceiving. This is helping the audience to come back to the influencers for more information. Moreover, AI strengthens cultural beliefs by prioritizing visual content and comments, which results in a reciprocal interaction between consumers (followers) and content. In this way, AI generates "performance" based on the expectations and interests of the followers (Goffman, 1959). According to Goffman, social interactions are prearranged according to "frontstage" and "backstage" actions. The influencer's Instagram posts, stories, and videos serve as the "frontstage" on social media, where they put a view by giving an informative message in an ideal form to their followers. AI simplifies this process by giving influencers knowledge about the wants and requirements of their followers as well as their audience. It enables them to "rehearse" content that appeals to them and builds their reputation. AI's feedback loop serves as a real-time audience feedback system that allows influencers to adapt their front-stage act according to what gets the most interaction. Side by side, the "backstage", in this context, includes the use of AI tools and content analytics by the pediatric influencers which helps them to keep track of what is effective and what is not effective content. They adjust their front-stage performance based on this backstage knowledge in order to improve their social media image. In this essence, pediatric influencers use AI-driven analysis of the followers' preferences and engagement patterns, which continuously control their impressions, improving their digital "selves" to meet the demands of their audience. These pediatricians are promoting child health literacy by prioritizing relevant topics—such as child vaccinations, nutrition, common health concerns, and myth-busting—based on the parents as followers actively engaging themselves in seeking health information from them. Therefore, by presenting child health information visually on Instagram in the form of videos, stories, and infographics, they make complex health concepts easier for parents to understand. So, Goffman's impression management theory applies here, where pediatric influencers are contributing to child health literacy by providing parents with easily accessible and comprehensible health information.

The findings highlight the need for healthcare professionals to embrace and optimize their digital presence, recognizing social media's powerful role in child health education and engagement. Observational netnography and interpretative analysis have proved to be a valuable tool for gaining insights into the impact of doctors' social media presence on child health literacy. Further research in this area could continue to shed light on the evolving dynamics of social media and its impact on health-related messaging.

REFERENCES

Agarwal, R. (2022). Predictive analysis in health care system using AI. In Garg, L., Basterrech, S., Banerjee, C., Sharma, T.K. (Eds.), *Artificial Intelligence in Healthcare* (pp. 117–131). Springer. https://doi.org/10.1007/978-981-16-6265-2_8

Bardhan, S. (2022). #Egyptiangirl and #Tunisiangirl: The (micro)politics of self-presentation on Instagram. *International Journal of Communication, 16,* 681–699.

Berger, J. (2008). *Ways of seeing*. Penguin UK (Original work published 1972).

Berkman, N. D., Sheridan, S. L., Donahue, K. E., Halpern, D. J., & Crotty, K. (2011). Low health literacy and health outcomes: An updated systematic review. *Annals of Internal Medicine, 155*(2), 97–107. https://doi.org/10.7326/0003-4819-155-2-201107190-00005

Borchers, N. S. (2019). Social media influencers in strategic communication. *International Journal of Strategic Communication, 13*(4), 255–277. https://doi.org/10.1080/1553118X.2019.1634075

Bourdieu, P. (1984). *Distinction: A social critique of the judgment of taste*. Harvard University Press.

Bourdieu, P. (1986). The forms of capital. In J. G. Richardson (Ed.), *Handbook of theory and research for the sociology of education* (pp. 241–258). Greenwood Press.

Boyd, D. M., & Ellison, N. B. (2007). Social network sites: Definition, history, and scholarship. *Journal of Computer-Mediated Communication, 13*(1), 210–230. https://doi.org/10.1111/j.1083-6101.2007.00393.x

Butdisuwan, S., Annamma, L. M., Subaveerapandiyan, A., George, B. T., & Kataria, S. (2024). Visualising medical research: Exploring the influence of infographics on professional dissemination. *Scientific World Journal, 2024,* 5422121. https://doi.org/10.1155/2024/5422121 PMID: 38919961; PMCID: PMC11199065.

Campbell, C., & Farrell, J. R. (2020). More than meets the eye: The functional components underlying influencer marketing. *Business Horizons, 63*(4), 469–479. https://doi.org/10.1016/j.bushor.2020.03.003

Chen, J., & Wang, Y. (2021). Social media use for health purposes: Systematic review. *Journal of Medical Internet Research, 23*(5), e17917. https://doi.org/10.2196/17917

Chen, W. (2015). Mediatizing the network model of cultural capital: Network diversity, media use, and cultural knowledge along and across ethnic boundaries. *Social Networks, 40,* 185–196. https://doi.org/10.1016/j.socnet.2014.10.003

Chowdhury, G. G. (2003). Natural language processing. *Annual Review of Information Science and Technology, 37*(1), 51–89. https://doi.org/10.1002/aris.1440370103

Das, A., & Faxvaag, A. (2014). What influences patient participation in an online forum for weight loss surgery? A qualitative case study. *Interactive Journal of Medical Research, 3*(1), e4. PMID: 24509408; PMCID: PMC3936279. https://doi.org/10.2196/ijmr.2847

Durkheim, E (1964). *The elementary forms of the religious life*. Allen & Unwin.

Erickson, B. H. (1996). Culture, class, and connections. *American Journal of Sociology*. https://doi.org/10.1086/230912.

Garcia, A. C., Standlee, A. I., Bechkoff, J., & Cui, Y. (2009). Ethnographic approaches to the internet and computer-mediated communication. *Journal of Contemporary Ethnography, 38*(1), 52–84. https://doi.org/10.1177/0891241607310839

Goffman, E. (1959). *The presentation of self in everyday life*. Knopf Doubleday Publishing Group.

Gupta, S. (2023). Social media marketing: Understanding social media algorithms through business growth. In A. Chakrabarti & V. Singh (Eds.), *Design in the Era of Industry 4.0, Volume 3* (pp. 91–103). Springer. https://doi.org/10.1007/978-981-99-0428-0_91

Hong, S., Jahng, M. R., Lee, N., & Wise, K. R. (2020). Do you filter who you are?: Excessive self-presentation, social cues, and user evaluations of Instagram selfies. *Computers in Human Behavior, 104,* 106159. https://doi.org/10.1016/j.chb.2019.106159

Hopkins, Z. H., Moreno, C., & Secrest, A. M. (2020). Influence of social media on cosmetic procedure interest. *Journal of Clinical and Aesthetic Dermatology, 13*(1), 28–31.

Househ, M., Borycki, E., & Kushniruk, A. (2014). Empowering patients through social media: The benefits and challenges. *Health Informatics Journal, 20*(1), 50–58. https://doi.org/10.1177/1460458213476969

Hussain, K., Khan, M. L., & Malik, A. (2024). Exploring audience engagement with ChatGPT-related content on YouTube: Implications for content creators and AI tool developers. *Digital Business, 4*(1), 100071. https://doi.org/10.1016/j.digbus.2023.100071

Inflact. (2024). Profile analyzer. Retrieved August 10, 2024, from https://inflact.com/tools/profile-analyzer/

Kickbusch, I. S. (2001). Health literacy: Addressing the health and education divide. *Health Promotion International, 16*(3), 289–297. https://doi.org/10.1093/heapro/16.3.289

Korda, H., & Itani, Z. (2013). Harnessing social media for health promotion and behavior change. *Health Promotion Practice, 14*(1), 15–23. https://doi.org/10.1177/1524839911405850

Kozinets, R. V. (2000). The field behind the screen: Using netnography for marketing research in online communities. *Journal of Marketing Research, 39*(1), 61–72. https://doi.org/10.1509/jmkr.39.1.61.18935

Kozinets, R. V. (2010). *Netnography: Doing ethnographic research online*. Sage.

Kozinets, R. V. (2015). *Netnography: Redefined* (2nd ed.). Sage Publications.

Kutner, M., Greenberg, E., Jin, Y., & Paulsen, C. (2006). *The health literacy of America's adults: Results from the 2003 National Assessment of Adult Literacy* National Center for Education Statistics. https://nces.ed.gov/pubsearch/pubsinfo.asp?pubid=2006483

Little-Wienert, K., Chang, T., Agarwal, R., Cramton, R., Hillenbrand, K., Panchal, A., Stubblefield, W., Mahan, J., Wright, M., Donato, L., & Chandran, L. (2021). Enhancing pediatricians' engagement on social media: The role of board style questions. *MedEdPublish, 10*(1). https://doi.org/10.15694/mep.2021.000177.1

Livingstone, S. (2009). On the mediation of everything: ICA Presidential Address 2008. *Journal of Communication, 59,* 1–18. https://doi.org/10.1111/j.1460-2466.2008.01401.x

Loughlin, M., Lewith, G., & Falkenberg, T. (2013). Science, practice, and mythology: A definition and examination of the implications of scientism in medicine. *Health Care Analysis, 21*(2), 130–145. https://doi.org/10.1007/s10728-012-0201-8

Martimianakis, M. A. T., Tilburt, J., Michalec, B., & Hafferty, F. W. (2020). Myths and social structure: The unbearable necessity of mythology in medical education. *Medical Education, 54*(1), 15–21. https://doi.org/10.1111/medu.13828

Mesko, B., Drobni, Z., Bényei, E., Gergely, B., & Gyorffy, Z. (2017). Digital health is a cultural transformation of traditional healthcare. *Mhealth, 3,* 38. https://doi.org/10.21037/mhealth.2017.08.07

Muniz, A. M., & O'Guinn, T. C. (2001). Brand community. *Journal of Consumer Research, 27*(4), 412–432. https://doi.org/10.1086/319618

Norman, C. D., & Skinner, H. A. (2006). eHealth literacy: Essential skills for consumer health in a networked world. *Journal of Medical Internet Research, 8*(2), e9. https://doi.org/10.2196/jmir.8.2.e9

Nutbeam, D. (2000). Health literacy as a public health goal: A challenge for contemporary health education and communication strategies into the 21st century. *Health Promotion International, 15*(3), 259–267. https://doi.org/10.1093/heapro/15.3.259

Salma, B., Fatima, T., Sara, A., & Merieme, B. (2024). Artificial intelligence in social media: From content personalization to user engagement. In Y. Farhaoui (Ed.), *Artificial intelligence, big data, IOT and block chain in healthcare: From concepts to applications* (Vol. 6). Springer. https://doi.org/10.1007/978-3-031-65018-5_5

Smailhodzic, E., Hooijsma, W., Boonstra, A., & Langley, D. J. (2016). Social media use in healthcare: A systematic review of effects on patients and on their relationship with healthcare professionals. *BMC Health Services Research, 16*(1), 442. https://doi.org/10.1186/s12913-016-1691-0

Sun, G., & Zhou, Y.-H. (2023). AI in healthcare: Handling potential and obstacles in digital communication. *Frontiers in Digital Health, 5,* 1291132. https://doi.org/10.3389/fdgth.2023.1291132

Tafesse, W., & Wood, B. P. (2020). Followers' engagement with Instagram influencers: The role of influencers' content and engagement strategy. *Journal of Retailing and Consumer Services, 58,* 102303. https://doi.org/10.1016/j.jretconser.2020.102303

Thompson, J. B. (2013). *Media and modernity: A social theory of the media.* Polity Press.

Väänänen, A., Haataja, K., Vehviläinen-Julkunen, K., & Toivanen, P. J. (2021). AI in healthcare: A narrative review. *F1000Research, 10,* 6. https://doi.org/10.12688/f1000research.26997.2

Vaca, E., Dorfman, R. G., Fine, N. A., & Schierle, C. F. (2017). Ethics and professionalism with Snapchat use in plastic surgery. *Plastic and Reconstructive Surgery – Global Open, 5*(9S), 44–45. https://doi.org/10.1097/01.GOX.0000526224.33465.92

Ventola, C. L. (2014). Social media and health care professionals: Benefits, risks, and best practices. *Pharmacy and Therapeutics, 39*(7), 491–499.

Wittgenstein, L. (1953). *Philosophical Investigations* (G. E. M. Anscombe, Trans.). Basil Blackwell.

Zarei, F., & Rakhshani, F. (2015). Netnography: A method to study health cultures and communities online. *Journal of Advanced Nursing Research, 3*(4). Retrieved from www.academia.edu/34273495

10 AI in Enhancing Social Services for Vulnerable Populations

Sheeba Khalid

10.1 INTRODUCTION

The term artificial intelligence (AI) was introduced in 1955 by John McCarthy, a professor at Stanford University. AI integrates computer science and data sets to emulate human intelligence and facilitate problem-solving across various situations. AI encompasses machine learning, which utilizes historical data to forecast and generate new outcomes. Generative AI denotes the production of images, videos, audio, text, and 3D models by employing learned patterns from existing data to create novel outputs. The European Commission (2019) has delineated AI as follows:

AI denotes systems that exhibit intelligent behavior by analyzing their surroundings and autonomously executing activities to attain certain objectives. AI systems may exist solely as software, functioning in the virtual realm (e.g., voice assistants, image analysis software, search engines, speech and facial recognition systems), or they can be integrated into hardware devices (e.g., advanced robots, autonomous vehicles, drones, or Internet of Things applications) (page 1). AI can manifest as expert systems, natural language processing, speech recognition, and machine vision. AI relies on algorithms to improve machine learning, reasoning, self-correction, and creativity. AI in healthcare has been employed to diagnose diseases, enhance patient treatment, automate repetitive operations, manage medical records, deliver customer support via chatbots, minimize dosage errors, offer robot-assisted services, analyze patient scans, and identify fraud (Reamer, 2023).

AI is progressively assuming a crucial role in improving social services for at-risk populations, providing novel avenues to tackle intricate societal issues. Marginalized populations, including the elderly, those with impairments, refugees, and those experiencing financial difficulties, sometimes face obstacles in obtaining needed services. Conventional social service systems, although beneficial, can be resource-demanding and ineffective, frequently failing to address the varied needs of various groups. AI technologies, such as machine learning, natural language processing, and predictive analytics, possess the capacity to transform these processes by facilitating more efficient, personalized, and proactive assistance. Utilizing AI, social services can enhance their responsiveness and customization to meet individual needs. AI-driven platforms can evaluate extensive data sets to forecast future threats, enabling social workers to intervene prior to crises arising. Virtual assistants and chatbots offer instantaneous

DOI: 10.1201/9781003560463-12

assistance and information to individuals, transcending linguistic obstacles and geographical constraints. The capacity of AI to discern trends in data can enhance resource allocation, ensuring that assistance is delivered more efficiently to the most vulnerable populations. The incorporation of AI in social services presents significant ethical issues, including the safeguarding of data privacy, the reduction of biases, and the preservation of human oversight. As AI advances, it serves as a potent instrument to enhance social services, rendering them more inclusive and adept at promptly and efficiently meeting the needs of vulnerable people.

AI is progressively utilized in social services to support at-risk groups, providing prospects for customized services and enhanced efficiency (Placeholder1). Nonetheless, its implementation prompts ethical concerns, especially with potential discrimination and the exclusion of disadvantaged consumers (Placeholder1; Reamer, 2023). Applications of AI in social work encompass risk assessments, crisis intervention, and result forecasting (Reamer, 2023). Researchers offer frameworks for the ethical integration of AI in social services to tackle ethical difficulties, including accessibility, interactive optimization, and enhancements in dynamic decision-making (Placeholder1) (Seniutis et al., 2024). Although AI presents potential advantages in public administration, including improved access and efficiency, it also poses dangers of marginalizing technology-deficient clients and insufficient accountability (Carney, 2021). The formulation of extensive ethical rules and meticulous evaluation of client attributes are essential for the effective integration of AI in social welfare systems (Seniutis et al., 2024; Carney, 2021).

There is a need to take all-inclusive social services focused on the vulnerable population in order to unite AI with the vision of creating new answers to the challenges of society. Among them, there are older people, disabled people, refugees, and those who are experiencing economic difficulties since they cannot always count on critical services at most times. However, despite the development of classical social service systems, such structures are always resource-intensive and ineffective, failing to notice the peculiar needs of the varied segments of populations. The efficient use of AI technologies, especially machine learning, natural language processing, and predictive analytics, may just enable such an innovation in practice: more efficient assistance, more relevant, and probationary.

Social services can be made more responsive and market their offerings toward meeting individualized needs with AI. Such AI-based technologies would analyze humongous data sets and hence make for strong predictions of possible future dangers, thus allowing researchers to intervene in the lives of social workers before emergencies occur. The virtual assistant and chatbot can be used to help users in a digital environment, having instant assistance and providing information without representatives, hence enabling people speaking several languages and located in geographically dispersed locations to communicate with one another.

With such rapid data pattern recognition from the AI, resource allocation is now more wisely channeled to the most severely threatened groups. Included are such important ethical concerns as data privacy, bias reductions, and human oversight in the use of AI in social services. These capabilities are highly powerful and enable the efficiency of social services in being comprehensive in yielding responses to meet the needs of at-risk populations in relatively quick and suitable manners. This is an attempt to critique the influence AI exerts on services provided in social work toward disadvantaged groups. In the current study, an effort will be made to trace and discuss the benefits emanating from the increased efficiency and accessibility of AI-facilitated social services mainly accruing to vulnerable groups.

The AI study further examines the ethical considerations of deploying AI, concerning issues of bias and the likelihood that such advantages will contribute to instances of marginalization in sub-populations. The study will critically look at current and future applications of AI in social work within risk assessments, crisis intervention, and outcome prediction, thus linking up with the ongoing debate regarding the ethical deployment of AI in services for social services. For example, several propositions by scholars have been advanced toward developing appropriate frameworks to enable the infusion of AI into social services with such ends in view. This includes improved accessibility, interactivity of optimization techniques, and improvement in dynamic decision-making. While AI promises increased accessibility and greater efficiency in public administration, there is a risk of excluding those with lower skills in information and communication technologies. Exclusion raises the specter of insufficient accountability on their part. Hence proper identification of clients, along with a strict ethical framework, constitutes the first two preconditions to the successful assimilation of AI in welfare systems. In this chapter, it will be argued that properly implemented and supervised by diligent and meticulous passion; AI could radically decrease disparities, empower community constituents, and create social inclusion that would improve welfare conditions for the most marginalized parts of society.

AI is transforming social services into more efficient, personal, and proactive assistance toward at-risk groups. Predictive analytics, machine learning, and natural language processing can be leveraged in AI technology to predict future dangers through immediate support and thereby augment resource allocation. However, there are also some adverse ethical issues in AI about data privacy and the existence of algorithmic prejudices. It intends to talk about how AI may help in social services by altering the risks and capabilities at a time when they require assistance through risk assessment, crisis intervention, and predictive modeling. Risks involved due to exclusion that can occur from unequal implementation of AI, excluding those not privileged enough to access technology, are further discussed.

This chapter deals with AI potential for social services when dealing with vulnerable populations as well as the ethical implications this will raise. Some examples of applications in areas such as risk assessment and crisis intervention and predictive modeling are discussed. We use new research to speak of frameworks for ethical incorporation and general principles of transparency, fairness, and accountability. Finally, this chapter would look at the risks of exclusion on people with limited access to technology and highlight strategies for equitable implementation of AI that can then reduce social and economic disparities and foster social inclusion among these groups.

AI in public administration: advantages and challenges public administration is fast-changing with benefits from AI. Still, as technology advances, the world governments realize that not only can I speed up things, save money, but also increase the quality of services delivered. It automates routine mailing and filing of documents, creating access and convenience for both administrators and citizens. Benefits include access to government services, enhancement of efficiency, data-driven decision-making, savings, and predictability of trends, including the area of public health, traffic patterns, or demands for social services.

However, AI brings related risks in public administration. Digital divide remains one of the biggest challenges because susceptible populations could be deprived of opportunities without access to technology or digital literacy and due to services brought about by AI. Poor design and accountability is an essential component of fairness and inclusion, and errors as well as redress mechanisms must be defined in order to protect citizens's rights.

Importantly, there are ethical considerations because the deployment of AI raises questions concerning surveillance, privacy, and consent. Policymakers should cut these complexities in order for public trust in AI initiatives to be rekindled. Public administration will not abate nor particularly intensify its relations with AI, since such complicated and promising integration will rather focus on the potential improvement of efficiency, accessibility, and decision-making, huge risks to be managed. A balanced approach emphasizing ethical questions alongside inclusivity will be critical in how this may go right (Figure 10.1).

10.2 THE NEED FOR AI IN SOCIAL SERVICES FOR VULNERABLE POPULATIONS

AI technologies can be used to tailor services to consumers's needs, but they also present ethical and operational challenges. This work aims to leverage AI to detect, serve, and empower vulnerable consumers by designing accessible services, optimizing customer experiences, and improving decision-making. The framework considers the implications for researchers, managers, consumers, and policymakers.

AI is transforming the social security sector by enabling proactive and automated delivery of social services. It is already providing relevant outcomes in areas such as error, evasion, and fraud, as well as developing effective automated solutions to customer's concerns. The applications of AI include medical care, robots carrying out dangerous activities, communication with insured people, and welfare benefit management. However, the rapid introduction of automated technological solutions poses significant challenges and potential risks. The increasing relevance of AI in social security is both inevitable and potentially beneficial for all parties involved. However, careful planning and a comparative assessment of the benefits and challenges of AI versus human labor are necessary. Many government agencies, including the UK, Australia, and Norway, are pursuing opportunities to utilize AI without a clear understanding of its costs, benefits, and risks for users. Since AI deployment is still in its infancy in the social security sector, public authorities and administrators may not be aware of the full range of AI application opportunities and related challenges. There is also little specific research on AI, which fails to provide an integrated view of AI applications and challenges for the social security sector (Benouachane, 2022).

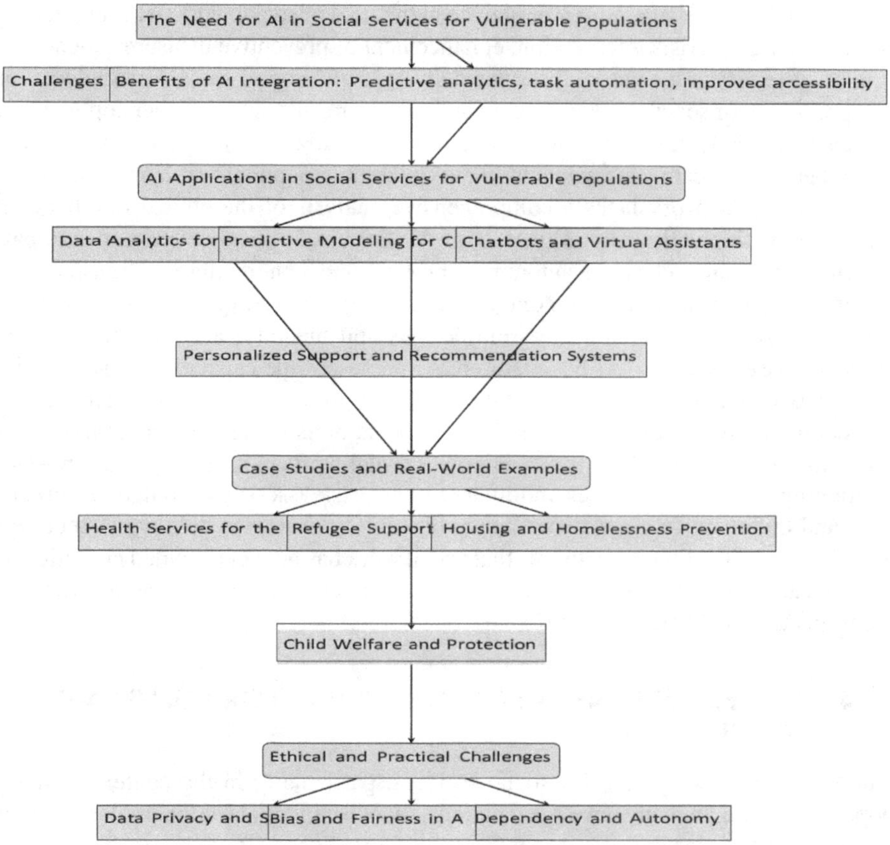

FIGURE 10.1 The needs for AI in social services for vulnerable populations.

10.3 CHALLENGES AND RISKS OF AI FOR SOCIAL SECURITY

AI) in social security presents substantial prospects for enhancing service delivery, efficiency, and fraud detection (Benouachane, 2022). Nonetheless, its execution also poses problems and hazards. Substandard design and hasty execution can disadvantage clients lacking technological resources and inadequately embody social welfare principles (Carney, 2020; Carney, 2021). The Australian experience reveals inconsistent outcomes attributable to these issues and a deficient comprehension of client characteristics and outsourced service dynamics (Carney, 2021). Notwithstanding these hurdles, the implementation of AI in social security is deemed inevitable and potentially advantageous (Benouachane, 2022; Carney, 2020). The International Social Security Association (ISSA) has established guidelines to promote the use of data-driven technology in social security (Ruggia-Frick, 2021). Effective execution necessitates meticulous preparation, weighing advantages against obstacles, and guaranteeing sufficient accountability and remedial measures (Benouachane, 2022; Carney, 2020).

AI is progressively gaining prominence in the field of social work. AI is utilized for risk assessments, crisis intervention, enhancement of preventive measures, identification of systemic biases in social service delivery, provision of social work education, and prediction of social worker burnout and service results, among other applications. A substantial body of literature exists regarding the utilization of AI by social workers and other human care professionals to aid vulnerable populations. However, the literature on social work lacks a comprehensive analysis of the ethical ramifications associated with practitioners utilization of AI. Principal ethical concerns encompass informed consent and client autonomy; privacy and confidentiality; transparency; client misdiagnosis; client abandonment; client surveillance; plagiarism, dishonesty, fraud, and misrepresentation; algorithmic bias and inequity; and the utilization of evidence-based AI tools. AI has matured. AI possesses the capacity to revolutionize social work and augment the professions efficacy in serving clients, organizations, and communities. AI can be utilized in clinical, administrative, advocacy, and policy domains within social work. It can empower social workers to rapidly examine data, facilitating significant services and interventions, risk assessment, outcomes prediction, and initiatives to combat systematic bias in social service delivery. A thorough examination of the literature reveals that social work has not yet provided practitioners and educators with a detailed analysis of significant ethical dilemmas associated with the utilization of AI (Reamer, 2023).

10.4 AI APPLICATIONS IN SOCIAL SERVICES FOR VULNERABLE POPULATIONS

Due to the increasing need for social services, particularly in the context of rising population aging in many developed nations, there is potential for the implementation of AI methodologies in this domain (Torras, 2015). The swift advancement of social robotics is primarily associated with the creation of robots for collaboration with individuals with disabilities (Taheri et al., 2019) and the elderly (Rodic et al., 2016), as well as robots intended to function as educational assistants (Cooney & Leister, 2019; Han et al., 2008; Vitanza et al., 2019) or caregivers, which are starting to transform social services in numerous nations. The current capacity to deliver social services through human labor and technological applications offers a novel perspective on social services. Technologies employed to administer, manage, and deliver social services encompass assistive technologies like mobile applications or screen readers (Wynne et al., 2016), as well as advanced assisted living technologies such as telecare, telehealth, smart homes, and social robots (Dunn, 2014; Wigfield, 2013).

A primary problem posed by current demographic trends is the aging population, necessitating innovative approaches to enhance the quality of life for the elderly. Significant advancements in mechanics, electronics, and computer technology have resulted in the creation of humanoid robots, which are intended for interaction with individuals in caregiving and the service industry, rather than in industrial (Torras, 2015). This technological progress in assistive robots provides many skills to facilitate long-term independent living, enhance healthcare, and mitigate social isolation among the elderly (Koceska et al., 2019). Support technologies, including e-care systems, proactive service systems, interactive home robots, and tracking systems for the elderly, are currently utilized in conjunction with nurses and family members to enhance the quality of life for older individuals through emotional and physical alleviation (Broadbent, 2009; Sasaki et al., 2007; Wada, 2003). Assistive robots for the elderly can be categorized as rehabilitative robots and socially assistive robots (Broekens, 2009; Feil-Seifer, 2005). Rehabilitation robots offer physical support, including artificial limbs and intelligent wheelchairs, while social assistance robots function as companion or service robots (Flandorfer, 2012; Torras, 2015). Nonetheless, the domain of social robotics has evolved beyond merely creating robots that adhere to fundamental care principles such as entertainment, safety, and healthcare. Geriatric medicine has challenges centered on enhancing the symptomatic identification of frailty (Torras, 2015). Robotics may therefore enhance social services and facilitate connections for users both at home and in medical facilities. Enhanced care and assistance for independent living is undoubtedly more economical and, crucially, signifies progress toward greater autonomy for the elderly. Social robotics encompasses the deployment of socially interactive robots to assist individuals with disabilities, and the collaboration of these robots with children diagnosed with autism is gaining significant traction. Specifically, it pertains to addressing deficits in social, emotional, and communicative domains. The creation of socially interactive robots, specifically imitator robots, is frequently employed to instruct children in basic coordinated behaviors through their imitative capabilities, benefiting not only children with autism but also those undergoing rehabilitation for complex developmental disabilities (Billard, 2007). These robots provide interactive games for several children, engaging them in a cooperative activity of alternating imitations of various stances. These games aim to enhance attention, improve turnover skills, and facilitate both verbal and non-verbal communication (Boccanfuso & O'Kane, 2011). Robots, autonomous systems, AI, networked data, and digital infrastructure can potentially revolutionize the delivery of social and healthcare services.

AI has the transformative potential within social services that provides innovative solution mechanisms for helping vulnerable population groups more efficiently and appropriately. Some of the uses of AI in social services include the following:

Data analysis for identification of need: AI could look across large, diverse data sets of health records or socioeconomic data to flag potential risks for problems like homelessness, domestic abuse, or a health crisis. Insights feed into targeted resource allocation so that social services can focus efforts at the right people who need help right now.

Predictive modeling for crisis intervention: AI can predict potential crises using predictive models prior to them getting too rough. It can identify certain patterns that may lead to a crisis in the future and alert social workers to take the necessary action. It can therefore be used in substance abuse, mental health crises, and chronic illness management.

Chatbots and virtual assistants: AI-powered chatbots provide instant assistance, offering emotional support, information, and guidance. This application is particularly valuable for individuals facing mental health challenges, as it enables access to services 24/7, overcoming barriers like language, geographic location, and social stigma.

Personalized support and recommendation systems: Social services can be tailored according to the unique needs, preferences, and history of an individual by AI. For instance, AI may suggest the right healthcare or educational resources that may make service provision more effective and user-centric. Real-time monitoring and alert the AI-powered monitoring systems can track real-time data, including the biometrics of the elderly and disabled, to immediately report anomalies to caregivers or emergency responders. This creates an environment of safety and provides instant response in emergencies, including health emergencies. It does more than optimize efficiency and increase accessibility and provides the possibility of freeing up the human service to perform more sophisticated work rather than ritualistic, including increasing data privacy and ensuring an opportunity for the process through the participation of the human senses for ensuring equitable results with respect to efficiency.

AI presents advantageous applications for assisting aging populations and at-risk consumers. AI can augment independent living for elderly individuals by tracking health metrics, assisting with daily tasks, and fostering social interaction (Czaja & Ceruso, 2022; Mhlanga, 2024). The technology can be utilized to identify and enhance services for vulnerable consumers by improving accessibility, facilitating engaging experiences, and refining decision-making (Placeholder1). The implementation of AI in senior care raises ethical dilemmas, such as privacy issues and the necessity for human-centered methodologies (Mhlanga, 2024). As nations encounter heightened demands on adult social care due to aging populations, AI and robotics may significantly contribute to meeting care requirements (Morgan, 2021). A multidisciplinary strategy that includes seniors, caregivers, healthcare experts, and AI developers is essential for the proper integration of AI, ensuring that technology aligns with the genuine requirements and preferences of older persons (Mhlanga, 2024).

10.5 AI-DRIVEN REMOTE HEALTH MONITORING: HEALTH SERVICES FOR THE ELDERLY

The aging population of Japan demands novel healthcare solutions. The principal market drivers are a diminishing workforce and escalating expenses for aged care. Remote monitoring is seen as a crucial element in providing excellent care for the Silver Generation. AI-enhanced health monitoring devices are essential to this change. These technologies allow older folks to oversee their health from the comfort and security of their homes, promoting autonomy and wellness. Additionally, they enable remote healthcare delivery by medical practitioners, particularly for individuals with chronic diseases. Partnerships between international and Japanese firms are expediting the advancement of these instruments. Opportunities in the Japanese industry for AI-driven geriatric care undoubtedly encompass this form of collaboration. AI technology integrated with these devices can provide real-time health data, facilitate early detection of potential health concerns, and deliver personalized health insights customized to individual requirements. The use of these modern technologies in Japan is crucial for addressing the particular healthcare needs of its elderly population.

AI-augmented health monitoring gadgets are transforming the methodology of senior care in Japan. These advanced tools enable elders to independently manage their health. They also offer a method for healthcare practitioners to administer remote care, check medication adherence and efficacy, and establish a direct communication channel for the elderly, enhancing their well-being and safety. The collaboration among leading firms generates a new array of tools. These AI-integrated gadgets deliver instantaneous health data, early indicators of health concerns, and personalized health insights. The incorporation of these technologies in Japan is crucial for meeting the specific healthcare requirements of its elderly demographic.

Recent studies highlight the growing importance of AI-driven remote health monitoring for the elderly. These systems integrate IoT, wearable devices, sensors, and AI to continuously monitor vital signs, emotional states, and cognitive conditions (Cantone, 2023; Awotunde, 2022). Such technologies enable early detection of health issues, particularly cardiac problems in seniors (Paraschiv, 2021). AI-powered remote patient monitoring devices like Dozee.ai and Qure.ai have proven effective in diagnosing COVID-19 patients and assisting healthcare personnel during critical times. These systems can reduce reliance on institutional care, allowing elderly individuals to maintain autonomy while receiving timely medical attention (Cantone, 2023). Despite their advantages, concerns remain regarding data privacy and security. Overall, AI-driven remote health monitoring shows promise in addressing the healthcare challenges posed by an aging population, potentially improving outcomes and reducing the workload on medical personnel (Awotunde, 2022).

10.6 AI-DRIVEN LANGUAGE TRANSLATION AND SOCIAL INTEGRATION FOR REFUGEES

A machine learning approach that efficiently and precisely pairs translator requests with volunteers aids in surmounting language hurdles for displaced individuals. Inhabitants of underdeveloped nations are disproportionately vulnerable to displacement due to humanitarian disasters. In times of crisis, linguistic challenges hinder humanitarian personnel from delivering services to displaced individuals. To cultivate resilience, such services must be adaptable and resilient to a variety of potential languages. Tarjimly seeks to eliminate obstacles by offering a platform that connects bilingual volunteers with displaced individuals or humanitarian workers who need translation services. Nonetheless, Tarjimly's extensive network of translators presents the difficulty of identifying the appropriate translation for each request. Recent studies underscore the substantial influence of AI-powered language translation and mobile applications on the integration of refugees (Agarwal et al., 2020). Recent studies underscore the capacity of AI and mobile technologies to enhance refugee assimilation. Martin-Gutierrez et al. (2020) present a framework that integrates AI technologies for user profile, skills alignment, and suggestions. Simao et al. (2018) present GeeBot, a robotic platform that provides translation and information services for refugees. Drydakis (2021) illustrates favorable correlations between mobile applications for immigrant integration (m-Integration) and enhanced integration levels, overall health, and mental well-being. AI-enhanced applications demonstrate heightened efficacy, with translation and voice assistant applications associated with a 4% rise in integration and a 0.8% enhancement in overall health. These technologies provide tailored services, instantaneous translations, and virtual communication, tackling diverse integration difficulties. The results underscore the significance of AI in improving outcomes and indicate that these technologies can substantially assist in surmounting language barriers and enabling access to vital services for refugees and immigrants.

10.7 HOUSING AND HOMELESSNESS PREVENTION: PREDICTIVE MODELS TO PREVENT HOMELESSNESS

Local governments and NGOs in Los Angeles, California, employ the AI models for working against homelessness. The predictive models employ data drawn from a variety of places such as health records, housing histories, and financial data to identify who among the users is prone to homelessness. For example, the Coordinated Entry System employs AI to rank people based on their risk scores so that minimal stockpiles of housing and resources can be effectively used. Predictive algorithms have demonstrated efficacy in identifying those at elevated risk of homelessness and in prioritizing scarce resources. Toros (2017) constructed a model utilizing administrative information to forecast high-cost homeless individuals, attaining a Receiver Operating Characteristic (ROC) curve of 0.83 and projecting potential cost savings of $19,000 per individual. Kithulgoda (2022) developed a predictive risk model that surpassed the VI-SPDAT in detecting customers susceptible to various adversities. Greer (2016) created models for individuals and families, enhancing accurate forecasts by 77% and decreasing unrecognized instances of subsequent homelessness by 85% relative to staff assessments. These findings illustrate that empirical models can markedly enhance the efficacy of homelessness prevention interventions by precisely identifying individuals most in need (Greer, 2016; Toros, 2017; Kithulgoda, 2022).

10.8 CHILD WELFARE AND PROTECTION: AI-POWERED CHILD WELFARE MONITORING SYSTEMS

UK child welfare agencies are testing AI applications in the observance and protection of high-risk children. For instance, in London Borough of Hackney, there has already been pilot testing of AI systems analyzing different kinds of risk factors, including family history, school attendance, health record, and so on. These tools provide an early warning system that case workers can take action upon to intervene before things worsen.

The application of predictive analytics in child welfare systems has elicited much apprehension among scholars. Although these tools intend to enhance decision-making and resource distribution, they may reinforce prevailing biases and disparities (Redden, 2020; Stapleton et al., 2022). Research indicates that machine learning models may moderately accurately identify adolescents at high risk of aging out of foster care; yet, racial differences in error rates remain (Ahn, 2021). Critics contend that the swift deployment of these technologies is devoid of sufficient public discourse and scrutiny, which may jeopardize openness and accountability (Redden, 2020; Glaberson, 2023). Stakeholders affected by child welfare systems voice concerns on the continuation of prevailing issues and propose alternative strategies, such as low-tech alternatives and community-oriented initiatives (Stapleton et al., 2022). Researchers underscore the necessity of meticulously addressing the basic inquiries posed by predictive analytics to prevent merely coding over the cracks in a defective system (Glaberson, 2023).

AI has been utilized across multiple areas, including healthcare, refugee assistance programs, housing and homelessness prevention, and child welfare. In Japan, AI-powered remote health monitoring systems are employed to track biometrics, facilitating early intervention and empowering the elderly. In Europe, AI solutions such as Tarjimly utilize machine learning to deliver instantaneous language translation and facilitate social integration for refugees, enhancing access to vital services and diminishing dependence on human interpreters. In Los Angeles, California, AI algorithms are employed to identify persons at elevated risk of homelessness, enabling social agencies to more effectively target early intervention initiatives. In the UK, AI-driven surveillance technologies are employed to oversee and safeguard at-risk youngsters. These AI tools offer early warning signals, allowing caseworkers to prioritize cases according to risk level. Notwithstanding issues like data privacy and potential bias, AI has demonstrated potential in mitigating harm through the facilitation of early interventions.

10.9 AI-DRIVEN SOCIAL SERVICES FOR VULNERABLE POPULATIONS IN INDIA

AI can markedly improve social services for vulnerable populations in India. AI-powered chatbots such as SnehAI create a secure environment for adolescents to engage in discussions on sensitive issues like sexual and reproductive health, delivering precise information and resources (Wang et al., 2022). AI technology can promote efficiency, facilitate data-driven decision-making, and improve overall resilience in case management for vulnerable migrants (Ediae, 2024). The social venture Haqdarshak employs AI to disseminate social welfare programs to underserved communities, thereby bridging the information gap (Ramdey & Bokhari, 2020). AI solutions can facilitate youth engagement in community life, as evidenced by the deployment of digital technology in Lebanese municipalities (Boustani et al., 2022). These case studies underscore AI capacity to enhance service delivery, foster inclusion, and assist vulnerable populations, while also recognizing the necessity of addressing ethical considerations and problems related to AI adoption.

10.10 ETHICAL AND PRACTICAL CHALLENGES

The introduction of AI in social services involves prospects and ethical dilemmas. AI may enhance the process of service delivery, assist in conducting risk assessments, support crisis situations, and identify the presence of systemic biases (Reamer, 2023).

Institutions of social security are applying AI to obtain proactive and automated service provisions, fraud clearance, as well as customer service (Benouachane, 2022). However, there exist ethical issues of possible discrimination and accountability. To solve these problems, authors suggest frameworks of responsible AI development in social services (Seniutis, 2024) and suggest relevance to citizens engagement in the innovation process (Kinder, 2023). The key requirements are openness, simplicity, and careful planning before the application of AI solutions. While AI poses great potential benefits to social services, the implementation of such requires harmonization to maximize positive impacts and minimize adverse ones through ethical review and consultation with stakeholders (Benouachane, 2022; Kinder, 2023).

A variety of ethics-based resources about social workers utilization of technology exist to assist practitioners in employing AI across many professional settings. This chapter addresses three interconnected domains: (1) practice standards, (2) regulatory and licensing standards, and (3) ethical code standards. In 2013, the board of directors of the Association of Social Work Boards (ASWB) established an international task group to formulate model regulatory requirements for technology in social work practice. The ASWB initiated the creation of new ethics-related standards in response to global regulatory authorities requests for guidance on the growing technological practices of social workers. The ASWB task group comprised representatives from leading global organizations in social work practice, regulation, and education. The task force established standards for social workers utilizing digital and electronic technology to disseminate information to the public, deliver client services, communicate regarding clients, manage confidential data and case records, and store and access client information. The group established model standards, encompassing comprehensive ethical guidelines concerning technology use, which address seven principal ethical concepts: practitioner competence; informed consent; privacy and confidentiality; boundaries, dual relationships, and conflicts of interest; records and documentation; collegial relationships; and social work practice across jurisdictional boundaries. The model standards, officially ratified in 2015, are currently shaping the formulation of licensing and regulatory legislation globally (Lopez Pelaez, 2023), in conjunction with other initiatives in other countries (Diez, 2023; Pascoe, 2023). In 2017, after exceptional collaboration among prominent social work organizations in the US the National Association of Social Workers (NASW), Council on Social Work Education (CSWE), ASWB, and Clinical Social Work Association (CSWA) the profession officially embraced new, comprehensive practice standards, which included detailed ethics guidelines addressing the utilization of technology by social workers and social work educators (Leyen, 2019). Endorsed by the boards of directors of these organizations, these comprehensive standards tackle a broad spectrum of significant ethical concerns regarding social workers utilization of technology for public information dissemination, service design and delivery, information collection, management, storage, and social worker education. In 2017, NASW implemented a revised code that incorporates comprehensive technology-related provisions regarding informed consent, competent practice, conflicts of interest, privacy and confidentiality, sexual relationships, service interruption, unethical conduct of colleagues, supervision and consultation, education and training, client records, and evaluation and research.

Social workers utilizing or considering the use of AI must adhere to established guidelines and address critical ethical issues concerning informed consent and client autonomy; privacy and confidentiality; transparency; client misdiagnosis; client abandonment; client surveillance; plagiarism, dishonesty, fraud, and misrepresentation; algorithmic bias and inequity; and the application of evidence-based AI tools. These fundamental ethical principles must be incorporated into guidelines that inform social workers utilization of AI.

Social work professionals are increasingly utilizing AI in various fields, including social work, to improve accessibility, personalization, and proactive addressing of various needs. However, there are significant ethical challenges and risks associated with AI, such as informed consent, privacy, transparency, misdiagnosis, surveillance, plagiarism, dishonesty, fraud, misrepresentation, algorithmic bias, and inequity. Social workers should implement proactive measures to safeguard clients and themselves, such as familiarizing themselves with AI protocols, examining ethical and practice norms related to AI usage, developing and executing training for personnel, including AI information in social work education programs, and conducting workshops and webinars for practitioners.

10.11 PROSPECTIVE STRATEGIES AND POLICY PROPOSALS FOR ARTIFICIAL INTELLIGENCE IN SOCIAL SERVICES

India is developing policies to leverage AI for economic advancement and societal benefit (Marda, 2018; Chatterjee, 2020). Nonetheless, the constraints and hazards associated with data-driven decision-making must be proactively addressed within policy frameworks (Marda, 2018). An extensive AI policy for India must prioritize security, privacy, and governance concerns (Chatterjee, 2020). The deployment of AI in India encounters obstacles including ethical dilemmas, socioeconomic ramifications, and infrastructural necessities (Bansal & Jain, 2023). The Indian government has launched several initiatives to tackle these difficulties, such as a national AI plan, AI centers of excellence, and foreign partnerships (Bansal & Jain, 2023). AI potential uses in India encompass multiple domains, including the legal system, where it might markedly enhance efficiency (Kumar et al., 2023). Policy considerations must include data governance, responsible AI development, regulatory norms, skills enhancement, and inclusivity (Bansal & Jain, 2023). Researchers suggest frameworks to utilize AI for identifying, assisting, and empowering disadvantaged populations (Hermann, 2023) and for addressing the technical constraints of AI systems in policy formulation (Hermann, 2023). Essential policy considerations encompass data governance, responsible AI, regulation and standards, skills development, and inclusivity (Bansal & Jain, 2023). India's AI plan emphasizes the creation of AI centers of excellence, the cultivation of a startup environment, and international collaboration (Chatterjee, 2020; Bansal & Jain, 2023). Robust regulations are essential to optimize advantages while mitigating potential hazards in AI deployment.

Recent study underscores the necessity for ethical and inclusive methodologies in the integration of AI inside social services. Moon (2023) promotes participatory government to guarantee that AI serves the common good. Seniutis et al. (2024) present a conceptual framework for the ethical development of AI in social services, highlighting the necessity of confronting ethical difficulties and advancements in digitalization. Zhang E. et al. (2024) present Social Environment Design, a paradigm for AI-driven policymaking that integrates voting on policy objectives to enhance governmental decision-making. Ho (2023) advocates for investment in infrastructure to oversee the AI supply chain and the establishment of essential AI standards to tackle the intricate difficulties of AI regulation. The papers jointly highlight AI potential to revolutionize social services and policymaking, while underscoring the necessity of ethical concerns, inclusive methodologies, and comprehensive monitoring mechanisms to guarantee the responsible development and use of AI in the social sector.

AI in social services is transforming how assistive services are rendered to the at-risk groups. Ethical AI calls for integration with key policy interventions such as data privacy, bias mitigation, transparency, and human oversight. Such policies must include openness, accountability, and equity in the implementations affecting disadvantageous populations.

AI in social services: an inter-disciplinary activity requires inclusiveness as in the involvement of the community and the participation of the hitherto excluded within the AI framework, establishing cross-sectoral collaborations with a two-way flow between legislators and technologists, ethics committees that critically assess social implications of AI projects as well as provide recommendations on its proper deployment. Training for social workers is also significant. Basic AI literacy courses should be implemented, and programs developed to address ethical implications of AI usage. There should be skill enhancement programs as well as continuous professional development whereby the professionals are well informed on the latest developments in AI and best practices. AI can bring changes that include making social services personalized, effective, and sympathetic care.

10.12 CONCLUSION

In brief, this chapter elaborates on how AI may change the mode of rendering social services to vulnerable groups. AI-based services will help bridge gaps in service delivery in areas that involve the elderly, refugees, and persons with disabilities, thus maximizing accessibility, efficiency, and personalization. This is how AI technologies have been deployed in such places as predictive modeling, virtual assistance, and real-time monitoring to address pressing social needs with timely support and better resource allocation. The positive impact of AI is in the following: it can anticipate a crisis, offer multilingual support, and give personalized care all of which empower people and social inclusion. However, while AI presents numerous benefits, it also raises complex ethical and practical concerns. Issues on data privacy, algorithmic bias, transparency, and accountability have critical challenges with the use of AI in social services. Balanced approaches, the chapter goes to say, must contain rigorous ethical frameworks and human-centered AI systems to address risks of exclusion, discrimination, and dependency on technology.

Emerging AI technologies are transforming social services by facilitating more proactive and automated service delivery (Benouachane, 2022). Technologies such as AI, blockchain, and big data are utilized in fields including healthcare, fraud detection, and consumer communication (Benouachane, 2022; Pombo, 2018). AI and data-driven innovations enable social security organizations to enhance processes, products, and organizational methodologies (Ruggia-Frick, 2021). The integration of AI in healthcare and social services can revolutionize service delivery and enhance quality of life (Stefanisinova, 2021). Nonetheless, challenges persist, particularly the necessity for meticulous planning and evaluation of advantages in relation to dangers (Benouachane, 2022). Governments must also account for the intricacies of integrating new technology while ensuring reliable operating operations (Ruggia-Frick, 2021). Notwithstanding these problems, the use of AI in social security is regarded as both unavoidable and potentially advantageous for all stakeholders (Benouachane, 2022).

AI will be ethically implemented, maintaining transparency and not only helping the social services but also doing it in all fairness with dignity, never neglecting a persons rights.

The future of research and policy initiatives in the use of AI should focus on design for ethical AI systems, with special care for vulnerable populations. If diverse stakeholders social workers, policymakers, technologists, and the communities themselves are involved in the design and deployment of solutions using AI, then a social environment may be fostered where AI actually does contribute toward social welfare. After all, this chapter proposes integration consonant with the principles of social equity, thus empowering disadvantaged groups and furthering a more inclusive society.

REFERENCES

Agarwal, D. B. et al. (2020). Accurate and Scalable Matching of Translators to Displaced Persons for Overcoming Language Barriers. *ArXiv, abs/2012.02595*.

Ahn, E. G.-H. (2021). Predicting Youth at High Risk of Aging Out of Foster Care Using Machine Learning Methods. *Child Abuse & Neglect, 117*, 105059.

Awotunde, J. A. (2022). Internet of Things with Wearable Devices and Artificial Intelligence for Elderly Uninterrupted Healthcare Monitoring Systems. *International Conference on Applied Informatics*.

Bansal, S. & Jain, D. (2023). A Comprehensive Study Assessing the Transformative Role of Artificial Intelligence in India\'s Governance Policy Framework. *International Journal for Research in Applied Science and Engineering Technology, 11(7)*, 1748–1756.

Benouachane, H. (2022). Artificial Intelligence in Social Security: Opportunities and Challenges. *The Journal of Social Policy Studies, 20(3)*, 407–418.

Billard, A. R. (2007). Building Robota, a Mini-Humanoid Robot for the Rehabilitation of Children with Autism. *Assistive Technology, 19(1)*, 37–49, https://doi.org/10.1080/10400435.2007.10131864

Boccanfuso, L. & O'Kane, J. M. (2011). CHARLIE: An Adaptive Robot Design with Hand and Face Tracking for Use in Autism Therapy. *International Journal of Social Robotics, 3(4)*, 337–347. https://doi.org/10.1007/s12369-011-0110-2

Boustani, N. S. et al. (2022). Strengthening Public Institutions and Social Inclusion of Vulnerable Groups in a Developing Country – Innovation in Organizations and Artificial Intelligence Implications. *Pacific Asia Journal of the Association for Information Systems, 14(3)*, 77–102.

Broadbent, E. S. (2009). Acceptance of Healthcare Robots for the Older Population: Review and Future Directions. *International Journal of Social Robotics, 1(4)*, 319–330. Retrieved from https://doi.org/10.1007/s12369-009-0030-6

Broekens, J. H. (2009). Assistive Social Robots in Elderly Care: A Review. *Gerontechnology, 8(2)*, 94–103.

Cantone, A. E. (2023). Enhancing Elderly Health Monitoring: Achieving Autonomous and Secure Living through the Integration of Artificial Intelligence, Autonomous Robots, and Sensors. *Electronics*, *12*, 3918

Carney, T. (2021). Artificial Intelligence in Welfare: Striking the Vulnerability Balance? *Social Science Research Network, 46(2)*, 1–29. https://doi.org/10.26180/13370369.v1,

Carney, T. (2020). Artificial Intelligence in Welfare: Striking the Vulnerability Balance? (Advance).

Chatterjee, S. (2020). AI Strategy of India: Policy Framework, Adoption Challenges and Actions for Government. *Transforming Government: People, Process and Policy, 14*, 757–775.

Cooney, M., & Leister, W. (2019). Using the Engagement Profile to Design an Engaging Robotic Teaching Assistant for Students. *Robotics, 8(1)*, 21.

Czaja, S. J., & Ceruso, M. (2022). The Promise of Artificial Intelligence in Supporting an Aging Population. *Journal of Cognitive Engineering and Decision Making, 16*, 182–193.

Diez, E. (2023). Artificial intelligence and social work: Contributions to an ethical artificial intelligence at the service of the people. In Esther Raya Diez (Ed.), *The Routledge international handbook of digital social work, Routledge* (pp. 368–381). Routledge.

Drydakis, N. (2021). Mobile Applications Aiming to Facilitate Immigrants' Societal Integration and Overall Level of Integration, Health and Mental health. Does Artificial Intelligence Enhance Outcomes? *Computers in Human Behavior, 117,* 106661.

Dunn, S., & Braddell, A. (2014). *Digital capabilities in the social care workforce: Rapid evidence review*. Skills for Care. 07/2014. Available at: www.skillsforcare.org.uk/Document-library/NMDS-SC,-workforce-intelligence-andinnovation/Research/Digital-literacy/Digital-Capability-Rapid-Evidence-Review-300714-FINAL.pdf.

Ediae, A. C. (2024). Leveraging AI in Case Management for Vulnerable Migrants: A Path toward Enhanced Resilience. *Computer Science & IT Research Journal*, *5(4)*, 985–1007.

Feil-Seifer, D. & Matari´c, M. J. (2005). Defining socially assistive robotics. *Proceedings of the 2005 IEEE 9th International Conference on Rehabilitation Robotics*, June 28–July 1, 2005, Chicago, IL, USA.

Flandorfer, P. (2012). Population Ageing and Socially Assistive Robots for Elderly Persons: The Importance of Sociodemographic Factors for User Acceptance. *International Journal of Population Research*, *1*, 2090–4029.

Glaberson, S. (2023). Coding Over the Cracks: Predictive Analytics and Child Protection. *SSRN Electronic Journal*, *46,* 307–363.

Greer, A. S. (2016). Targeting Services to Individuals Most Likely to Enter Shelter: Evaluating the Efficiency of Homelessness Prevention. *Social Service Review*, *90*, 130–155.

Han, J. J. et al. (2008). Comparative Study on the Educational Use of Home Robots for Children. *Journal of Information Processing Systems, 4(4)*, 159–168. Retrieved from https://doi.org/10.3745/JIPS.2008.4.4.159

Harper, C. (1994). *An assessment of vulnerable groups in Mongolia: Strategies for social policy planning.* World Bank Discussion Papers No. 229. Washington, D.C.: World Bank.

Hermann, E. W. (2023). Deploying Artificial Intelligence in Services to AID Vulnerable Consumers. *Journal of the Academy of Marketing Science*, *52*, 1–21.

Ho, A. (2023). Future-Proof: Monitoring the Development, Deployment, and Impacts of Artificial Intelligence. *Journal of Science Policy & Governance*, *22(3)*, 3–13.

Kinder, T. S. (2023). Local Public Services and the Ethical Deployment of Artificial Intelligence. *Government Information Quarterly, 40*, 101865.

Kithulgoda, C. V. (2022). Predictive Risk Modeling to Identify Homeless Clients at Risk for Prioritizing Services using Routinely Collected Data. *Journal of Technology in Human Services, 40*, 134–156.

Koceska, N. K. et al. (2019). A Telemedicine Robot System for Assisted and Independent Living. *Sensors, 19(4)*, 834 https://doi.org/10.3390/s19040834

Kumar, N. K. et al. (2023). The Indian Justice System Significantly Improved By Using Big Data Utilization and Artificial Intelligence Techniques. *Journal of Nonlinear Analysis and Optimization.* https://jnao-nu.com/Vol.%2014,%20Issue.%2001,%20January-June%20:%202023/30.1.pdf

Leyen, U. v. d. (2019). *A Union that strives for more : my agenda for Europe : political guidelines for the next European Commission 2019-2024*. Publications Office. European Commission: Directorate-General for Communication. https://data.europa.eu/doi/10.2775/018127

Lopez Pelaez, A. (2023). *The Routledge international handbook of digital social work* Routledge.

Marda, V. (2018). Artificial Intelligence Policy in India: A Framework for Engaging the Limits of Data-Driven Decision-Making. *Philosophical Transactions of the Royal Society A: Mathematical, Physical and Engineering Sciences, 376*, 2133. https://doi.org/10.1098/rsta.2018.0087

Martín-Gutiérrez, D. H.-P. et al. (2020). A Novel Multi-Modal Framework for Migrants Integration Based on AI Tools and Digital Companion. *IEEE International Conference on Multimedia & Expo Workshops (ICMEW)*, pp. 1-16.

Mhlanga, D. (2024). Artificial Intelligence in Elderly Care: Navigating Ethical and Responsible AI Adoption for Seniors. *SSRN Electronic Journal.* Available at SSRN: http://dx.doi.org/10.2139/ssrn.4675564

Moon, M. (2023). Searching for Inclusive Artificial Intelligence for Social Good: Participatory Governance and Policy Recommendations for Making AI More Inclusive and Benign for Society. *Public Administration Review*, *83(*6), 1496–1505.

Morgan, J. (2021). Artificial intelligence and the challenge of social care in aging societies: Who or what will care for us in the future? In Carrigan, M. & Porpora, D., (Eds.), *Post-human futures: Human enhancement, artificial intelligence and social theory. The Future of the Human* (pp. 92–116). Routledge.

Paraschiv, E. I. (2021). Underpinning Improved Outcomes through Preventative Patient Care Models Based on Remote Monitoring and AI. *2021 International Conference on e-Health and Bioengineering (EHB)*, pp. 1–4.

Pascoe, K. (2023). Considerations for Integrating Technology into Social Work Practice: A Content Analysis of Nine Professional Social Work Associations' Codes of Ethics. *International Social Work, 66*, 298–312. https://doi.org/10.1177/0020872820980833

Pombo, C. G. (2018). *Servicios sociales para ciudadanos digitales: Oportunidades para América Latina y el Caribe*. Inter-American Development Bank.

Ramdey, K. & Bokhari, H. (2020). ICT4D in Channelling Welfare Services in India: Case Study of Haqdarshak. In Singh, S. K., Singh, A. K., Singh, V. P., Singh, J. P., & Singh, S. N. (Eds.), *Advances in intelligent systems and computing* (pp. 835–843). Springer Nature

Reamer, F. (2023). Artificial Intelligence in Social Work: Emerging Ethical Issues. *International Journal of Social Work Values and Ethics*, *20(2)*, 52–71.

Redden, J. (2020). Predictive Analytics and Child Welfare: Toward Data Justice. *Canadian Journal of Communication, 45*, 101-111.

Rodić, A., Vujović, M., Stevanović, I., Jovanović, M. (2016). Development of Human-Centered Social Robot with Embedded Personality for Elderly Care. In Wenger, P., Chevallereau, C., Pisla, D., Bleuler, H., Rodić, A. (Eds.), *New Trends in Medical and Service Robots. Mechanisms and Machine Science*, vol 39 (pp. 233–247). Springer. https://doi.org/10.1007/978-3-319-30674-2_18

Ruggia-Frick, R. (2021). Applying emerging data-driven technologies in social security. Country experiences and ISSA guidelines. *Ubezpieczenia Społeczne. Teoria i praktyka.*

Sasaki, J., Yamada, K., Tanaka, M., & Funyu, Y. (2007). Experiments Of Life Monitoring Systems for Elderly People Living in Rural Areas. *International Journal of Systems Applications, Engineering & Development, 1(4)*, 120–124. Retrieved from https://citeseerx.ist.psu.edu/viewdoc/download?doi=10.1.1.613.4131&rep=rep1&-type=pdf

Seniutis, M. G. (2024). Conceptual framework for ethical artificial intelligence development in social services sector. *Human Technology*.

Seniutis, M. G. et al. (2024). Conceptual Framework for Ethical Artificial Intelligence Development in Social Services Sector. *Human Technology.*

Simão, H., Avelino, J., Duarte, N., & Figueiredo, R. (2018). GeeBot: A robotic platform for refugee integration. In *Companion of the 2018 ACM/IEEE International Conference on Human-Robot Interaction* (pp. 365–366).

Stapleton, L., Lee, M. H., Qing, D., Wright, M., Chouldechova, A., Holstein, K., Zhiwei, S. W., & Zhu, H. (2022). Imagining new futures beyond predictive systems in child welfare: A qualitative study with impacted stakeholders. *Proceedings of the 2022 ACM Conference on Fairness, Accountability, and Transparency.*

Štefanišinová, N. J. (2021). Implementation and Application of Artificial Intelligence in Selected Public Services. *Hrvatska i komparativna javna uprava, 21(4)*, 601–622,

Taheri, A. M. et al. (2019). Teaching Music to Children with Autism: A Social Robotics Challenge. *Scientia Iranica, 26(1)*, 40–58.

Toros, H., & Flaming, D. (2017). Prioritizing Which Homeless People Get Housing Using Predictive Algorithms. *Econometrics: Mathematical Methods & Programming* eJournal. http://dx.doi.org/10.2139/ssrn.2960410

Torras, C. (2015). Social Robots: A Meeting Point between Science and Fiction. *Mètode Science Studies Journal, 5*, 111–115. Retrieved from, https://doi.org/10.7203/metode.82.3546

Vitanza, A. R. et al. (2019). Robot Swarms as an Educational Tool: The Thymio's Way. *International Journal of Advanced Robotic, 16(1)*, https://doi.org/10.1177/17298814188251

Wada, K., Shibata, T., Saito, T., & Tanie, K. (2003). Effects of Robot-assisted Activity for Elderly People and Nurses at a Day Service Center. *Proceedings of the IEEE, 92*, 1780–1788.

Wang, H., Gupta, S., Singhal, A., Muttreja, P., Singh, S., Sharma, P., & Piterova, A. (2022). An Artificial Intelligence Chatbot for Young People's Sexual and Reproductive Health in India (SnehAI): Instrumental Case Study. *Journal of Medical Internet Research, 24*(1), e29969. https://doi.org/10.2196/29969.

Wigfield, A. W. (2013). Assisted Living Technology in Social Care: Workforce Development Implications. *Journal of Assistive Technologies, 7(4)*, 204–218.

Wynne, R., McAnaney, D., MacKeogh, T., Stapleton, P., Delaney, S., Dowling, N., & Jeffares, I. (2016). Assistive technology/equipment in supporting the education of children with special educational needs – what works best? Retrieved from http://ncse.ie/wp-content/uploads/2016/07/NCSE-Assistive-Technology-Research-Report-No22.pdf

Zhang, E., Zhao, S., Wang, T., Hossain, S., Gasztowtt, H., Zheng, S., Parkes, D. C., Tambe, M. & Chen, Y. (2024). "Social Environment Design," Papers 2402.14090, arXiv.org, revised Jun 2024.

11 AI-Driven Disaster Management

Case Studies and Best Practices

Sidharth Verma and Durgesh Tripathi

ABBREVIATIONS

AI	Artificial Intelligence
BP	Back Propagation
CD	Change Detection
CNN	Convolutional Neural Network
DRR	Disaster Risk Reduction
IFRC	International Federation of Red Cross and Red Crescent Societies
IOU	Intersection Over Union
ML	Machine Learning
MLP	Multilayer Perception
MLR	Multiple Linear Regression
NLP	Natural Language Processing
PTSD	Post Traumatic Stress Disorder
SAR	Synthetic Aperture Radar
TC	Tropical Cyclone
VH	Vertical-Horizontal Polarization
VV	Vertical-Vertical Polarization
XAI	Explainable Artificial Intelligence

11.1 INTRODUCTION

The most threatening impacts of humanity in the new world, regardless of their cause being man-made or natural, are disasters (Ogie et al., 2018a,b). It unfolds social structures, causes significant damage to vital infrastructure, and has a wide range of socio-economic as well as environmental implications. The scale and stringency of such disasters have manifoldly increased with climate change, rapid urbanization, and environmental degradation. In the last decade alone, 2007–2017, 3,469 natural disasters, including floods, earthquakes, wildfires, storms, landslides, and volcanic eruptions, have taken place. These shocked 1.4 billion people; injuries amounted to 1.2 million, deaths to 597,287, and total damages topped $1.6 billion (Ogie et al., 2018a,b).

DOI: 10.1201/9781003560463-13

In this ever-growing backdrop, disaster risk reduction (DRR) has now gained relevance at the international level. DRR refers to the coordinated efforts toward discovering, assessing, and reducing disaster risks with state-of-the-art proactive measures to reduce vulnerabilities and enhance resilience. Contrasting the conventional reaction after a disaster, this approach encompasses measures of preparedness, prevention, and recovery to minimize loss of life and the economy and accelerate progress toward sustainable development.

DRR is no longer imperative since disasters are no longer local but cross geographical, economic, and social boundaries. With the abovementioned rationale, disaster impact mitigation strategies have been the strategic agenda for all governments, international agencies, and communities to embrace. Global frameworks such as the Sendai Framework for DRR 2015–2030 established comprehensive guidelines to reduce disaster risks and enhance local, national, and international preparedness (NDMA, n.d.). These frameworks underscore the need to build resilient societies through collaborative action among stakeholders across disciplines and sectors. DRR, therefore, combines science, technology, policy, and community involvement to address immediate risk factors and underlying vulnerabilities such as poverty, inequality, and poor urban planning that make disasters more catastrophic.

Considering the constantly changing face of disaster management, technology has become the backbone of an effective strategy for DRR. Artificial intelligence (AI) is, in fact, that transformative tool that has transformed the process of disaster prediction, management, and mitigation. The technologies involved in AI are a broad range, covering even machine learning (ML), natural language processing (NLP), and data analytics. Systems can learn from data and make intelligent decisions through ML and other methodologies. Such abilities will have profound implications for disaster management. For example, this AI-based early warning system may monitor large data sets in real time and raise timely alarms for an impending flood, cyclone, or earthquake (Varsha et al., 2024). All these data streams come from satellites, sensors, and social media; AI will give precise forecasts so that vulnerable people can be evacuated and resources can be appropriately accommodated.

One of the most important contributions that AI can give to DRR is the development of a preparedness stage during disasters. By analyzing historical data on disasters using ML algorithms, patterns and trends can be identified and predicted to inform policymakers and communities of the likelihood of future events. At the same time, predictive models powered by AI are used to simulate disasters so better planning and response strategies can be achieved. For example, AI models can be developed in seismically vulnerable areas to estimate infrastructure vulnerability so policymakers know which areas need retrofitting or reinforcement. It saves lives and has an enormous impact on reducing the cost of economic reconstruction activity after disasters.

AI is also very critical during and after the occurrence of the disaster. Immediately after the disaster, the most probable quick and efficient resource allocation is necessary to minimize loss of life and ensure the welfare of affected populations. AI-based systems can quickly run real-time analyses of data that help them determine the areas that have been hit the worst and prioritize relief efforts accordingly. For example, after the hurricane has swept through a region, aerial drones can fly over the destroyed landscape and use AI algorithms to map and survey impacted regions, determine the effects on infrastructures, and locate stranded people to enable rescue teams to respond more accurately. Similarly, AI-based tools also exist that optimize supply chain logistics to deliver food, water, and medical supplies to affected areas post-disaster.

While AI's potential in DRR is massive, there are still challenges. The application of these AI technologies depends on substantial access to quality data, which is often unavailable or inconsistent in disaster-prone regions, especially in low-income countries. Other things being equal, ethical concerns over the use of anything from data privacy to algorithmic bias to the digital divide must be considered so that these technologies are only deployed with thought to social responsibility and equity. For example, over-reliance on AI systems will leave behind or disadvantage various communities or minorities-whose data needs to be suitably represented in training datasets. Building capacity, encouraging cooperation, and establishing ethical guidelines for governing the conduct of AI will be a few tough challenges to overcome.

The other challenge with AI integration into DRR strategies is governance and accountability. With more autonomy for AI tools, it will be crucial to determine who is trustworthy and transparent about the underlying decision-making process. Policy must provide some form of regulatory framework to define the roles and responsibilities of stakeholders: governments, private sector entities, and technology developers in AI-driven disaster interventions. Interagency collaboration is of prime importance in filling gaps in many areas so that effective yet inclusive AI tools can be developed and deployed.

However, numerous real-world case studies of success stories of AI-based applications in disaster management exist. There's a case in India where early warning systems based on AI enabled the anticipation of cyclones, so the authorities evacuated millions of people and saved thousands of lives. The spread of the COVID-19 virus was tracked by AI-driven tools, tracing hotspots and giving much-needed input for crucial policy decisions. Such examples demonstrate what AI can do for complex and dynamically changing problems in the realm of disaster management and offer valuable lessons for its broader application in DRR.

The multi-tasking role of AI in DRR will be explored through a critical analysis of applications, benefits, and challenges. Through a combination of literature review, case studies, and secondary data analyses, contributions will be made toward an argument about how AI can strengthen disaster forecasting, early warning systems, and emergency response efforts. Such leading issues to be discussed in this chapter include considerations on the use of AI in terms of ethics, technology, and policy, alongside best-practice insights and future directions. The discussion below presents the nuances with which AI can unfold transformative potential in the reduction of risks of disasters and building resilient societies. As disasters become more frequent and intense, harnessing AI for DRR is no longer optional but something that could prevent loss of lives and livelihoods and ecosystems' destruction for many generations ahead.

11.2 THE ROLE OF ARTIFICIAL INTELLIGENCE IN DISASTER RISK REDUCTION

AI has become a core technology that can help address the growing complexities of DRR. Its capabilities add value to creating better readiness, response, and rehabilitation to disasters by scrutinizing extended datasets, recognizing patterns, and supporting automation toward responses. This section will discuss its role in disaster prediction, early alerting mechanisms, emergency management, and recovery initiatives, all informed by practical illustrations and underpinned by academic validation.

11.2.1 AI in Disaster Forecasting and Early Warning Systems

AI transforms disaster management through the alertness of early warning systems, real-time monitoring and post-disaster recovery. It can identify disasters when they are just about to happen by assessing satellite imagery, sensor data, and historical records. This includes predicting flooding by processing the amount of rainfall, water levels in rivers, and data related to terrain or being able to detect pre-shock signs of earthquakes via seismic analysis. It will analyze real-time data sources about damage assessment, relief priorities, and displaced populations, including drones, satellites, and social media. AI optimizes resource allocation: it allocates supplies, supply networks, and transportation – so they arrive as soon as possible and medical resources are distributed effectively. In communications, AI allows automated emergency alerts to be disseminated and guided in real time via chatbots. After the disaster hits, AI comes into play in the reconstruction plan of projects by prioritizing according to assessments of the damage and economic impact analyses in making recovery strategies. AI is an essential component in reducing disaster-related losses and accelerating recovery by improving the accuracy of prediction, response efficiency, and long-term recovery planning.

11.2.1.1 Flood Forecasting and Monitoring

Floods are the most common natural disaster globally. Flooding is the most common form of natural disaster in the world, affecting millions of lives and causing nearly $10 billion in damages every year (Nevo, 2020) and causing between 6,000 and 18,000 fatalities every year of which 20% are in India (INDIAai, 2020) and accounting for between 6,000 and 18,000 deaths annually-in which 20% take place in India. Reliability in early warnings has been demonstrated to prevent a significant portion of deaths and economic loss but many people don't have access to these types of warning systems. In the year 2022, worldwide, there were 176 documented flood events. With sea levels increasing with climate change and extreme events becoming more intense and frequent, it is expected that at 1.5°C global warming, the population exposed to flooded hazards is to have surged by 24% (Alves, 2024).

The Google Flood Forecasting Initiative allows for using Google's advanced infrastructure and ML capabilities to provide affected communities with accurate real-time flood forecasting and alerts by combining AI with physics-based modeling to deliver scalable, accurate inundation models. These models are further enriched with reliable data from ground sources collected by government agencies, allowing Google's flood forecasting system to better predict the timing, location, and intensity of potential flood events. In this sense, the project will deliver convenient and actionable knowledge on preparedness and response in flood-prone areas (Nevo, 2020). Advanced modeling of morphological flooding simulates the flow and spreading of water over land surfaces, incorporating impacts caused by changes in terrain and changing landscape patterns during the flood forecasting and floodplain management process (Hassaballah et al., 2020). The definition describes morphology as "the form and structure of the land's surface, including its elevation, slope, and features such as rivers, floodplains, and coastal areas." Therefore, such modeling does not merely concern itself with nature but also how alterations in the landscape, such as erosion or sedimentation or human interventions like levees or dams, would influence flooding dynamics apart from the inundation pattern.

Recently, Google developed a new paradigm for modeling inundation, termed the morphological inundation model. These models are physics-based models associated with ML, increasing the accuracy and scalability of inundation models. According to the morphological model, accuracy has increased by 3% compared to classical physics-based models, which will help significantly enhance the forecast for large areas but permit much faster model development by reducing manual modeling and correction (INDIAai, 2020).

This flood forecasting initiative, launched in 2018 in Patna, Bihar, and scaled across the whole country by 2020, now reaches 200 million people across over 250,000 square kilometers. Alerts are targeted and come with three layers of flood risk; they're also available in multiple formats, including visual presentations for local languages like Hindi and Bengali. Google.org is partnering with the International Federation of Red Cross and Red Crescent Societies (IFRC) to build regional networks that provide disaster warnings to individuals without smartphones. Over the past year, improvements to the forecasting model have increased the lead time for many alerts by doubling it, allowing governments and communities to prepare an extra day before floods hit.

In the work of (Garg et al., 2024), the flood detection accuracy has been improved by 25% in terms of intersection over union (IOU) and 16% as F1 scores with the integration of pre-flood and post-flood Sentinel-1 synthetic aperture radar (SAR) imagery. Vertical-vertical polarization (VV) performs better than vertical-horizontal polarization (VH), with an increase of 29% in IOU. The combination of amplitude and coherence data improves the task of flood mapping by 20% in terms of IOU and 10% in F1 scores. Critical features for good accuracy in mapping involve amplitude change detection (CD) and co-flood coherence in VV polarization. The results demonstrate that the models based on Sentinel-1 can generalize across regions.

Hence, these provide fast, scalable solutions for global flood monitoring. Demonstrates that integrating pre-flood and post-flood Sentinel-1 SAR imagery significantly improves flood detection accuracy in arid regions, with a 25% increase in IOU and 16% in F1 scores. VV polarization outperforms VH polarization, showing a 29% increase in IOU. Combining amplitude and coherence data enhances flood mapping, yielding a 20% improvement in IOU and 10% in F1 scores. Critical features for accurate mapping include amplitude CD and co-flood coherence in VV polarization. The findings suggest that Sentinel-1-based models can generalize across regions, offering faster, scalable solutions for global flood monitoring.

11.2.1.2 Cyclone and Hurricane Tracking

Cyclones, hurricanes, or typhoons are the most frequent natural disasters worldwide (Chen et al., 2004; Duan et al., 2020). Cyclones originate from pressure disturbances in the atmosphere that feature quick and, at times, destructive airflow. India is in the northern Indian Ocean, where tropical cyclones (TCs) pose a very high threat. On average, India faces two to three landfalling TCs a year, of which one develops into a severe TC or even more robust (Babu, 2022).

Time is of the essence for forecasting a TC: the more advanced warning possible for a community means that its members will be better prepared at its landfall (Derouin, 2024). Satellite imagery is equally crucial in the weather forecasting model. Techniques that use deep learning for automatic image processing require large datasets to be utilized for training by tapping various properties. The accuracy of weather forecasts is quite high when data have relatively high temporal resolution (Vyavahare et al., 2021).

Deep learning applications in the intensity estimation and forecasting of typhoons have multiplied. For example, Wimmers et al. (2019) utilized convolutional neural networks (CNNs) to forecast typhoon intensity from cloud imagery patterns. The application of ANNs (artificial neural networks) was used by Wang et al. (2015) for the prediction of the intensity of TCs. Baik and Paek (2000) optimized the back propagation (BP) algorithm in a multilayer perceptron network utilizing the given forecast model to predict TC intensity from 12 to 72 hours over the Northwest Pacific. Using the multidimensional predictors applied in multiple linear regression (MLR), the multilayer perception (MLP) can reduce the average error by 7%–16% compared to the MLR model. Apart from that, Chaudhuri et al. (2013) utilized MLP networks to predict the intensity of typhoons; its inputs include central pressure, maximum sustained wind speed, pressure drop, total atmospheric ozone, and sea surface temperature.

Research by (Zhou et al., 2022) proposed an AI-based innovative cyclone monitoring and forecasting system. This system has crucially advanced the automation of identifying, locating, and determining the intensity of TCs. There are deep learning models in the identification of vortices, classification of intensities, and rapid intensification identification. High-resolution images, real-time tracking and intensity assessment results were intriguing and would be helpful for typhoon forecasting.

Remote sensing provides real-time data on the scale and impact of disasters. This data allows organizations to understand the situation, prioritize actions, rebalance resources, and decide what is needed. Continuous monitoring of weather and possible dangers ensures proactive preparation for disaster response through remote sensing. This also provides a basis for the location and extent of damage to essential infrastructure like roads, bridges, communication networks, and works, hence enabling organizations to channel effective efforts toward restoring the most critical services (Velev et al., 2023).

Deep learning algorithms thus form an essential means to process big data, such as satellite imagery and sensor data, and facilitate the management of disasters and recoveries. For example, the interpretation of satellite images may indicate probable objects of interest, such as buildings and roads, making damage estimation due to a disaster more straightforward and resource allocation easier for quicker recovery (Hong et al., 2022).

Cloud computing is important during disaster response as it gives access on-demand to computing resources, data storage, and applications (Hurson & Wu, 2021).

It gives scalable, flexible, and secure data storage vital in dealing with large volumes of data after a disaster. High-performance computing and AI allow for real-time analysis, which is always helpful in making decisions and modeling scenarios. It ensures collaboration and information sharing between the organizations, and also enhances coordination, resource pooling, and effective decision-making to meet the needs of the affected communities.

11.2.1.3 Earthquake Prediction and Early Warning

Earthquakes are difficult to predict and respond to because of their lack of predictability and the immense damage they cause. The coincidence of earthquakes with other natural phenomena, such as rainstorms and floods, adds to the complexity. Lousy planning for early warning systems, inadequate distribution of emergency supplies, and bad logistics planning could endanger the effectiveness of earthquake disaster response. In this connection, an Earthquake Emergency Response Network coordinates numerous organizations and agencies for effective disaster management (Biswas et al, 2024).

Applying the first-ever AI in earthquake prediction minimizes the damage caused by earthquakes to human lives and economies. Through seven months of trials conducted in China, researchers from The University of Texas at Austin successfully created a new algorithm using AI, which can predict 70% of the earthquakes that will occur a week before they happen in reality. Trained to recognize statistical patterns in real-time seismic data correlated with past earthquake records, the AI made 14 correct predictions of earthquakes located within 200 miles and magnitude but missed one event and caused eight false alarms. Though it has not been tested outside the region, it is an essential milestone toward AI-driven earthquake forecasting. Applications in already existing seismic networks, such as California, Japan, and Texas, would thus create a base for future improvement in predictions. Future plans include setting up model physics for AI-based regions with scarce seismic data. Hence, some convergence of these two approaches could result in a holistic model applicable at the global level. This research is instrumental in improving earthquake preparedness and lessening human and economic losses globally (University of Texas at Austin, 2024).

11.3 AI IN EMERGENCY RESPONSE AND RESOURCE ALLOCATION

Disaster response operations, inherently complex and demanding quick decisions on accurate resource distribution under immense pressure, constitute highly stressful circumstances. The fluid nature of emergencies-inadequate information, changing conditions drastically, and the most basic human needs-throws an overwhelming impetus on the traditional system. Emergent AI has genuinely become revolutionary for these tasks, offering best-in-class analytical tools in real-time data analysis, prioritized intervention, and resource optimization. AI will employ combined ML and predictive analytics with geospatial technologies to expedite emergency operations at much higher accuracy levels.

One of its greatest strengths is the processing and interpreting of large volumes of data created during disasters. For instance, in real-time satellite imagery, aerial drone footage, and social media feeds, responders can determine the worst damage and assess how much damage is done. AI-driven tools include computer vision algorithms that detect stranded individuals, blocked roads, and collapsed buildings, which may become actionable intelligence to ground teams. Predictive models can forecast the needed resources like medical supplies, food, or evacuation transportation and may arrange fair and timely distribution to stricken areas.

AI integration is also helpful in streamlining the logistics in crises. Systems that ensure dynamic resource management through AI adapt in real-time conditions while helping agencies determine areas where much need exists yet not waste resources, In the 2023 Turkey-Syria earthquakes, mapping tools that integrate AI were of the utmost importance in pinpointing rescue hotspots and coordinating relief responses across treacherous terrain. As the effects of disasters worsen and intensify in a changing climate, urbanization, and increased occurrence, AI-driven emergency response systems are necessary.

11.3.1 Aerial Reconnaissance with Drones

The symbiotic connection between AI and drone technology ushered in a new frontier for opportunities to alter the landscape of various initiatives and sectors (Obiuto et at., 2024). The fusion of AI and drone technology is much more profound than automation and uncrewed flight; it is simply an essential convergence of innovation (Alqahtani & Kumar, 2024). A paradigm has shifted in drones, also known as unmanned aerial vehicles or unmanned aerial vehicles (UAVs), which are fitted with AI. Thus, they have changed from remote-controlled devices to sophisticated, autonomous systems (Johnson, 2020).

Perhaps the most significant application of AI in drones is autonomous navigation, which is achieved through AI-based flight path planning. Unlike the already programmed flight paths that fail to change contingent upon environment, mission, and safety conditions, AI allows the drones to function using dynamic flight paths contingent upon real-time adjustments due to changes in the environment, objectives of the mission, or safety concerns (Nguyen et al., 2021).

AI promises to be revolutionary in how Australians think about their devastating wildfires and response mechanisms. One of the AI models developed can enable cube satellites to spot fires in space 500 times faster than how imagery through the ground was traditionally processed. In that respect, the AI analyzes the "hyperspectral imagery" captured by the satellite, identifying smoke before a fire produces heat, thereby being able to facilitate quicker detection and response. This results in early detection, enabling ground crews to respond quickly, thus possibly reducing the impact of losses regarding human life and property. AI technology in the Kanyini mission also saves space for storing data and energy consumption, making it more efficient than any other similar fire detection systems (University of South Australia, 2024).

Although drones are presently gaining much publicity for their role in modern warfare, they are excellent tools for disaster relief situations. After the 2023 devastating earthquake in Turkey and Syria, snapping pictures for a vivid depiction of extreme devastation has been greatly helpful to emergency relief operations in getting a more accurate idea of the situation before them. Spain sent drones to Turkey as part of its humanitarian package. In addition to aerial imagery, drones could have installed various sensors for increased accuracy in the damage assessment process, such as detecting high methane levels in the air, suggesting that gas lines were ruptured and are liable to explode (Sky News, 2023).

In July 2024, heavy monsoon precipitation triggered a tremendous landslide in Wayanad, Kerala. The landslide wrecked several houses in the Mundakkai locality and displaced hundreds of villagers. Drones have proven helpful in the new science of disaster management by rapidly entering afflicted regions, providing high-definition imagery, and imparting real-time information. This has been well projected in the devastating Wayanad landslide tragedy in Kerala, India. Thermal imaging drones were a real game-changer, as they could find heat signatures from those trapped in the rubble, further reducing response times and increasing survival rates. In addition, drones were used for communication in areas where regular links had become dysfunctional; they acted as satellite relays between relief teams. Increased autonomy, improved sensors, and swarm technology will likely enhance the efficiency and scalability of drone operations, allowing for better assessments of damage, search and rescue efforts, and extensive monitoring during disasters. As the technology for drones keeps advancing, its capacity to save lives, prevent destruction, and improve disaster response efforts is likely to become more vital to global approaches toward the management of disasters (Robotics Innovation and Research, 2024).

11.3.2 Optimizing Relief Distribution

Optimizing relief distribution is critical in disaster management because affected populations are provided with assistance on time and in reasonable amounts. Relief traditionally suffers logistical inefficiency and resource misallocations and cannot reach more remote areas promptly. Such factors cause critical delays and do not help in proper recovery. AI can make the distribution of aid much better by using predictive analytics to identify which regions are the most exposed, making the delivery of resources even more prioritized.

Furthermore, ML algorithms can optimize supply chains by predicting demand and delivering routes faster to avoid delays and waste. Moreover, AI can provide better response team communication and coordination by giving real-time updates on each action. Organizations might automate their decision-making through AI tools, making processes faster in terms of human intervention time. With advances in AI technology, its role in disaster management progresses from throughput efficiency and responsiveness to the ultimate payoff of saving more lives by getting help to those needing it much quicker.

AI can be extensively used to develop a response to disasters by improving communication, decision-making, and resource allocation to increase the effectiveness and efficiency of relief efforts. From artificially intelligent drones and satellite imagery, responders can get an immediate sense of affected regions, including inaccessible ones. Algorithms in computer vision analyze visual data to discern the range of damage and highlight critical hotspots. AI regarding wildfires can pinpoint the most active zones and predict possible spread so that firefighting teams can plan their actions forward.

Similarly, AI algorithms can forecast areas that might be submerged in the event of flooding, helping governments streamline rescue operations and deployment while minimizing resources. AI also smooths out resource allocation by looking into issues related to logistics embedded in disaster management. ML models analyze data from previous events, population density, and transportation networks to make supply delivery smoother with food, water, or medical aid. AI-based chatbots can enhance the impacts of disaster response by providing affected communities with timely information, triggering alerts, giving evacuation instructions, and answering key questions. Such technologies free up human responders to focus on more complex interventions, improving disaster operations' overall effectiveness (Tewari, 2024).

In scenarios where land or conventional air transport cannot be accessed due to infrastructure damage, drones may bring communication equipment, mobile healthcare units, and temporary shelters. The doctors can collect blood samples from home-bound patients and then forward them to a lab for tests through drones. In addition, drugs such as antibiotics and many other treatments can be given directly to patients' homes, further improving home care for the elderly staying in nursing homes and empowering the elderly population. Drones can also monitor patients suffering from chronic conditions such as dementia or feed those who cannot cook independently. The involvement of drones in healthcare operations facilitates emergency response at a faster speed and an increased provision of medical services, significantly to patients in far-flung regions or home-based care. The most significant advantage of drones in healthcare is covering immense distances at high speeds, especially when transporting blood products and laboratory samples. Unlike land-based vehicles, which are subject to breakdowns and accidents, unmanned aerial vehicles make the healthcare supply more reliable and efficient (Papyan et al., 2024).

During May 2020, amidst the global COVID-19 pandemic, a medical drone delivery trial was conducted by SOARIZON across the Scottish Highlands. Personal protective equipment and COVID-19 test kits were flown out, among other deliveries. The trial ultimately showed that the journey could be completed in 15 minutes using a drone instead of up to 6 hours by road and ferry. As heavy rain pounded parts of northern and eastern India, including Himachal Pradesh, West Bengal, Sikkim, Arunachal Pradesh, Assam, and Meghalaya, Uttarakhand entered a red alert for the month of July 2023. The UAV solution company from Tamil Nadu, Garuda Aerospace, sent their logistics and delivery drones to Assam and Himachal Pradesh immediately after seeing that foreseen situation to assist the NDRF (National Disaster Response Force) in its rescue missions and support any ensuing supplies during the emergency (Ministry of Communication, Government of India, 2024).

The earthquake that occurred in Nepal on 25 April killed more than 8,500 people making it the deadliest disaster in Nepal's history. On the other hand, the disaster marked a new record for disaster management. It was the first instance of extensive use of AI in managing data accrued through mobile phones, satellites, and social media. AI proved particularly crucial in helping the search teams locate victims, estimate relief requirements, and move across the rugged terrain, which enormously increased the efficiency and effectiveness of the relief work (Meier, 2015).

11.3.3 Social Media Analytics for Real-Time Insights

Social media sites have emerged as an absolute exigency for getting the word out quickly in disaster situations (Velev, 2023). They act as focal points for information sharing, requests for help, and resource mobilization in an intervention. Analysis of different posts, tweets, images, and videos by AI algorithms may be used to identify the affected areas and the severity of a disaster and prioritize the efforts to respond first. People increasingly use social media sites like Facebook and Twitter during disasters and crises. Empirical research has demonstrated how data obtained in disaster relief efforts, situational awareness, actionable intelligence extraction, and so much more are highly effective in many humanitarian missions. Social media will also help to alleviate one part of the information scarcity problem when a sudden onset disaster occurs, namely, during those early hours when credible sources of information are in very short supply (Imran et al., 2018).

Recent work conducted by the University of Alabama in Huntsville has led to the development of a framework designed to apply AI techniques for improvement in the usage of social media data related to disaster situations which investigated the ability of AI to glean essential insights from social media interactions, thereby enabling immediate response strategies. For this project, the researchers analyzed 1.7 million tweets produced during two critical times during the COVID-19 pandemic: March to April 2020 in the United States and May to June 2021 in India. The states affected during these periods were subjected to extreme supply chain disruptions that resulted in a lack of several provisions, including face masks, medication, and ventilators. The authors used the techniques of AI and ML applied in analyzing and interpreting social media to find critical supply chain disruption indicators and classify all the tweets into "imperative" (urgent calls for help) or "non-imperative" (informational items). The framework also enabled the prediction of the geographical locations of critical tweets without geo-tag information for better coordination of the response efforts.

Furthermore, automating most of the content analysis identifies real-time urgent needs and gives decision-makers actionable information as soon as possible. This methodology optimizes the entire process of managing disruptions within supply chains but also suggests the great promise of AI in disaster management. It establishes rapid analysis of voluminous data on social media that makes responses to crises more efficient and effective. The strategies developed here can be varied according to various disaster scenarios, which stands as testimony to the flexibility of AI in handling a range of problems in real time (Lewis, 2024).

AIDR, or AI Digital Response, can handle massive amounts of data, extract critical humanitarian information for rapid decision-making by organizations involved in disaster response, save time to act, and enhance the effectiveness of providing aid through automated analysis of posts carrying texts and images (Ghaffarian et al., 2023). NLP can be used to analyze unstructured data from social media, articles, and online forums to detect affected areas during disasters, track the spreading of misinformation, and strengthen communication and information transmission among response and recovery organizations (Campesato, 2020).

11.4 AI IN POST DISASTER RECOVERY

Post-disaster recovery is one of the most complicated phases after a calamity since it includes rebuilding infrastructure, supporting livelihoods, and addressing psychological and social impacts. Recovery in a holistic manner would effectively imply some strategic planning on the distribution of resources and how their resilience will be boosted to minimize future disaster impacts. AI has emerged as a mighty tool in post-disaster recovery, offering solutions in streamlined processes, improved decision-making, and community resilience.

11.4.1 Damage Assessment and Reconstructing Planning

Another critical application of AI in handling natural disasters is its support to the response teams in obtaining real-time data that would be fundamental to correct decision-making and resource allocation. Primarily, AI systems can offer real-time damage assessments and monitor people in distress, thus assisting in relieving some of the resources to the affected most efficiently. The speed of response will accelerate with AI; aid will arrive in the affected area much sooner, and frontline personnel will have the information they need to make rational decisions. This especially concerns the area of natural disasters wherein emergency communication channels through calling numbers tend to get clogged by distress calls, and therefore, critical details should be addressed while routing through human operators. AI technologies, particularly in NLP and speech-to-text abilities, can manage massive communication volumes in record time. NLP algorithms extract contextual information needed for these calls, enabling a better understanding of the nature of an emergency and the caller's exact location. In this concern, AI can do many things to improve the efficiency of operations in emergency response and timely deploy the most critical resources with precision and urgency (Vina, 2024).

Research by Ghaffarian et al. (2023) is applied to explain the use, challenges, and opportunities of explainable artificial intelligence (XAI) in disaster risk management (DRM). Most notably, according to the study, in 2022, usage of XAI accounted for markedly increased, meaning increased transparency in a disaster setting. The chapter outlined core areas for further research; these comprise delineating the multi-hazard integration into risk assessments, explicability of disaster early warning frameworks, and applying causal inference techniques in decision-making (Ghaffarian et al., 2023).

11.4.2 Mental Health Support

The American Psychiatric Association defines post-traumatic stress disorder (PTSD) as "a psychiatric disorder that can occur in people who have experienced or witnessed a traumatic event such as a natural disaster, a serious accident, a terrorist act, war/combat, rape or other violent personal assault" (Bertl et al., 2022). Individuals with PTSD are always characterized by recurrent and intrusive recollections of the traumatic events, severely interfering with their everyday lives. The lifetime prevalence of PTSD is about 12.5% (Spottswood et al., 2017).

AI is mainly known for helping disaster management by predicting and minimizing physical harm during terrible events like floods, hurricanes, and earthquakes. However, the psychological impacts of those events should be addressed. Mental health issues, such as anxiety, depression, and similar disorders, have become increasingly prevalent following these calamities, affecting the community at its deepest end. Although a living area for applying AI in treating psychological distress in climate-related disasters is still pretty activated at this point, "the potential of AI in enhancing psychological well-being" receives continuously growing attention from scientific circles and media. Other studies, such as Shatte et al. (2019), relate research on AI to mental health and highlight that AI can better help diagnose and improve outcomes through the study of speech and facial expressions as well as behavioral patterns. Diagnostic errors can be significantly reduced with AI technologies, thus increasing the timeliness and accuracy of interventions. The integration of AI within the frameworks of disaster response can help them deal with not only the physical effects of impacts on communities but also their possible psychological effects (Pieters & Nagabhatla, 2023).

11.5 CHALLENGES AND ETHICAL CONSIDERATIONS

AI transforms DRR and opens up a new frontier in risk management and disaster impact reduction. Significant factors contributing to enhanced preparedness and response and recovery systems during a disaster were processing enormous quantities of data and providing real-time insights to build better decision-making mechanisms. However, along with advantages, challenges and ethical dilemmas accompany the inclusion of AI in DRR. That is where the hopes of AI lie in saving lives, optimizing resource allocation, and predicting disasters occurring. But it will be realized only when headway against such large blocks of obstacles is made.

Issues arise primarily from data quality, accessibility, and interoperability. AI systems rely heavily on high-quality, robust, reliable datasets that hold promise for accurate predictions and actionable insights. However, in many regions including developing countries data often is incomplete or outdated, or the selection is biased. This downplays the trustworthiness of AI tools significantly. The digital divide also exacerbates this problem because underserved areas lack the infrastructure and resources to procure advanced technologies. Beyond technical problems, ethical considerations pose formidable obstacles. Issues of fairness, transparency, and accountability persist as central concerns. Algorithmic biases embedded within AI models may lead to discriminatory treatment of vulnerable communities, while the opaque "black box" nature of most AI systems feeds fears related to decision-making transparency and overall trustworthiness. Since gathering and scrutinizing personal information usually occurs without adequate protection measures, privacy and data security concerns obtrude.

Given that AI is being infused into emergency management, these systems must be developed to embody transparency and explainability. This enables them to build trust between emergency managers, the public, and the technology itself by putting stakeholders in the position of understanding why an AI makes certain decisions. For instance, during a crisis, while using AI systems for automating resource allocation, there is a need to explain how these decisions are made to maintain public confidence and ensure fairness in dispensing aid.

Equitable access to AI technologies is an ethical challenge in disaster management. This approach becomes a must so that the already underserved and marginalized communities are not undermined. AI-driven systems as early warning mechanisms for disasters would be implemented and accessible to individuals with disabilities and those in remote areas. Such an approach would ensure fair benefits to be provided through AI to all demographics during a crisis.

The prospect of AI replacing human beings in emergency management raises enormous ethical issues of workforce displacement. Its potential has raised questions about which jobs should be reinstated with support from retraining to match evolving AI practices. For instance, predictive analytics will alter many job roles through AI application virtues. This implies the need for proactive measures and a just transition to avoid disruption in the workforce. AI systems will have unintended impacts on disaster response. Extensive reliance on AI systems with less human control will tend to increase errors of critical consequence. These risks suggest suitable testing, validation, and cooperation between human operators and AI systems.

Accountability in the choices of the AI should be done for the sake of fairness and to prevent potential injury. Thus, there's a continuous tracking and auditing of AI systems to have unjust biases or outcomes. For example, an AI system that provides resources following a calamity must constantly be audited to avoid inadvertently favoring certain groups or regions. Holding these ethical concerns in perspective and elaborating on them with vivid examples, policymakers and organizations can take steps to implement AI responsibly in disaster management disasters, hence attaining the ethics and efficiency of disaster response (Visave, 2024).

The existence of AI in emergency management raises several issues that need careful consideration to use them responsibly like

Algorithmic Bias Mitigation: Full evaluation processes and measures must ensure that AI systems are fair and don't discriminate against anyone. If AI systems are not constantly monitored for biases, biased outcomes are always possible. Training data must be diverse and representative to ensure that the final solutions developed with them are fair.

Protection of Privacy and Data Security: Ethical considerations and governance frameworks must protect personal data in AI-driven emergency systems. Data processing policies and secure storage measures must be respected while ensuring that all privacy rights are safeguarded and practical steps are taken to minimize breaches.

Accountability and Transpicuity: AI systems used in emergency management need to be transparent and clearly explain the decisions made by AI algorithms. This will be trusted by the public and ensure that they can understand why AI-driven interventions were undertaken for their needs.

Inclusion and Involvement: Community involvement in developing and implementing AI-driven emergency services is essential for fostering diversity in opinion and needs. Properly including stakeholders in the process can help pinpoint potential ethical issues. In this way, any AI solution launched would resonate well with community values and priorities. Proper safety measures must undoubtedly be taken in designing AI-based emergency management systems because risks need to be factored in, and people need to be protected in times of crisis. Fail-safe mechanisms, system audits, and disaster response simulations are examples of testing the reliability and effectiveness of AI intervention.

11.5.1 Balancing Efficiency with Ethics: AI for Emergency Management

It has proven helpful in augmenting operations at emergency responses, particularly in mass care and disaster management. Yet, whenever this is implemented in emergency management, it subtly integrates efficiency enhancement and critical ethical considerations. First, AI holds tremendous potential to improve the response efforts. For instance, it can complement highly advanced warning systems concerning natural disasters. Still, many ethical issues have to be taken with precaution. One of the most significant concerns is the dependability and validation of the predictions provided by AI. For instance, the use of AI in earthquake prediction and fire management calls for strict validation to avoid false alarms and threats due to missed warnings.

Another ethical issue is that algorithms developed and deployed for responding to disasters should be fair and unbiased. AI systems must be designed to avoid biases that eventually tilt toward vulnerable communities or ecosystems. There's also a matter of careful inclusion of AI in healthcare emergencies concerning patient privacy and the risk of algorithmic bias in decision-making, which will, no doubt, maintain stern data security measures and treat all individuals fairly. Therefore, in humanitarian aid and disaster response, AI must be used with respect for affected populations' dignity, autonomy, and rights. It cannot add up to or even create inequalities and power imbalances. Therefore, clear ethical guidelines and governance frameworks must be established to help guide proper integration, especially now that AI is coming into such contexts. Issues are dealt with using a broad-ranging, transparent approach that calls for a high degree of accountability, fairness, and judicious use of decisions. This way, AI maximizes the efficacy of emergency management systems while maintaining trust and equity with the public.

11.6 CONCLUSION

AI has developed capabilities to enhance preparedness, response, and recovery systems to give a new makeover to the DRR landscape. The high capacity of AI in processing vast volumes of data in real time, with an ultimate motive to support the decision-making processes, enables AI significantly to enhance disaster management practices. AI effectively reduces disasters' effects and promotes accelerated recovery efforts through predictive analytics, resource optimization, and enhanced early warning systems. Serious problems include data quality, accessibility, and interoperability, and in a developing region, incomplete, outdated, or biased datasets can easily "break the power of AI." Another critical ethical issue is how an algorithm might produce bias, perform operations without clarity, and even disregard personally identifiable information, especially in emergency management. Enhancing the transparency and the explainability of the AI decision processes is relevant, mainly in the context of when AI-based systems make essential decisions, including resource allocation, during a crisis between management officials and the general public. There is also the need for fair access to AI not to exacerbate existing inequalities, particularly for marginalized and underrepresented groups whose members often face the highest risks when disasters strike. Furthermore, the possibility of substituting human labor with AI in emergency management processes brings forth concerns regarding adjustment to the workforce. AI presents many benefits when applied to disaster management and must be independent of such technologies; otherwise, significant blunders and missed opportunities by human evaluation and involvement will be present. Continuous evaluation, validation, and collaboration between humans and technology significantly reduce risks and ensure reliability in AI-enabled interventions. These will also raise issues of privacy and data security. Therefore, excellent governance frameworks are to be developed to protect sensitive data without violating citizens' rights through AI systems used in disaster management. Real-time performance is required in disaster response; rapid processing of data by AI improves situational awareness that supports more effective interventions. Conclusion Thus, AI harmonization in disaster management frameworks should be made only on a balance between operational effectiveness and ethical implications to respect the dignity, autonomy, and rights of affected populations. For that reason, continuous monitoring, auditing, and active engagement of the community in establishing and implementing AI systems are required to uphold fairness, accountability, and transparency. The responsible and inclusive use of AI would maximize disaster response effectiveness, maintain public confidence, and ensure a fair outcome for all communities. Success in AI for DRR, therefore, depends on the challenges developed as holistic frameworks that could ensure that application into practices about disaster management is ethical, effective, and inclusive.

REFERENCES

Alqahtani, H., & Kumar, G. (2024). Machine learning for enhancing transportation security: A comprehensive analysis of electric and flying vehicle systems. *Engineering Applications of Artificial Intelligence*, *129*, 107667.

Alves, B. (2024, September 26). *Floods – Statistics & Facts*. Statista.

Babu, S. A. (2022, December 2). *5th World Congress on Disaster Management: Volume II* (p. 262). Taylor & Francis. ISBN 978-1-000-82747-7.

Baik, J. J., & Paek, J. S. (2000). A neural network model for predicting typhoon intensity. *Journal of the Meteorological Society of Japan*, *78*, 857–869.

Bertl, M., Metsallik, J., & Ross, P. (2022). A systematic literature review of AI-based digital decision support systems for post-traumatic stress disorder. *Frontiers in Psychiatry, 13*, 923613.

Biswas, S., Kumar, D., Hajiaghaei-Keshteli, M., & Bera, U. K. (2024). An AI-based framework for earthquake relief demand forecasting: A case study in Türkiye. *International Journal of Disaster Risk Reduction, 102*, 104287.

Campesato, O. (2020). *Artificial Intelligence, Machine Learning, and Deep Learning*. Mercury Learning and Information.

Chaudhuri, S., Dutta, D., Goswami, S., & Middey, A. (2013). Intensity forecast of tropical cyclones over North Indian Ocean using multilayer perceptron model: Skill and performance verification. *Natural Hazards, 65*, 97–113.

Chen, L., Luo, Z., & Li, Y. (2004). Research advances on tropical cyclone landfall process. *Acta Meteorologica Sinica, 62*, 541–549.

Derouin, S. (2024). *Physics Meets Machine Learning for Better Cyclone Predictions*. Prevention Web.

Duan, Y. H., Fang, J., Cheng, Z. Q., Xu, J., Li, Q. Q., Zhan, R. F., et al. (2020). Advances and trends in tropical cyclone research and forecasting: An overview of the ninth world meteorological organization international workshop on tropical cyclones (IWTC-9). *Acta Meteorologica Sinica, 78*(3), 537–550.

Garg, S., Dasgupta, A., Motagh, M., Martinis, S., & Selvakumaran, S. (2024). Unlocking the full potential of Sentinel-1 for flood detection in arid regions. *Remote Sensing of Environment, 315*, 114417.

Ghaffarian, S., Taghikhah, F. R., & Maier, H. R. (2023). Explainable artificial intelligence in disaster risk management: Achievements and prospective futures. *International Journal of Disaster Risk Reduction, 98*, 104123.

Hassaballah, K., Mohamed, Y., Omer, A., & Uhlenbrook, S. (2020). Modelling the inundation and morphology of the seasonally flooded Mayas Wetlands in the Dinder National Park-Sudan. *Environmental Processes, 7*(3), 723–747.

Hong, T. P., Serrano-Estrada, L., Saxena, A., & Biswas, A. (Eds.). (2022). *Deep Learning for Social Media Data Analytics*. Springer.

Hurson, A. R., & Wu, S. (2021). AI and cloud computing preface. *AI and Cloud Computing, 120*, XI–XII.

Imran, M., Alam, F., Ofli, F., & Aupetit, M. (2018, March). Artificial intelligence and social media to aid disaster response and management. In Qatar Foundation Annual Research Conference Proceedings (Vol. 2018, No. 3, p. ICTPD1030). Qatar: HBKU Press.

INDIAai. (2020, December 4). *Using AI to Predict Floods and Save Lives*. https://indiaai.gov.in/case-study/using-ai-to-predict-floods-and-save-lives

Johnson, J. (2020). Artificial intelligence, drone swarming and escalation risks in future warfare. *The RUSI Journal, 165*(2), 26–36.

Lewis, S. (2024, May 3). *Tapping Social Media and AI for Faster Disaster Response*. Illinois Tech.

Meier, P. (2015, June 1). *Virtual Aid to Nepal Using Artificial Intelligence in Disaster Relief*. Foreign Affairs.

Ministry of Communications, Government of India. (2024). (publication). Use Cases of Drones in Disaster Management (pp. 01–39). New Delhi, Delhi.

NDMA. (n.d.). Sendai Framework. https://ndma.gov.in/Global/sendai-framework

Nevo, S. (2020, September 3). The technology behind our recent improvements in flood forecasting. *Google Research*. November 20, 2024.

Nguyen, D. D., Rohacs, J., & Rohacs, D. (2021). Autonomous flight trajectory control system for drones in smart city traffic management. *ISPRS International Journal of Geo-Information*, *10*(5), 338.

Obiuto, N. C., Festus-Ikhuoria, I. C., Olajiga, O. K., & Adebayo, R. A. (2024). Reviewing the role of AI in drone technology and applications. *Computer Science & IT Research Journal*, *5*(4), 741–756.

Ogie, R. I., Forehead, H., Clarke, R. J., & Perez, P. (2018a). Participation patterns and reliability of human sensing in crowd-sourced disaster management. *Information Systems Frontiers*, *20*, 713–728.

Ogie, R. I., Rho, J. C., & Clarke, R. J. (2018b, December). Artificial intelligence in disaster risk communication: A systematic literature review. In 2018 5th International Conference on Information and Communication Technologies for Disaster Management (ICT-DM) (pp. 1–8). IEEE.

Papyan, N., Kulhandjian, M., Kulhandjian, H., & Aslanyan, L. (2024). AI-based drone assisted human rescue in disaster environments: Challenges and opportunities. *Pattern Recognition and Image Analysis*, *34*(1), 169–186.

Pieters, R., & Nagabhatla, N. (2023, September 18). *Unlocking Solutions: AI for Disaster Management and Psychosocial Well-Being*. United Nations University.

Robotics Innovation and Research. (2024, August 27). *The Role of Drones in Disaster Management: A Case Study of the Wayanad Landslide*. https://hcrobo.com/the-role-of-drones-in-disaster-management-a-case-study-of-the-wayanad-landslide/

Shatte, A. B., Hutchinson, D. M., & Teague, S. J. (2019). Machine learning in mental health: a scoping review of methods and applications. *Psychological Medicine*, *49*(9), 1426–1448.

Sky News. (2023, February 7). *Turkey-Syria Earthquake: How Drones and a NASA Heartbeat Detector Could Transform Disaster Search.* https://hcrobo.com/the-role-of-drones-in-disaster-management-a-case-study-of-the-wayanad-landslide/

Spottswood, M., Davydow, D. S., & Huang, H. (2017). The prevalence of posttraumatic stress disorder in primary care: A systematic review. *Harvard Review of Psychiatry*, *25*(4), 159–169.

Tewari, V. (2024, October 11). *AI in Disaster Response: Predicting and Managing Crises.* StatusNeo.

University of South Australia. (2024, June 5). Fighting Fires from Space in Record Time: How AI Could Prevent a Repeat of Australia's Devastating Wildfires. Prevention Web.

University of Texas at Austin. (2024, August 20). *Artificial Intelligence Predicts Earthquakes with Unprecedented Accuracy*. SciTechDaily.

Varsha, V. R., Naganandini, S., & Hariharan, C. (2024). Utilizing AI and machine learning for natural disaster management: predicting natural disasters with AI and machine learning. In *Internet of Things and AI for Natural Disaster Management and Prediction* (pp. 279–304). IGI Global.

Velev, D., & Zlateva, P. (2023). Challenges of artificial intelligence application for disaster risk management. *The International Archives of the Photogrammetry, Remote Sensing and Spatial Information Sciences*, *48*, 387–394.

Vina, A. (2024, September 25). *AI in Natural Disaster Management*. Ultralytics.

Visave, J. (2024). AI in emergency management: Ethical considerations and challenges. *Journal of Emergency Management and Disaster Communications*, *4*(3), 1–19.

Vyavahare, S., & Khanuja, H. K. (2021). Fidelity of AI Model in Representing the Different Types of Cyclone Track Estimation. www.irjet.net/archives/V8/i8/IRJET-V8I848.pdf

Wang, X., Wang, C., Zhang, L., & Wang, X. (2015). Multidecadal variability of tropical cyclone rapid intensification in the western North Pacific. *Journal of Climate*, *28*(9), 3806–3820.

Wimmers, A., Velden, C., & Cossuth, J. H. (2019). Using deep learning to estimate tropical cyclone intensity from satellite passive microwave imagery. *Monthly Weather Review*, *147*(6), 2261–2282.

Zhou, G., Fang, X., Qian, Q., Lv, X., Cao, J., & Jiang, Y. (2022). Application of artificial intelligence technology in typhoon monitoring and forecasting. *Frontiers in Earth Science*, *10*, 974497.

Part III

AI Technology and Data Systems

12 The Evolution of Modern Data Platforms

Powering the AI-Driven World

Vaishali Mishra, Harshal Darade, A.V. Senthil Kumar, and B. Gayathri

12.1 INTRODUCTION

In the context of the digital revolution culminating in big data and artificial intelligence (AI), businesses are forced to reassess and redefine their business processes in terms of how information is used. This includes its most well-known iterations, machine learning (ML), natural language processing (NLP), and computer vision, which allow machines to process and analyze data in a human-like manner. Such tools enable organizations to automate decision-making, optimize operations, and improve user experiences by learning from large datasets (Russell & Norvig, 2020).

Big Data is a term used for extremely large and complex datasets which is beyond the scope of traditional data processing systems. Big Data is frequently described by the "three Vs": volume (the large amounts of data), velocity (the rate of data generation), and variety (multiple types and formats of data) (Laney, 2001). Nowadays, companies generate data from a wide range of sources such as social media, IoT devices, customer interactions, and online transactions. Such rapid data growth calls for advanced platforms for its processing, management, and analysis (Jagadish et al., 2014). AI and Big Data are understood not as disparate technologies and are becoming relatively complementary tools (Figure 12.1) to solve complex problems, provide automation, and deeper insights (Hu et al., 2014; Kaur & Saxena, 2022).

12.2 COMPONENTS OF CURRENT DATA PLATFORM

A data platform is an interconnected system of technologies that collects, stores, manages, and analyzes structured, semi-structured, and unstructured data. These platforms act as a nerve center for organizations to centralize and standardize their data-oriented toward analytics and decision-making. Such workloads are crucial for allowing advanced analytics, AI, and ML, enabling companies to mechanize workflows, improve performance, and assist decision-making (Sun & Zhang, 2021). And data lakes, flexible data storage and processing solutions tailored for big data, enable advanced analytics and support big data processing workflows (Schatzberg, 2019).

DOI: 10.1201/9781003560463-15

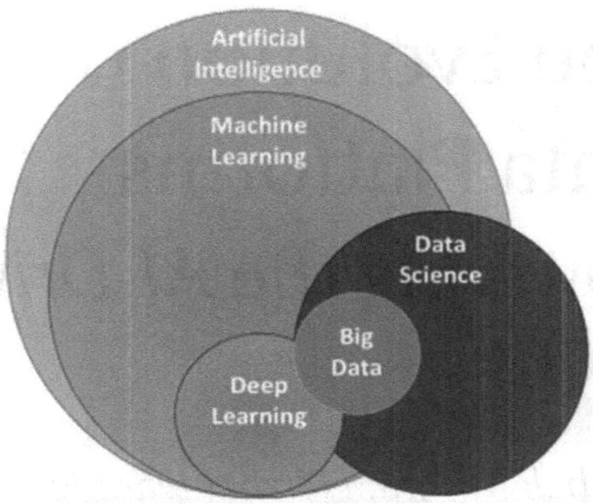

FIGURE 12.1 The Venn-diagram showing AI-related terminologies overlap (Thakur, 2020).

12.2.1 Traditional Data Warehouses

They are designed for structured data but have difficulty handling unstructured or semi-structured data in many of today's applications, resulting in high latency in modern data platforms. Cloud-based solutions like Snowflake, Google BigQuery, and Azure Synapse Analytics centralize data and offer real-time analytics. Their cloud-native designs enable organizations to grow storage and computing resources in accordance with the requirements, optimize costs while ensuring their real-time data availability. This flexibility is crucial for AI projects, which often require rapid processing of large datasets to feed into ML models (Smith, 2023).

12.2.2 AI and Machine Learning Integration

With the growing dependency on AI for insights and automation, Modern data platforms now package AI and ML technologies together. Numerous platforms have integrated AI/ML tools to facilitate the end-to-end flow of data from data pipeline building to model deployment. This consolidation helps minimize the complications of operating AI infrastructures and enhances productivity (Anderson, 2023a). AI models need real-time access to data, and with features such as parallel processing, distributed storage, and in-memory computing, lag times are dramatically minimized, enabling AI algorithms to scale (Davis, 2022).

12.2.3 Created by Big Data Scalability

As the Big Data explosion and AI-powered projects are on the rise, platforms are required to scale better. Scalability means increasing resources (disk space, computing power, etc.); performance does not deteriorate. These modern platforms correlate and can adapt to increasing data volumes, providing companies with tailored data operations and high-performance analytics and decision-making in the face of evolving data (Johnson, 2021). This scaling is necessary in AI initiatives when they involve real-time data processing and analysis (Mishra et al., 2024).

12.2.4 Effectiveness for Big Data

The efficiency is equally important. Such responsive platforms enable organizations to move from data to insight faster than ever before. Distributed computing too where tasks are broken down into smaller chunks and processed in parallel across multiple nodes, delivers speed (Smith, 2023). Modern platforms are built to leverage scalability and efficiency that allow organizations to handle Big Data for AI-oriented projects, while real-time analytics and innovation can provide deep insights into useful data.

By investing in horizontal, efficient data platforms, organizations will unlock the full potential of AI and Big Data for ongoing innovation and growth (Sun & Zhang, 2021).

12.3 EVOLUTION OF DATA PLATFORMS

The world Big Data analytics market is valued at $350 billion currently and is predicted to grow to about $655 billion by 2029 (Dean & Ghemawat, 2004). Data platforms have evolved to meet the needs of modern enterprises as the scale, speed, and complexity of data have continued to increase exponentially. The evolution of data platforms, from on-premise solutions to cloud-based architectures to the emergence of open-source platforms and modern data warehouses, showcases the industry's constant quest for scalability, flexibility, and performance (Tomar & Agarwal, 2022) (Mishra et al., 2024) (e.g., as a topic or high-level full-text source). We will discuss the limitations of traditional systems, the rise of cloud-based platforms, the adoption of open-source compute frameworks (Hadoop, Spark, etc.), data lakes and data warehouses, and the phases of this evolution.

12.3.1 Legacy Data Platforms

Traditional data platform implementations were primarily built on top of Relational Database Management Systems (RDBMS), which created, stored, and managed structured data in a unidimensional matrix comprising rows and columns inside a predefined schema. Examples of this include Oracle, Microsoft SQL Server, and IBM DB2, which are commonly used systems for managing transactional workloads (Smith, 2021). These platforms were frequently deployed on-site and demanded significant hardware resources such as dedicated servers, storage solutions, and networking devices, all of which required routine upkeep and monitoring (Johnson & Miller, 2022). These systems were critical in maintaining data consistency and integrity through ACID (Atomicity, Consistency, Isolation, Durability) properties and offered transactional guarantees necessary for economic sectors such as finance and manufacturing.

Pros of Traditional Data Platforms

- Exceptionally Processes of Structured Data: In summary, traditional RDBMS structures were empirically designed to structure a collection of data in relational tables with the organized approach of SQL support to manage itself for query and transaction processes (Smith, 2021).

- Reliability and Consistency: A guarantee of data consistency and durability of transactions was ensured with the enforcement of ACID properties (Johnson & Miller, 2022).
- Indexing and Query Optimization: Query optimizers and advanced indexing techniques (like B-trees and hash indexes) further enhanced the speed and performance of data retrieval (Chen et al., 2020).

Shortcomings of Conventional Data Platforms

- Scalability Limitations: Traditional platforms had difficulties scaling horizontally (adding additional servers) because of data center limitations. They default to vertical scaling, i.e., adding more power (CPU, memory, storage) to the existing hardware. As businesses became larger, this process became costly and tedious (Chen et al., 2020).
- Reduced Data Range: These systems were designed to work with structured data and had difficulty dealing with semi-structured or unstructured data such as documents, multimedia files, or social media data. As a result, supporting these types of data often involved complex and inefficient workarounds (Lee, 2023).
- Slow Data Processing: Traditional systems often rely on batch processing, which collects data over a period and processes it at prescribed times. This resulted in a lag that created challenges for industries needing to process real-time data, such as e-commerce and finance, to make timely decisions (Davemport & Ronanki, 2018).
- Inordinate Owning Costs: Compared to other platforms, they needed a lot of real estate for hardware as well as for electrical power and cooling, and frequent, costly software updates. The downtime due to upgrade and maintenance may bring business disruption (Liu & Peng, 2021; Zhou et al., 2022).

12.3.2 Cloud-Based Data Platforms

Because of the limitations of older on-premise systems, cloud-based data platforms were the way forward. Cloud computing services (e.g., Amazon Web Services (AWS), Microsoft Azure) revolutionized data management by providing them with scalable on-demand resources, ranging from storage to analytics (Tomar & Agarwal, 2022). Recent developments have caused a pivot from this, as cloud computing provided the required flexibility, scaling, and efficiency as legacy systems accrued heavy costs with little return (Harris, 2021). Cloud platforms turn capital expenditure into operational expenditure, allowing for real-time data processing necessary for modern business (Jain & Kapoor, 2021).
Advantages

- Scalability: With their elastic scalability, cloud platforms allow organizations to scale their storage and computing resources up or down depending on demand, leading to better cost control (Wang, 2023). They help avoid costly infrastructure—freeing up capital expenses and turning costs into flexible operating expenses (Harris, 2021).

- Advanced Analytics: Built-in analytics and ML tools, such as the ones proposed by Blanco and McKinney (2021), can simplify the requirements for businesses to carry out data-driven actions, refreshing the process of real-time business models.

Security and Compliance: Cloud services deliver robust security measures, including encryption and identity management, enabling organizations to adhere to regulations such as General Data Protection Regulation (GDPR) and Health Insurance Portability and Accountability Act (HIPAA), especially important for sectors such as healthcare and finance (Gandomi & Haider, 2015; AZTech IT Solutions, 2018).
Limitations

- Internet Dependence: Cloud platforms need stable internet access; outages can lead to downtime and delays (Harris, 2021).
- Data Privacy Issues: Security may provide some assurances, but storing sensitive data with third parties and off-site poses compliance and privacy risks (Harris, 2021).
- Long-Term Costs: Using cloud services creates long-term costs that can add up quicker than the one-time investment associated with traditional systems (Harris, 2021).

12.3.3 Open-Source Platforms

Apache Hadoop Introduction Open-source frameworks like Apache Hadoop and Apache Spark have revolutionized data management in the cloud. Hadoop, which emerged in the mid-2000s, provided a foundation for large-scale distributed dataset storage and processing with its core components, the Hadoop Distributed File System (HDFS) and MapReduce. HDFS provides high access throughput, fault tolerance for large file storage across nodes, and MapReduce enables task processing to be carried out in parallel. A lot of organizations require affordable solutions for their big data problems, which is how Hadoop's design came into play, as it works on commodity hardware to reduce costs. Apache Spark then came up as an in-memory processing framework, which added more power to Hadoop with faster, real-time data processing. Spark caches this intermediate data in memory, which speeds up computations and allows for various workloads such as batch processing and streaming analytics, making it a suitable platform for ML.
Pros:

- Cost-Efficiency: Organizations can handle large sets of data without the need to invest in costly proprietary systems (Shvachko et al., 2010).
- Community Innovation: Open source naturally promotes collaborative growth through community contributions (Apache Software Foundation5, 201).

- Interoperability: Built to integrate with a multitude of storage services, like Amazon S3 and Google Cloud Storage, to avoid vendor lock-in (Delaet, 2014).

Cons:

- Complexity: The configuration and maintenance of these frameworks may need a high level of technical knowledge (McWherter, 2014).
- Support Limitations: Organizations do not have access to direct vendor support, unlike proprietary systems (Shvachko et al., 2010).
- Performance Overheads: Given certain scenarios, open-source solutions may not deliver the same performance level as proprietary ones (Zaharia et al., 2016).

12.3.4 Data Lakes

The rising volumes of data have led to the emergence of centralized storage solutions capable of accommodating various data types. Data lakes and modern data warehouses are now vital components of the data platform landscape.

A data lake is a centralized repository that allows businesses to store all types of structured and unstructured data at scale (Harris, 2021). Data lakes, unlike traditional data warehouses, not only can accept raw data from different sources; they also can store data in its native format without having to rearrange its structure before the ingestion (Blanco & McKinney, 2021). Especially if an organization deals with a diversity of data types like video files, IoT sensor data, and social media feeds, this flexibility is a boon (Lee, 2023). Data lakes enable both batch and real-time processing, as well as ML, predictive analytics, and bypassing business intelligence (Tomar & Agarwal, 2022).

But working with data lakes is not without its challenges. Without proper governance and data management, unstructured and poorly organized data becomes a "data swamp" that is hardly useful for analysis (Chen et al., 2020). Organizations gradually adopt data lake management tools that improve data cataloging, metadata management, and data quality controls to mitigate these issues (Gandomi & Haider, 2015). In contrast, the contemporary data warehouse—e.g., Amazon Redshift, Snowflake, or Google BigQuery—has been designed for both structured and semi-structured data. Typically, cloud data warehousing platforms are a hybrid between the performance and scalability of traditional warehouses and cloud-native flexibility (Harris, 2021). The system is optimized to execute analytical queries for which the response is often on a large dataset, thus returning very fast once the data is stored (Johnson & Millar, 2022).

The main difference between the data lake and the data warehouse can be found in the way they store and process the data. Data warehouses usually use a schema-on-write method of data structuring, where the data is organized and structured before being stored. While data warehouses leverage the schema-on-write principle, data lakes adopt schema-on-read, where raw data is stored and schema is applied only during the querying of the data (Zhou et al., 2022). Each has its advantages, so many organizations have adopted hybrid systems that combine data lakes with data warehouses that take advantage of both types of platforms (Davenport & Ronanki, 2018).

12.4 FEATURES OF MODERN DATA PLATFORMS

Modern data platforms have changed the way we find the systems that capture and store big data and analytics from different sources, as the big data market becomes more competitive with the demand to create scalable, elastic, and flexible infrastructures designed to store and process big amounts of data. Modern platforms support distributed computing, real-time data processing as well as multi-format data storage. Moreover, technologies such as containerization have become a game changer, allowing for quicker, more efficient processes than ever before. Here, you can read the main attributes that define modern data platforms and their application when it comes to processing data in our data-first world.

12.4.1 Scalability, Elasticity, and Flexibility

The ability to scale resources as required drives the asset of modern data platforms. Traditional systems would run into limitations with scaling as they primarily depended on the upgrading of physical hardware or vertical scaling where more powerful hardware was needed to handle bigger workloads (Cohen et al., 2009). By contrast, modern platforms—especially those that run in the cloud—allow us both horizontal and vertical scaling (we can both add more servers and make existing servers more powerful). As data continues to increase with numerous digital interactions, IoT devices, and others generating data rapidly, services such as AWS and Microsoft Azure make it very easy for businesses to scale up storage and compute resources (Papadopoulos & Xiang, 2022). For instance, Netflix uses AWS to scale its infrastructure for streaming services during peak periods, such as when a new show is released.

Elasticity allows the resources to automatically adapt to changing workloads, avoiding the costs of over-provisioning as well as the performance problems caused by under-provisioning. Airbnb, for example, utilizes Google Cloud Platform (GCP) to manage resource spikes during busy travel seasons (Pal & Kumar, 2021).

Flexibility is a critical characteristic of modern data platforms that allows organizations to execute various workloads and easily accommodate multiple data formats—structured, semi-structured, and unstructured—without rigid limitations. To handle such data types, platforms such as Google Cloud BigQuery and Snowflake might serve IoT and social media analytics because they easily manage the above-mentioned data types (Kim & Park, 2022). This approach enables seamless collaboration between different tools, frameworks, and business models, making it versatile for complex analytics tasks. Beyond that, microservices architecture like Spotify facilitates the separate development, deployment, and scaling of components to accommodate different workload requirements (Hale & Burton, 2020). Also, the microservices architecture of modern platforms enables each component to be developed, deployed, and scaled independently.

12.4.2 Distributed Computing Architecture

Distributed computing architectures have become the mainstay of modern data platforms, allowing organizations to easily scale to growing datasets. These systems spread tasks between servers and facilitate parallel processing, which enables much faster completion of data operations than would be possible on a single machine (Mohamed Ghazala, 2023). This paradigm allows for optimal performance in both data storage and computation.

One of the major concepts of distributed computing comes from Apache Hadoop, which introduced the notion of "distributing both data and computation over clusters of commodity hardware" (Smith, 2020). Its HDFS and MapReduce processing model were instrumental in facilitating the storage and processing of large amounts of data across distributed clusters. However, technologies such as Apache Spark have also since improved upon Hadoop by offering quicker in-memory processing and more flexibility (Johnson & Kosslyn, 2018).

You can access your applications without having to worry about latency (speed), fault tolerance, and high availability. Furthermore, since data is replicated between different machines, if one fails, the system can stay running (Hernandez & Miller, 2021). Big-scale data analyzing platforms such as Amazon Redshift and Google BigQuery incorporate these architectures for data management and include auto-scaling, load balancing, and other features to enhance resilience and support real-time processing of big datasets (Kaur, 2020; Davenport, 2014; Sitaram & Manjunath, 2012).

12.4.3 Data Processing/Application in Real Time and Streaming Analytics

Advancement in real-time ML has witnessed a growing demand for instant decision-making. Today's data platforms can support real-time analytics that enable organizations to process and analyze data the moment it's created. This is a significant step up from batch processing, which incurs the latency before insights can be extracted (Cohen et al, 2009).

Apache Kafka, Apache Flink, and Amazon Kinesis are examples of streaming platforms that allow real-time data processing by taking events from multiple data streams and processing them in real time (Xu & He, 2023). For example, banks can conduct real-time transaction monitoring to identify potential fraud and initiate instant alerts. Retailers can adapt their marketing efforts or stock based on live customer actions (Pal & Kumar, 2021).

In IoT applications, for example, new devices continuously create massive data streams in near-real time, which need to be analyzed in real time. For instance, real-time analytics is vital for autonomous vehicles, processing sensor data and making rapid decisions for safe operation (Kim & Park, 2022; Hale & Burton, 2020).

Additionally, these platforms can connect to traditional data stores, providing organizations with a consolidated view by merging real-time analytics with historical data. This hybrid strategy becomes crucial for businesses that require a comparison between real-time trends and historical data in order to make informed decisions (Mohamed Ghazala, 2023).

12.4.4 Support for Structured, Semi-Structured, and Unstructured Data

Today's organizations operate across various data types, from structured relational data to semi-structured data like JSON or XML and unstructured data like videos, images, and social media posts (Kaur, 2020). Modern data platforms facilitate the management of different data types, providing tools for ingestion, storage, processing, and analysis.

Traditional relational databases are specially designed for structured data but do not work well for semi-structured and unstructured data (Smith, 2020). Modern platforms bypass this restriction through schema-on-read techniques, as are commonly seen in data lakes, where the data is saved in an original format and structure is only applied when needed during analysis (Johnson & Kosslyn, 2018). Such flexibility is essential for organizations handling diverse data sources. In contrast, utilizing technologies such as Apache Hadoop and Spark allows users to store raw data and implement a schema at the time of analysis, enabling support of semi-structured and unstructured data (Hernandez & Miller, 2021).

Furthermore, current data warehouses such as Snowflake and Google BigQuery offer support for both structured as well as semi-structured data, so organizations can query JSON, Parquet, and Avro data types natively without transforming them into tables (Davenport, 2014). Such versatility in handling structured, semi-structured, and unstructured data allows organizations to interact with a range of data sources, whether from transactional databases to multimedia sources or the output of IoT sensors (Garrison et al., 2012). This wide-ranging support helps businesses leverage AI and ML models across the whole of their data ecosystem, enabling richer insights and smarter decision-making (Moore & Clark, 2023).

12.4.5 Containerization (Kubernetes, Docker, etc.)

Containerization plays a crucial role in modern data platforms and enables developers to run applications in isolated environments, thus eliminating the need to manage and maintain infrastructure in serverless architectures. It enables developers to build applications without any concern about underlying infrastructure (Kim, 2022). Containerization is used in serverless platforms, including AWS Lambda and Google Cloud Functions, to automatically scale applications according to demand, improving scalability and flexibility (Davenport, 2014).

Docker "gathers applications with all dependencies and offers the same performance in development, testing and production" (Moore, & Clark, 2023). Kubernetes abstracts over these containers, making automation of management easy, handling scaling, load balancing, and self-healing tasks to effectively run on a large scale (Broderick & Ellis, 2019) in a multi-cloud or hybrid environment.

Docker makes it easier to develop and deploy applications, especially when using a microservices architecture, where each separate service runs as a container. That's helpful for data platforms with complicated parts such as data ingestion and ML models (Thomas, 2022). Furthermore, the version control offered by Docker also guarantees reproducibility, making it essential for collaborative projects (Garrison et al., 2015). Kubernetes takes these features and adds the necessary automation to efficiently maintain large-scale containerized environments and enable the dynamic scaling needed for serverless platforms (Kim, 2022; Davenport, 2014).

12.5 AI-DRIVEN DATA PLATFORMS

Modern data platforms equipped with AI and ML have revolutionized the way data is collected, stored, processed, and analyzed. As enterprises continue their pursuit of value from large datasets, AI and ML capabilities enable automation of processes, predictive modeling, and data-oriented decision-making. The application of AI and ML to data platforms is also featured in this section, exploring how these technologies can automate data preparation and feature engineering, enable predictive and prescriptive analytics, and ultimately improve decision-making.

12.5.1 Data Platforms with AI/ML Capabilities Integration

Big data is the momentum driven by big data, which is a necessity for any platform, not only to store the data but also helps in deriving design by going through the generated data. As a result, AI and ML are now foundational elements of these platforms, allowing for automated data analysis and enhanced pattern identification, once tedious and error-ridden when done manually (Garrison et al., 2012).

For example, one of the major breakthroughs is the deployment of AI/ML models directly to data platforms, making it possible for users to reach complex ML tasks without transferring data across systems. The integration of analytics close to the sources of information allows for a high reduction in latency and the complexity of the proposed methods driven by AI (Davenport, 2014). For instance, one relevant Google Product is Google BigQuery ML, which enables users to create and run ML models in Google's cloud data warehouse using SQL commands to create, train, and deploy models (Moore & Clark, 2023). Snowflake has similar functionalities through Snowpark, which is a programming and data processing environment that allows users to write code in Python or Java and run ML capabilities in the Snowflake environment (Kim, 2022).

These platforms are intrinsically highly scalable, leveraging cloud computing to process terabyte-sized datasets. Native integration of AI and ML also helps enable data scientists, engineers, and analysts to iterate quickly on models and deploy them in real-time environments. This capability lowers the barrier of do-it-yourself ML infrastructure while democratizing advanced analytics beyond only data science expertise (Kaur, 2020).

These platforms are intrinsically highly scalable, leveraging cloud computing to process terabyte-sized datasets. Native integration of AI and ML also helps enable data scientists, engineers, and analysts to iterate quickly on models and deploy them in real-time environments. This capability lowers the barrier of do-it-yourself ML infrastructure while democratizing advanced analytics beyond only data science expertise (Kaur, 2020).

12.5.2 Automated Data Preparation, Data Wrangling, and Feature Engineering

Data preparation and cleaning are one of the most time-consuming tasks in data science. Data must be cleansed, transformed, and often blended from multiple sources before any real analysis can occur. In conventional data wrangling, the process is heavily reliant on manual scripting, where substantial human intervention is necessary and prone to error, resulting in a bottleneck in the overall analytical process (Smith, 2020).

AI data platforms automate large parts of data preparation and feature engineering and reduce the time and effort involved. Databricks' Delta Lake and Azure Synapse Analytics are examples of these tools that offer automated data wrangling capabilities, automatically spotting patterns, missing values, and outliers without human interaction (Johnson & Kosslyn, 2018). In order to streamline the data cleaning and standardization process, these platforms harness ML algorithms to automatically recommend transformations and adjustments (Hernandez & Miller, 2021).

"Feature engineering, which is another important but time-consuming step in the data science process, is the process of converting raw data into a format that is accepted for the machine learning algorithms." This process can be tedious, but AI-driven data platforms can automate it by utilizing algorithms for feature selection to identify the most significant variables and from a dataset, such solutions will help to maximize time for model development (Davenport, 2014). For example, Google BigQuery ML and Snowpark running on Snowflake allow automatic feature selection and transformation, which shortens the time to go from raw data to predictive model (Xu & He, 2023).

The recent generation of AI-powered data catalogs, such as those by Alation and Collibra, also provides a combination of data governance with automated data profiling and wrangling capabilities (Pal & Kumar, 2021). Such catalogs not only organize and classify the data, but also they use AI to suggest relevant data sets for a task and recommend transformations to clean and improve data quality (Kim, 2022).

12.5.3 Predictive and Prescriptive Analytics with AI

Building, training, and deploying models at scale have reshaped predictive and prescriptive analytics with the help of AI-powered data platforms (Figure 12.2). Predictive analytics focuses on analyzing historical data to predict future outcomes, and prescriptive analytics builds on predictive analytics by providing recommendations on specific courses of action (Garrison et al., 2012). These capabilities combine to enable organizations to transition from descriptive reporting to proactive, actionable insight.

Data platforms powered by an AI and ML stack have democratized predictive analytics, empowering business users, analysts, and data scientists to apply sophisticated models without deep ML knowledge (Kaur, 2020). For example, Google BigQuery ML offers the ability to write simple SQL queries to train and deploy predictive models such as regression, classification, and time-series forecasting (Moore & Clark, 2023). This helps to avoid unnecessary data moving between environments and makes the model deployment process much easier.

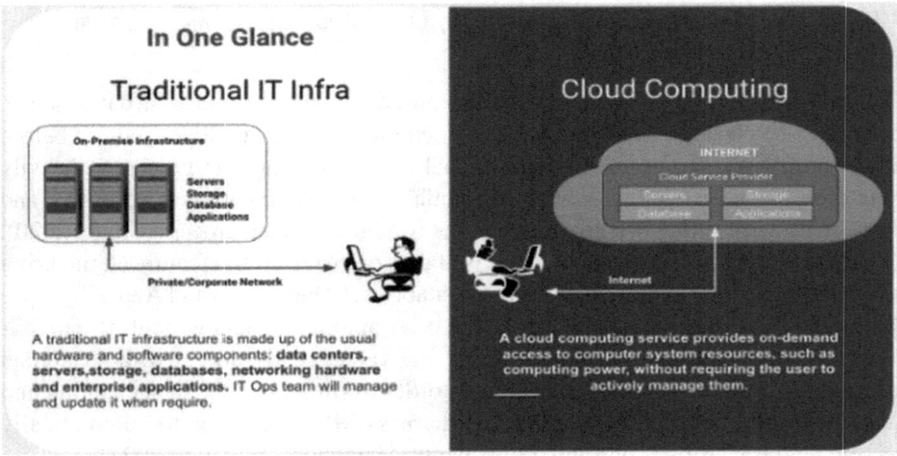

FIGURE 12.2 Using descriptive, predictive, and prescriptive analytics to make decisions (Hayes, 2018).

Snowflake enables predictive analytics as well, using its Snowpark and UDF (user-defined functions) to allow users to run Python or Java code natively in the platform (Kim, 2022). This lowers latency and promotes collaboration between data engineers and data scientists, enabling models to be built and deployed directly in the platform.

Prescriptive analytics simply predicts outcomes; prescriptive analytics goes one step further and recommends which course of action to take (Davenport, 2014). In contrast, platforms powered by AI, like Azure Synapse Analytics, merge ML models and optimization algorithms to recommend actionable strategies based on data (Xu & He, 2023). As an illustration, in retail, prescriptive analytics may advise on changing inventory levels according to estimated demand, and in healthcare, it can prescribe personalized treatment plans based on patient data (Pal & Kumar, 2021).

An even larger factor has been the introduction of reinforcement learning, a kind of AI that learns by interacting with the environment around it. Second, reinforcement learning algorithms can enhance prescriptive analytics, as they are capable of learning and adapting their decisions based on the changes in the environment and feedback from the users (Hale & Burton, 2020). They are Dynamic capabilities that allow organizations to adapt, pivot, and change their strategies on the fly as conditions change.

12.6 AI-DRIVEN DATA SOLUTIONS: EXAMPLES

Here are a few AI-driven platforms that demonstrate the strength of integrated AI/ML capabilities within new data architectures:

- Google BigQuery ML: Users can build and deploy ML models using SQL statements in BigQuery, making AI-driven analytics accessible without needing to set up specialized ML infrastructure (Google Cloud, 2023; Veeramachaneni et al., 2021).

- Azure Synapse Analytics: Azure Synapse unifies big data and data warehousing in the same solution, empowering users to perform prescriptive and predictive analytics using AI and ML in real time (Microsoft Azure, 2023).
- Snowflake with Snowpark: Snowpark enables developers and data scientists to create ML models in Snowflake, providing a scalable environment to perform AI-driven analytics without moving data in and out of the platform (Snowflake Inc., 2023).

12.7 SECURITY, PRIVACY, AND DATA GOVERNANCE

In the age of AI and big data, where data gives rise to innovative solutions and decision-making processes, governance, security, and privacy measures have become key areas of focus. As data ecosystems become larger in size and more extensively distributed, adequate frameworks and tools are needed to ensure that data is processed safely and on legal grounds. Also, cover the importance of governance frameworks, data security challenges in distributed environments, ensuring data privacy in AI-powered analytics, and the key tools and frameworks used for ensuring governance and compliance.

12.7.1 Governance Frameworks Matter

Data governance is the processes and policies instituted to manage the availability, usability, integrity and security of the data in the enterprise systems (Abraham et al., 2019) Data governance frameworks play a vital role in AI and big data ecosystems, helping to ensure that data is accurate, consistent, and managed responsibly throughout its lifecycle. The lack of a proper governance framework can lead to AI models producing biased, partial, or inaccurate outputs that drive flawed business decisions (Duan et al., 2019).

- Governance frameworks lay the groundwork for compliance with regulatory requirements like the GDPR in the European Union and the California Consumer Privacy Act (CCPA) in the United States.
- They assist organizations in tracking data lineage, which guarantees that the data being used in analytics is from trusted sources and hasn't been corrupted.
- Data governance frameworks can help in bridging the technical and ethical aspects of AI by ensuring that data is properly handled and that businesses are accountable for their use of data (Otto, 2020).

A key issue in AI governance is the provision of high-quality training data, as well as data that is representative enough to avoid biased outcomes. With the ongoing advancement of AI, and its increasing integration into decision-making, organizations must implement governance practices that match the fluid nature of AI-driven analytics (Tannenbaum & Wingfield, 2020) (Mishra et al., 2024).

12.7.2 Issues with Security of the Automation Data

With organizations gradually moving toward distributed and cloud-based data platforms, new security challenges emerge. Conventional security paradigms that were developed for on-premises systems are often ill-suited for the unique complexities of cloud environments (Grobauer et al., 2011). Cloud data can travel through several locations and devices, raising the risks of breaches, and unauthorized access and/or data loss.

Cloud providers (like AWS, Azure, Google Cloud) provide their own security controls, but ultimately, with a shared responsibility model, organizations are responsible for protecting their own data. Security aspects in distributed environments concern access control management, data encryption both at rest and in transit, and defense against advanced persistent threats (Chen et al., 2019).

Because of their dependence on extensive datasets, AI-powered systems can be especially sensitive to data poisoning attacks, in which attackers insert contaminated entries into the training data to compromise the models (Cao et al., 2020). In addition, centralized monitoring is often absent in distributed environments, inhibiting organizations from keeping an eye on their data in multiple nodes (Singh et al., 2021).

Another major hurdle facing these organizations is compliance with regulatory regimes like GDPR, which requires organizations to maintain data with strict security protocols to prevent unauthorized access and breaches. Ignoring these requirements can result in severe penalties, and as a result, organizations must take a serious consideration toward building strong security mechanisms in their distributed systems.

12.7.3 Data Privacy

AI-driven analytics also raises important issues about data privacy, as these models often process large amounts of personal information. AI and ML systems often need access to sensitive data, including health records, financial information, or personal preferences, in order to deliver accurate insights. But if a careless or ineffective effort is made to accomplish data anonymization and security, this presents serious privacy risks (Acquisti et al., 2016).

A clear point of contention in AI is the ability (or inability) of ML models to pull sensitive and confidential data from seemingly non-concrete data points (Shokri et al., 2017). To counteract this threat, privacy-preserving methods (including but not limited to differential privacy, homomorphic encryption, and federated learning) are increasingly integral to AI systems. For example, differential privacy adds noise to the data, such that individual records cannot be traced, which allows AI approaches to operate around sensitive information (Dwork & Roth, 2014).

Both GDPR and CCPA impose stringent data privacy regulations, albeit the context being regulatory. GDPR, for instance, empowers individuals by granting them the right to view, amend, and remove their personal information. Data processing activities carried out by organizations that utilize AI models for analyzing data must comply with these regulations to prevent any legal consequences (Voigt & Bussche, 2017).

12.8 GOVERNANCE AND COMPLIANCE TOOLS AND FRAMEWORKS (GDPR, CCPA, ETC.)

A range of tools and frameworks have arisen to assist with data governance, security, and privacy compliance. Here are some of the biggest ones:

12.8.1 General Data Protection Regulation (GDPR)

Implemented in the European Union, GDPR is one of the most extensive global data protection laws. It regulates the way organizations gather, store, and process personal information and requires organizations to put stringent security measures in place to safeguard (Voigt & Bussche, 2017). AI platforms should comply with GDPR and must fulfill several important legal obligations.

12.8.2 CCPA (California Consumer Privacy Act)

CCPA gives California consumers rights to access, delete, and opt out of the sale of their personal information. The impact on organizations using AI relates to the need to ensure that AI systems do not unintentionally process information about people in a manner that violates questions of privacy (Gellman & Dixon, 2019) in a manner that is not CCPA compliant.

12.8.3 ISO/IEC 27001

An international standard that determines the elements of managing sensitive information of the company It has specifications for establishing an information security management system (ISMS), ensuring data confidentiality, integrity, and availability (Humphreys, 2013).

12.8.4 NIST Privacy Framework

This framework, created by the United States National Institute of Standards and Technology, assists organizations in integrating privacy into their data processing activities. It provides a risk-based approach focused on use and accountability, making it particularly well-suited for AI-enabled platforms that utilize personal data (NIST, 2018).

It offers continuous security assessments of resources and compliance with standards like GDPR and ISO/IEC 27001, helping organizations stay aware of their data security posture.

Such tools and frameworks make certain that AI-based platforms stay compliant with the developments of data privacy and security regulations. It is essential to note that adherence is not a one-time activity, and organizations need to keep themselves updated as well as audit their practices with the most recent laws and standards.

12.9 INDUSTRIAL CASE STUDIES

12.9.1 Machine Learning and AI

The Future of Modern Data Platforms Modern data platforms are paving the way for the increasing impact of AI and big data analytics across varying industries, providing scalable infrastructure, near real-time data processing, seamless AI, and ML capabilities integration. Here, we share in-depth case studies from healthcare, finance and retail showcasing how these platforms are revolutionizing operations and driving significant business results. The detailed case studies capture the technical sophistication of AI-driven analytics, demonstrating its power in diverse sectors.

12.9.2 Mayo Clinic

For instance, the Mayo Clinic has utilized Google Cloud's BigQuery and AI-based analytics to process and analyze enormous volumes of patient information, driving diagnostic accuracy and operational efficiency. The clinic created predictive algorithms that leveraged various data from patients, including unstructured medical records, images, and genomic sequences, by integrating AI models into their workflows. More specifically, TensorFlow (i.e., Google's ML platform) was instrumental in building reliable deep learning models for medical image analysis and has greatly improved the performance of diagnostic activities like cancer detection (Castellanos, 2021; Goedert, 2019).

The application of AI-powered predictive analytics led to a 30% decrease in diagnostic errors along with better patient outcomes that resonate with previous studies highlighting the ability of AI to enhance diagnostic quality (Ghassemi et al., 2018). In addition, AI-powered scheduling and predictive analytics models improved operating room (OR) utilization by freeing up 12% more ORs and decreasing patient wait times. This holistic method not only enhanced patient experience but also resulted in a 15% reduction in patient readmissions (Ghassemi et al., 2018; Topol, 2019). This partnership between Mayo Clinic and tech giant Google demonstrates the transformative power of AI and big data analytics in the healthcare landscape, showcasing how the data platforms of today can facilitate major strides in patient care.

12.9.3 JPMorgan Chase

The reimagined, labor-intensive function of contract review demonstrates the transformative potential of AI and applies it at scale, as evidenced by financial giant JPMorgan Chase's Contract Intelligence (COiN) platform. COiN applies NLP algorithms on an Apache Hadoop and Spark infrastructure to parse and analyze legal documents such as credit default swaps and International Swaps and Derivatives Association (ISDA) agreements. Using NLP techniques like named entity recognition (NER) and topic modeling, the platform can identify and extract relevant information from contracts in a fraction of the time it would take to analyze the documents manually, significantly speeding up the document review process. For instance, COiN can review 12,000 contracts in mere seconds, in contrast with manual processes that would take upward of 360,000 hours to accomplish (Nasir et al., 2020).

The COiN platform contributed to significant cost savings, with 80% less spent on legal and compliance operations. In addition, the mobile integration with live data pipelines developed by Apache Kafka ensures that contracts are always up to date based on the ability to quickly analyze live data pipelines. By analyzing documents and extracting information in real time, this ability allows JPMorgan

Chase to improve its compliance with regulations and efficiency in operations (Nasir et al., 2020).

12.9.4 Amazon

Amazon has advanced AI and big data techniques in its recommendation engine, which has a share of 35% in total sales. Single-site data processing: Run on AWS and use distributed frameworks like Apache Spark and Hadoop for processing large-scale data. Using real-time data analysis, the company's personalized product recommendations are built from information gained from clickstream data, purchase history, and product metadata, among various other factors.

Kinesis Data Streams were used to process customer activity in combination with AWS SageMaker for the collaborative filtering algorithms and deep learning models that power the engine. Also, continuous A/B testing (between various upsell sets) results in a multiplication of an algorithm performance (Smith, 2019) where user engagement increases by 10% and cart abandonment rates decrease by 15%.

Amazon uses predictive analytics for demand forecasting in supply chain management to optimize inventory and logistics. The company can adjust what they are stocking based on customer tastes after real for example gives a look at how you might do that using Kinesis and AWS Lambda to real-time process your supply chain data. This increase in efficiency has resulted in a 20% decrease in stockouts and faster average delivery times (Kohavi et al., 2020).

12.9.5 Uber

Uber utilizes AI-based analytics platforms to better its global ride sharing business. Its data architecture, built on Apache Hadoop and Apache Spark towers, handles huge scale, real-time data on user demand, driver availability and traffic regional traffic patterns. Based on predictive analytics through ML models, Uber manages the rider-driver match as well as optimizes pricing through surge pricing techniques.

Kafka powers the real-time streaming architecture and processes millions of ride requests, driver locations, and traffic data points per second. Apache Flink processes this data and manifests real-time changes in pricing and dispatching. This has led to a 12% reduction in idle times, a 5% increase in routes taken and annual savings of $1 billion on fuel and operational expenses (Zhang et al., 2019).

Moreover, Uber uses geospatial data analytics to also improve the routing algorithm via reinforcement learning, whereby the system learns from historical trip data. The results were a 10% reduction in variance of trip time and greater customer satisfaction due to shorter wait times (Chen & Hsieh, 2020; Miller & Lemoine, 2019).

12.9.6 Netflix

One of the most prominent examples of AI and big data analytics in action is Netflix. The recommendation engine, which is based on Apache Kafka and Hadoop, analyzes large volumes of viewing history, search queries, and video metadata to provide customized recommendations. Using collaborative filtering techniques, such as matrix factorization and deep neural networks, Netflix successfully predicts users' preferences.

The real-time data pipeline handles over a trillion zillion events a day in Apache Kafka and directly hooks into Content Delivery Network (CDN), ensuring that recommendations are updated in real time based on user actions. Driven by such personalized designed, 80% of content consumption on the platform is guided by the recommendation's engine, which helped reduce customer churn by 10% and improved user retention by 20% (Amatriain & Basilico, 2012).

Additionally, Netflix uses AI-powered compression algorithms to ensure the best video quality out there. Netflix dynamically optimizes its video encoding using ML models of network states and viewer preferences based on massive datasets. This strategy saves 25% data consumption but delivers great video quality, especially in poor Internet scenarios (Kohavi et al., 2020).

12.10 DATA PLATFORM CHALLENGES

Modern Data Brains: The Growth of Enterprise data platforms fueled by new AI, IoT and big data technologies mean organizations can now have unfettered access to the information they need to make real-time informed decisions. But making these platforms a reality at scale brings its own challenges—especially when it comes to integrating with legacy systems and keeping up with new trends such as edge computing and quantum technologies. In this part we talk about major technical challenges and future directions in terms of convergence of AI, IoT and big data.

12.10.1 Integrating and Interoperating Data

Integrating data between different hybrid environments is still a significant challenge. Organizations typically handle data from legacy relational database management systems (RDMS) (e.g., Oracle, MySQL) as well as new age NoSQL (e.g., MongoDB, Cassandra) database systems, requiring complex transformation of data flows to enable seamless interaction (Sun & Zhang, 2021). Tools such as Apache NiFi and Kafka are key to managing real-time streaming pipelines; however, mismatches in the schema can lead to inconsistencies. Data virtualization solutions (Denodo) and serialization frameworks (Avro, Parquet) can abstract and standardize data across systems (Johnson, 2021).

12.10.2 Scalability and Performance

For increasingly larger volumes of data, elastic scaling is done via distributed processing systems such as (but not limited to) Apache Hadoop, Spark, and cloud-native services. Nonetheless, situations including data skew and expensive shuffle operations in distributed environments can lead to such performance degradation (Anderson, 2023b). These issues are alleviated through vertical scaling via autoscaling mechanisms (e.g., AWS Auto Scaling, Kubernetes) and in-memory processing (e.g., Apache Ignite).

12.10.3 Governance, Compliance and Security

Data governance and compliance are of utmost importance, especially with stringent regulations such as GDPR and CCPA in place. Governance frameworks like Apache Atlas (Smith, 2023) with capable of retrieving relevant metadata in terms of lineage are needed to manage operational metadata, while multi-tenant level security is ensured by encryption protocols (TLS (Transport Layer Security), Advanced Encryption Standard (AES-256)). Similar solutions such as AWS Key Management Service (KMS) and Attribute-Based Access Control (ABAC) frameworks provide improved security and regulatory compliance (National Institute of Standards and Technology, 2018).

12.10.4 Cost Management

Cloud-based platforms have flexible pricing, but unpredictable workloads can create cost overruns. Auto-scaling allocates and releases resources as needed, but there are still challenges in optimizing cluster size and handling spot instances (Kothapalli, 2023). Some AWS services, such as AWS Cost Explorer, can be used to help monitor and control costs, and query optimization techniques can minimize computational overhead (Anderson, 2023b).

12.11 FUTURE TRENDS

12.11.1 Edge Computing

In the context of IoT, edge computing allows for local data processing, minimizing latency and bandwidth consumption by sending only pertinent data to the cloud. This is supported by platforms such as AWS Greengrass and Azure IoT Edge, which have great applications ranging from real-time monitoring to autonomous driving (Bonawitz et al., 2019).

12.11.2 Quantum Data Platforms

Quantum computing will revolutionize data platforms by solving complex problems in shorter amounts of time compared to classical systems. Platforms such as IBM Quantum and Google Quantum AI hold potential across a wide range of applications, including encryption, ML, and big data processing (Grover, 1996; Shor, 1994).

12.11.3 Serverless Architectures

With serverless computing platforms like AWS Lambda and Azure Functions, automatic scaling and resource allocation mean low-cost, event-driven workloads without needing to manage servers, making it perfect for bursty or intermittent data needs (Baldini, 2017).

12.12 CONCLUSION

Trends in AI and Big Data have also brought about the rise of modern data platforms that support ML, automation, and real-time analysis. They offer the scalability and efficiency required to enable more complex AI models and larger datasets, maximizing the potential of these technologies.

- Moving from legacy, on-premise systems into cloud-based, distributed architecture has transformed the management of data so that organizations can deal with ever-larger amounts of data in a more elastic and cost-efficient way. With technologies such as Hadoop, Spark, and containerization (with Kubernetes and Docker), organizations are faced with the ability to process this data in a scalable, in-flight manner, which is critical in a business environment that evolves quickly.
- One of these trends is embedding AI and ML directly into data platforms themselves, automating tasks such as data preparation and model training. While initiatives to bring AI behind the scenes—through platforms such as Google BigQuery ML, Azure Synapse Analytics, and Snowflake with Snowpark—provide a foundation for analytics that is more efficient and accurate.
- Modern data platforms, while beneficial, can be fraught with challenges, including technical complexity and integration with legacy systems. Cutting-edge technologies like edge computing and quantum data platforms might do the trick, from fabricating faster systems to boasting even more processing power.
- In conclusion, Modern data platforms are the foundation for businesses to succeed in today's AI/Big Data space, providing the necessary infrastructure to manage more extensive and complex data needs. In the future, advancements in edge computing and quantum systems will continue to enhance their capabilities, establishing benchmarks for performance and scalability, and helping organizations remains a market leaders in a data-driven world.

REFERENCES

Abraham, R., Schneider, J., & vom Brocke, J. (2019). Data governance: A conceptual framework, structured review, and research agenda. *International Journal of Information Management*, 49, 424–438.

Acquisti, A., Brandimarte, L., & Loewenstein, G. (2016). Privacy and human behavior in the age of information. *Science*, 347(6221), 509–514.

Amatriain, X., & Basilico, J. (2012). Netflix Recommendations: Beyond the 5 Stars. In *Proceedings of the Fifth ACM Conference on Recommender Systems*, 1–2. Association for Computing Machinery.

Anderson, J. (2023a). *AI and Machine Learning in Modern Data Platforms*. Tech Innovators Press.

Anderson, L. (2023b). Tackling scalability issues in distributed systems. *Cloud Computing Review*, 15(2), 45–58.

Apache Software Foundation. (2015, June 2). *The Apache Software Foundation Annual Report: 2014–2015 fiscal year*. The Apache Software Foundation.

AZTech IT Solutions. (2018). Key Differences between Cloud Computing vs. Traditional. Aztech IT Solutions. Retrieved from www.aztechit.co.uk/blog/topic/cloud

Baldini, I., Castro, P., Chang, K., Cheng, P., Fink, S., Ishakian, V., Mitchell, N., Muthusamy, V., Rabbah, R., Slominski, A., & Suter, P. (2017). Serverless computing: Current trends and open problems. *Research Advances in Cloud Computing*, 4, 1–20.

Blanco, I., & McKinney, W. (2021). *Python for Data Analysis: Data Wrangling with Pandas, NumPy, and IPython*. O'Reilly Media.

Bonawitz, K., Eichner, H., Grieskamp, W., Huba, D., Ingerman, A., Ivanov, V., Kiddon, C., Konecny, J., Mazzocchi, S., McMahan, H. B., Overveldt, T. V., Petrou, D., Ramage, D., & Roselander, J. (2019). Federated learning: Challenges, methods, and future directions. *Journal of Machine Learning Systems*, 37(3), 50–60.

Broderick, T., & Ellis, J. (2019). Kubernetes orchestration and management. *Journal of Cloud Computing*, 15(2), 45–60.

Cao, X., et al. (2020). Data Poisoning Attacks on Machine Learning Models. *Proceedings of the ACM SIGSAC Conference on Computer and Communications Security*, 27–40.

Castellanos, H. (2021). Mayo Clinic and Google Cloud partner to improve patient outcomes. *Fierce Biotech*. Retrieved from www.fiercebiotech.com

Chen, D., et al. (2019). Security issues and challenges for cloud computing. *Future Generation Computer Systems*, 86, 1017–1029.

Chen, L., Mao, S., & Liu, Y. (2020). Big data: A survey on data storage and processing. *Journal of Cloud Computing: Advances, Systems and Applications*, 9(1), 1–17.

Chen, M., Mao, S., Zhang, Y., & Leung, V. C. M. (2014). Big data: Related technologies, challenges and future prospects. *SpringerPlus*, 3(1), 1–18.

Chen, T. D., & Hsieh, Y. C. (2020). The impact of Uber on traffic congestion: A case study of Austin, Texas. *Transportation Research Part A: Policy and Practice*, 138, 1–17. https://doi.org/10.1016/j.tra.2020.05.006

Cohen, A., Gibbons, P., & Matias, Y. (2009). Real-time analytics: The future of decision making. *Journal of Data Systems*, 5(3), 12–24.

Davenport, T. H. (2014). *Big Data at Work: Dispelling the Myths, Uncovering the Opportunities*. Harvard Business Review Press.

Davenport, T. H., & Ronanki, R. (2018). Artificial intelligence for the real world. *Harvard Business Review*, 96(1), 108–116.

Davis, R. (2022). *Optimizing AI Workloads with Data Platforms*. Data Science Quarterly.

Dean, J., & Ghemawat, S. (2004). MapReduce: Simplified data processing on large clusters. *Communications of the ACM*, 51(1), 107–113.

Delaet, T., Joosen, W., & Vanbrabant, B. (2014). Critical review of vendor lock-in and its impact on adoption of cloud computing. *Journal of Cloud Computing: Advances, Systems and Applications*, 3, Article 20.

Duan, Y., et al. (2019). Artificial intelligence for decision making in the era of big data–evolution, challenges and research agenda. *International Journal of Information Management*, 48, 63–71.

Dwork, C., & Roth, A. (2014). The algorithmic foundations of differential privacy. *Foundations and Trends in Theoretical Computer Science*, 9(3–4), 211–407.

Gandomi, A., & Haider, Z. (2015). Beyond the hype: Big Data concepts, methods, and analytics. *International Journal of Information Management*, 35(2), 137–144.

Garrison, R., Kim, H., & Lee, J. (2012). Containerization in collaborative environments. *Journal of Software Engineering*, 7(1), 100–115.

Garrison, G., Kim, H. J., & Wakefield, R. L. (2015). Success factors for implementing cloud-based information systems. *Journal of Information Technology Management*, 26(3), 45–56.

Gellman, R., & Dixon, P. (2019). California's CCPA and the new global privacy era. *Journal of Law, Technology & Policy*, 2020(1), 21–34.

Ghassemi, M., Naumann, T., Schulam, P., Beam, A. L., & Chen, I. Y. (2018). Opportunities in machine learning for healthcare. *Journal of the American Medical Informatics Association*, 25(10), 1417–1426. https://doi.org/10.1093/jamia/ocy112

Goedert, J. (2019). Mayo Clinic, Google partner to use AI, patient data to catapult research. *Health Data Management*. Retrieved from www.healthdatamanagement.com/news/mayo-clinic-google-partner-to-use-ai-patient-data-to-catapult-research

Google Cloud. (2023). *Introduction to BigQuery ML*. Retrieved from https://cloud.google.com/bigquery/docs/bqml-introduction BigQuery ML Documentation. Retrieved from https://cloud.google.com/bigquery/docs/bqml-introduction

Grobauer, B., Walloschek, T., & Stocker, E. (2011). Understanding cloud computing vulnerabilities. *IEEE Security & Privacy*, 9(2), 50–57.

Grover, L. K. (1996). A Fast Quantum Mechanical Algorithm for Database Search. *Proceedings of the 28th Annual ACM Symposium on Theory of Computing*.

Hale, J., & Burton, K. (2020). Microservices in modern platforms: Scaling flexibility. *Journal of Software Engineering*, 7(2), 112–127.

Harris, K. (2021). *Cloud Data Warehousing: The Definitive Guide. O'Reilly Media*.

Hayes, B. (2018). *Using Predictive Analytics and Artificial Intelligence to Improve Customer Loyalty*. Business over Broadway.

Hernandez, L., & Miller, R. (2021). Distributed systems and fault tolerance. *Data Systems Review*, 11(3), 67–82.

Hu, Y., Wen, Y., Chua, T., & Li, X. (2014). Toward scalable systems for Big Data analytics: A technology tutorial. *IEEE Access*, 2, 652–687.

Humphreys, E. (2013). Implementing the ISO/IEC 27001 information security management system standard. *Computers & Security*, 31(3), 201–207.

Jagadish, H. V., Gehrke, J., Labrinidis, A., Papakonstantinou, Y., Patel, J. M., Ramakrishnan, R., & Shahabi, C. (2014). Big Data and its technical challenges. *Communications of the ACM*, 57(7), 86–94.

Jain, P., & Kapoor, S. (2021). *Cloud Computing: Concepts, Technology, and Architecture*. Prentice Hall.

Johnson, A., & Miller, B. (2022). Understanding relational database management systems. *Database Systems Journal*, 13(4), 19–34.

Johnson, P. (2021). *Scalability in Big Data and AI: The Role of Modern Platforms*. IT Solutions Review.

Johnson, S., & Kosslyn, S. (2018). Apache Spark: An improvement on Hadoop. *Data Science Innovations*, 9(1), 44–59.

Kaur, P. (2020). Distributed computing in cloud platforms. *Journal of Cloud Computing*, 8(4), 120–135.

Kaur, H., & Saxena, S. (2022). Integration of AI and Big Data: Opportunities and challenges. *Journal of Emerging Technologies*, 9(2), 34–45.

Kim, S. (2022). Containerization trends in data platforms. *Data Science Review*, 8(3), 32–44.

Kim, S., & Park, H. (2022). Flexibility in cloud-based data platforms. *Journal of Data Engineering*, 14(2), 98–112.

Kohavi, R., Longbotham, R., Sommerfield, D., & Henne, E. (2020). Controlled experiments on the web: Survey and practical guide. *Data Mining and Knowledge Discovery*, 18(1), 140–181.

Kothapalli, M. (2023). Cost optimization strategies for cloud infrastructure. *Journal of Artificial Intelligence & Cloud Computing*, 2(1), 1–4.

Laney, D. (2001). *3D Data Management: Controlling Data Volume, Velocity, and Variety*. META Group.

Lee, J. (2023). The future of data management: Trends and challenges. *Journal of Data Management*, 10(2), 25–40.

Liu, S., & Peng, W. (2021). A survey on data management in cloud computing. *IEEE Access*, 9, 124234–124250.

McWherter, J. (2014). *Professional Mobile Application Development*. Wrox.

Microsoft Azure. (2023). *What Is Azure Synapse Analytics?* Microsoft Learn. Retrieved from https://learn.microsoft.com/en-us/azure/synapse-analytics/overview-what-is

Miller, E. J., & Lemoine, C. (2019). The impact of ride-sharing on urban transportation. *Transportation Research Part C: Emerging Technologies*, 100, 250–259. https://doi.org/10.1016/j.trc.2019.02.003

Mishra, V., & Darade, H. (2024). Harnessing Big Data: An Expert Review of Trends, Challenges, and Future Directions with a Focus on Data Quality. In *Proceedings of the Third International Conference on Artificial Intelligence and Data Science (AIDeS 2024)*. IEEE. https://doi.org/10.1109/AIDeS.2024.10957681

Mishra, V., et al. (2024). Navigating the Evolution and Management of AI-Driven Data Platforms. *An In-Depth Review & Expert Analysis, ICACTCE.*

Mohamed Ghazala, S. (2023). *Modern Data Platforms for Sustainable Data-Driven Transformation*. Data Insights Publishing.

Moore, D., & Clark, A. (2023). Version control in containerized environments. *Data Platform Technologies*, 9(4), 67–79.

Nasir, M., Awan, M. U., & Shah, M. A. (2020). Artificial intelligence for document review: A case study of JPMorgan chase COiN platform. *Journal of Banking and Finance Technology*, 4(3), 263–275. https://doi.org/10.1007/s42786-020-00037-x

National Institute of Standards and Technology. (2018). Guide to attribute based access control (ABAC) definition and considerations (NIST Special Publication 800-162).

NIST. (2018). Guide to attribute based access control (ABAC) definition and considerations (NIST Special Publication 800-162).

Otto, B. (2020). Data governance in the digital age: Unlocking the value of data. *Data & Knowledge Engineering*, 124, 101812.

Pal, R., & Kumar, S. (2021). Real-time scaling in cloud platforms. *Cloud Solutions Journal*, 16(3), 56–73.

Papadopoulos, A., & Xiang, L. (2022). Scalability and elasticity in cloud computing. *Journal of Cloud Architecture*, 10(2), 77–93.

Russell, S., & Norvig, P. (2020). *Artificial Intelligence: A Modern Approach* (4th ed.). Pearson.

Schatzberg, E. (2019). *Technology: Critical History of a Concept*. University of Chicago Press.

Shokri, R., et al. (2017). Membership Inference Attacks against Machine Learning models. *Proceedings of the IEEE Symposium on Security and Privacy*, 3–18.

Shor, P. W. (1994). Algorithms for Quantum Computation: Discrete Logarithms and Factoring. *Proceedings of the 35th Annual Symposium on Foundations of Computer Science.*

Shvachko, K., Kuang, H., Radia, S., & Chansler, R. (2010). The Hadoop distributed file system. *Proceedings of the 2010 IEEE 26th Symposium on Mass Storage Systems and Technologies* (MSST), 1–10.

Singh, A., et al. (2021). Cloud computing security: Privacy and data protection. *Journal of Cloud Computing*, 10(1), 1–17.

Sitaram, D., & Manjunath, G. (2012). Chapter 3 – Platform as a Service. In D. Sitaram & G. Manjunath (Eds.), *Moving to the Cloud* (pp. 73–152). Syngress.

Smith, J. (2019). How Amazon uses AI to drive sales. *Journal of Business Insights*, 5(2), 12–19.

Smith, J. (2021). The rise of relational databases in the age of Big Data. *Journal of Database Management*, 32(3), 14–29.

Smith, L. (2023). *Real-Time Data Processing in Cloud Architectures*. Analytics Today.

Smith, R. (2020). *Hadoop for Data Storage and Processing: A Comprehensive Guide*. O'Reilly Media.

Sun, Y., & Zhang, W. (2021). Kubernetes in hybrid cloud environments. *Journal of Cloud Architecture*, 11(6), 120–134.

Tannenbaum, R., & Wingfield, C. (2020). AI and the evolving role of data governance. *Harvard Data Science Review*, 2(4), 1–19.

Thakur, N. (2020). The Differences between Data Science, Artificial Intelligence, Machine Learning, and Deep Learning. APA: Plain English. (n.d.). In *Data Science vs. Artificial Intelligence vs. Machine Learning vs. Deep Learning*. AI Plain.

Thomas, L. (2022). Microservices architecture with containerization. *Modern Software Development Journal*, 6(8), 90–105.

Tomar, D., & Agarwal, R. (2022). Cloud data management: Techniques and challenges. *Future Generation Computer Systems*, 124, 28–42.

Topol, E. J. (2019). *Deep Medicine: How Artificial Intelligence Can Make Healthcare Human Again.* Basic Books.

Veeramachaneni, K., et al. (2021). Google BigQuery ML: Bringing machine learning to data warehousing. *Proceedings of the VLDB Endowment*, 13(12), 3209–3222.

Voigt, P., & von dem Bussche, A. (2017). *The EU General Data Protection Regulation (GDPR): A Practical Guide*. Springer.

Wang, Y. (2023). Elastic scaling in cloud data platforms: A review. *Journal of Cloud Computing: Advances, Systems and Applications*, 10(1), 11–23.

Xu, D., & He, M. (2023). Streaming platforms and the evolution of real-time analytics. *Journal of Data Engineering*, 14(1), 20–35.

Zaharia, M., Chowdhury, M., Franklin, M. J., Shenker, S., & Stoica, I. (2016). Apache Spark: A unified engine for big data processing. *Communications of the ACM*, 59(11), 56–65.

Zhang, H., Wang, Y., & Zhao, D. (2019). The application of big data analytics in transportation: A review. *Transportation Research Part C: Emerging Technologies*, 102, 145–161. https://doi.org/10.1016/j.trc.2019.03.017

Zhou, M., Pan, S., & Wang, Y. (2022). Cost-effectiveness of cloud vs. traditional data management systems. *International Journal of Information Management*, 62, 102446.

13 AI-Driven Data Excellence

Enhancing Governance, Quality, and Business Outcome

Vaishali Mishra, Harshal Darade, A. V. Senthil Kumar, and Namita Mishra

13.1 OVERVIEW OF DATA EXCELLENCE

Data excellence is the ability of an enterprise to effectively manage and leverage data as a strategic asset. The process looks at more than just data management, focusing on quality, governance, and security to allow a company to effectively use data for decisions and operational efficiency (Johnson, 2023). With organizations relying more on data for competitive advantage, achieving data excellence is critical to harnessing the full value of data assets. Data has been described in the digital economy as the "new oil"—a critical resource for innovation, growth, and competitive insertion (Mishra & Darade, 2024a). Collecting a lot of data is no guarantee of strategic oversight. The data must be correct, accessible, and in line with the business goals. The emphasis in data excellence frameworks is on cleansing, validation, and integration, so that all data is usable across the enterprise, consistently (Mishra & Darade, 2024b). As a result, it leads to better decisions, enhanced customer experiences, and greater operational efficiency.

13.2 ARTIFICIAL INTELLIGENCE IN DATA TRANSFORMATION

Data management is being transformed with the automation of complex activities, for example, data cleansing, validation, and anomaly detection, because of artificial intelligence (AI) (Davis, 2022). AI tools improve data accuracy and facilitate real-time analytics, allowing organizations to glean insights to make better decisions in a timely manner.

AI also assists a lot in data governance, such as automating compliance checks and data usage monitoring in accordance with regulations such as the General Data Protection Regulation (GDPR) or California Consumer Privacy Act (CCPA) (Anderson, 2023). AI models identify anomalies in data usage, correcting potential problems before they grow into larger issues, helping maintain a high standard of data quality while remaining compliant with legal regulations.

DOI: 10.1201/9781003560463-16

Moreover, through analyzing historical data, AI empowers predictive analytics to anticipate emerging patterns. This allows the companies to not fall prey to market fluctuations, customer tastes, or operational nightmares (Prasad, 2020).

13.2.1 Purpose of the Chapter

In this chapter, we delve into AI's role in data excellence—from data governance, quality, and enterprise data strategy. You will learn about the main pillars of data excellence, how AI is transforming traditional data practices, and how AI enables organizations to become data-driven in multiple fields. This chapter will also discuss how AI enhances efficiency, accuracy, and innovation in data management, and address challenges in implementing AI-based data strategies. Finally, it will show how AI-powered data excellence enables organizations to meet market needs, adhere to regulations, and stay ahead of the competition.

13.2.2 The Role of Data Management and AI: A Historical View of Evolution

Practices of Data Management Evolution

- *Issues and Constraints with Conventional Data Management*

While data management has always been an essential function for organizations, traditional practices were often manual and fragmented, and came with a host of challenges. During the initial years, businesses mostly depended on manual processes to store, organize, and retrieve data. Such processes were slow, error-prone, and not scalable at all while dealing with thousands of data points. One of the major hurdles that businesses faced was siloed data systems—data was scattered amongst various databases or even physical records, making it difficult to access, integrate, and analyze across the organization (Watson, 2008).

Data governance and quality control were also inconsistent, and they were magnified by issues of scalability and accessibility. Data governance, which is the corollary framework of rules, processes, and policies to maintain data accuracy and consistency—was often weak or nonexistent. Most organizations didn't have a standard way to process data, leading to errors, redundancies, and inconsistencies. Flawed decision-making, inefficiencies, and a lack of trust in data-driven processes were often a consequence of poor data quality. This led to obstacles in innovation and operational efficiency, as decision makers had to make decisions based on incomplete or incorrect data (Redman, 1996).

These older techniques also suffered from data security issues. As more and more data were collected and stored, data breaches and unauthorized access to sensitive information also became more prevalent. Unfortunately, organizations did not always have the necessary security protocols in place to prevent or lessen the effects of such an event. This led to the unsetting of traditional data management systems as they were not designed to cater to the burgeoning demand for sophisticated data security in the digital epoch.

- *The Pre-AI Era of Data Governance, Quality Control, and Enterprise Management*

Prior to the advent of AI, organizations were dependent on human intervention and basic software tools for data governance and data quality troubleshooting. Data governance frameworks were typically not well defined and proved to be most often a set of high-level guidelines on the processing of data for the organization, ultimately lacking a mechanism to enforce consistency across the organization. And data quality control involved a lot of manual checks, which were time-consuming and inefficient. For instance, data validation processes were done by teams of data professionals who would inspect datasets manually for errors, outliers, and inconsistencies. These approaches were neither scalable nor efficient for the vast datasets that contemporary organizations handled (Codd, 1970).

Enterprise Data Management (EDM) practices were equally constrained. Businesses generally store data in relational databases by querying and manipulating data using structured query languages (SQL). Relational databases had a way to organize data into structured formats but were rigid and unable to support the management of unstructured data (text, images, video, etc.), which was increasingly important in the modern data landscape (Chen et al., 2014). Moreover, legacy systems struggled with real-time data analytics, resulting in insights produced too late to guide timely business decisions.

This included drawing focus on these constraints that exposed the need to search for advanced, scalable, and computerized information, adding resonances that were able to deal with the increasing complexities and volume of venture information.

- AI in Data Management

The embodiments of AI in data management practices are enabled by a few major technology revolutions that occurred over the previous two decades. Among the earliest technologies that opened the path for AI-driven data management was the emergence of big data analytics. As organizations started to generate large volumes of data, traditional storage and processing methods were no longer adequate (Mishra et al., 2024a). Also, big data technologies (e.g., Hadoop, NoSQL databases) made it possible for organizations to store and process massive unstructured datasets, which paved the way for the development of AI applications (Dean & Ghemawat, 2004).

Machine learning (ML) and natural language processing (NLP) advancements that allow the execution of data management activities through AI also contributed significantly. ML is a core concept of AI that helps systems develop advanced algorithms that learn from a system and its data patterns and improve with time; NLP allows a data-based system to interpret human language. This allowed some of the tasks that were previously needed by humans to be performed automatically (for example, selection or cleaning of the data, anomaly detection, statistical techniques, and predictive analytics efforts) (LeCun et al., 2015).

Another significant development was the emergence of cloud computing. For organizations, cloud platforms offered the flexibility to scale data storage and processing services in real time, eliminating the need for costly on-premises solutions (Mishra et al., 2024a). This was facilitated due to the nature of such technologies, which are heavily reliant on computation that can quickly lead to the increased use of a data center (Armbrust et al., 2010) and therefore ease for organizations in deploying such products (Garg et al., 2013).

13.2.3 Initial Applications of AI for Data Management

AI was one of the first use cases that came to mind for data management, particularly when it came to automating data cleaning processes. ML algorithms were implemented to detect and fix errors in large datasets, such as missing values, duplicates, and inconsistencies. This had dual advantages as it not only enhanced the quality of the data but also decreased the time and efforts needed for data preparation immensely. IBM's Watson, as well as tools like Google's TensorFlow, were some of the first to automate numerous data management processes, such as data integration and transformation, that allowed data workflows to be increasingly automated (Chen & Lin, 2014).

AI was used to improve data security as well. ML Algorithms-Based Errors: ML algorithms were able to identify anomalous behavior within data access patterns, enabling organizations to receive real-time alerts of some potential security attacks. AI systems, for instance, could identify abnormal login activity or unauthorized access to sensitive data, resulting in quicker reactions to security threats (Shin, 2020). This proactive approach to data security was a departure from traditional methods, which were often reactive and slow to find fixes for security vulnerabilities.

Predictive analytics was another early use case, with AI systems analyzing historical data and creating forecasts about future trends. AI applications in areas like finance, healthcare, and retail use predictive analytics to provide insights on customer behavior, market trends, and operational efficiencies, enabling organizations to make better decisions (Davenport & Ronanki, 2018).

13.2.4 The Latest Trends in AI-Based Data Management

***AI Data Trends: How The Augmentation of *AI is Reshaping the Data Landscape*

In recent years, AI has turned the world of data management and utilization upside down. The automation of data management tasks is one of the most important trends. AI-driven tools can automatically manage the processes involved in data ingestion, cleaning, and transformation, streamlining the elaborate process of making data ready for the analytical process considerably. Organizations have been able to use this automation to be able to process the data in real time for faster insights and more flexible decision-making (Zhou et al., 2019).

Another major trend involves adopting AI for predictive analytics. It allows organizations to make proactive, data-driven decisions because of ML algorithms analyzing historical data to predict trends and outcomes going forward. This functionality is especially useful in industries like finance, where distinguishing between false positives and true threats can greatly influence investment plans, and health care, where predictive analytics should direct resource management and treatment plans (Topol, 2019).

AI has enabled data processing via more intelligent analytics methods like deep learning and neural networks. These techniques allow organizations to extract useful information from formats that were once too unruly for conventional processing methods, such as text, images, and video. AI has opened a new realm of data analysis, by extracting value from unstructured data (Goodfellow et al., 2016), gaining deeper insights and providing better applications in domains such as NLP and computer vision.

In conclusion, organizations that embrace AI technologies for data management will not only benefit from increased efficiency but also gain a competitive edge by unlocking new avenues of value creation in the era of digital transformation.

13.3 DATA GOVERNANCE WITH AI

Excerpt: The Definition and Significance of Data Governance

• Effective Data Governance Principles and Elements

Data governance is the set of policies, practices, and processes to ensure the data is well managed, accessible, and secured. Data accurately, timely, and correctly available for decision-making in business is its objective. The main principles in accordance with data governance are accountability, transparency, integrity, stewardship, and compliance (Abraham et al., 2019).

A successful data governance program includes a few key elements:

- Data Stewardship: Assigning data stewards—individuals responsible for data assets. They make sure that data is appropriately formatted, available, and secure (Psychoula et al., 2021).
- Qualitative Management of Data: The process of accuracy, consistency, and completeness of data is required for operational decision-making. Good governance frameworks have controls in place to routinely audit and cleanse data to eliminate errors and inaccuracies.
- Data Lifecycle Management: Making data retention, archiving, and deletion policies part of data management ensures that organizations meet legal requirements while avoiding the costs and risks associated with outdated or unnecessary data (Brous et al., 2019).
- Metadata Management: This includes information about the source, structure, and use of data. In this respect, without correct metadata management, no data will be interpreted (Otto, 2011).
- Data Protection and Privacy: Organizations need to ensure measures are in place to protect sensitive data against unauthorized access, breach, and misuse. This is particularly important in sectors that deal with personal or financial data, including health care and financial services (Cavoukian, 2012).
- Regulatory Compliance: Compliance with data privacy legislation, including the GDPR and the CCPA, is a pillar of data governance (Vanberg & Ünver, 2017).
- Compliance with Regulations (CCPA and GDPR): These make up the rules of data governance and legality. Laws such as the GDPR in Europe and the CCPA in the U.S. introduce significant requirements about how organizations will collect, store, and use personal data. For instance, the GDPR requires

organizations to ask for the explicit consent of individuals to process their personal data, and empowers individuals to access, correct, or have their information deleted (Voigt & Von dem Bussche, 2017). Failure to comply with these regulations could lead to hefty fines and damage to an organization's reputation.

Similarly, the CCPA grants California residents the right to know which personal data is being collected, to demand its deletion, and to opt out of data sales. Both regulations, therefore, reflect the broader goals of data governance (Weber, 2020) in their mutual aim at safeguarding individuals and their data as well as at keeping transparency about their use. In an increasingly challenging, complex regulatory environment spanning continents, compliance monitoring and robust enforcement are becoming more difficult—and here, AI can be transformational.

13.4 HOW AI CAN STRENGTHEN DATA GOVERNANCE

13.4.1 AI-Driven Compliance Oversight and Policy Enforcement

The Rise of AI in Data Governance: Improving Compliance Monitoring and Policy Enforcement Manual audits and periodic reviews have long been applied to keep compliance in check, but these can be time-consuming and often error-prone. Regulatory agencies can leverage AI solutions like ML algorithms to automate this monitoring by continuously reviewing big datasets to assess compliance with regulatory policies and detecting cases of non-compliance. Such as detecting when sensitive data is accessed without authorization, data retention periods exceeded, or customer data is being used for purposes that violate privacy laws (Khatri & Brown, 2010).

AI systems, for instance, can analyze real-time data access patterns, identifying anything unusual that hints at unauthorized access. When trained to learn the normal behaviors of users in an organization, ML models can identify anomalies that may signal a breach or misuse of information. Not only do these systems enhance the efficiency and precision of compliance enforcement, but they also help to mitigate the risk of data breaches, ultimately shielding organizations from potential fines and reputational damage (Fennelly & Kaspersky, 2019).

13.4.2 Data Lineage and Metadata Management Automation

AI plays a crucial role in automating data lineage and metadata management, both of which are essential elements of data governance. Data lineage is the ability to document the flow of data from the point of origin through different systems in an organization. Data lineage is crucial for maintaining data integrity, locating issues, and enhancing transparency. Historically, tracking data lineage has been manual and costly. On the other hand, AI-based software can assess data flows and system logs and automatically map and visualize data lineage, offering a clearer and more precise representation of how data is used and transformed throughout the organization (Heinrich et al., 2018).

The management of automated metadata is subject to similar improvements using AI. AI tools automate the process of creating and updating this metadata, making sure data remains correctly labeled, classified, and organized. NLP algorithms can evaluate the information and derive metadata from it, thereby enabling discovery, accessibility, and utilization of the information. The use of metadata also enhances data discoverability and enables decision-making when users can more readily identify datasets of interest and understand their context (Wang et al., 2019).

13.4.3 Artificial Intelligence for Data Privacy and Ethical Usage

AI is key to protecting data and enabling ethical use. Data privacy laws such as the GDPR in Europe and the CCPA in the USA have set clear guidelines for organizations on how to collect, store, and process personal data, providing individuals with rights such as access to their data and the ability to have it deleted. AI systems can aid organizations in complying with these regulations by automatically identifying personal or sensitive data within large datasets and applying appropriate protections, such as encryption or anonymization (Schiff et al., 2020).

Furthermore, AI algorithms can help to establish policies that restrict the usage of data to ethical purposes, preventing data from being exploited for biased or discriminatory outcomes. AI, for example, is used to analyze decision-making processes to determine if data is being leveraged in ways that could lead to certain groups of individuals being treated unfairly based on race, gender, or socioeconomic status.

13.5 REAL-LIFE CASE STUDIES OF AI-DRIVEN DATA GOVERNANCE

- *Use Cases in Finance, Public Sector, and Healthcare*

AI-driven data governance is already being widely leveraged across finance, healthcare, and the public sector. AI in finance is helping in adhering to the regulatory standards like Know Your Customer (KYC) and Anti-Money Laundering (AML) requirements. The AI algorithms can automatically verify customer identity, monitor transactions for any suspicious activity, and help ensure that financial data is treated in accordance with industry regulations. As an example, banks like JPMorgan Chase have adopted AI-based systems that have been used to analyze trading patterns to identify potential cases of either fraud or market manipulation (Bose & Leung, 2018). Not only do these AIs minimize the chances of non-compliance, but they also automate exceedingly complex fraud detection processes.

In the healthcare domain, privacy and data security hold a prime place due to the sensitive nature of patient data. By monitoring access to patient data and ensuring that it is shared only with authorized personnel, AI-driven solutions enable hospitals and healthcare providers to comply with regulations, including the Health Insurance Portability and Accountability Act (HIPAA). For instance, the Cleveland Clinic is using AI to monitor the usage of medical data and ensure adherence to privacy policies. AI tools also perform data quality management by highlighting incomplete or inaccurate patient records so that healthcare providers can make decisions based on reliable information (Reddy & Aggarwal, 2020).

In the public sector, we have seen the use of AI to strengthen data governance in government agencies that manage sensitive citizen data. Example: U.S. Department of Homeland Security implements AI-based tools to oversee and address data privacy concerns within its departments. AI systems provide facilities to automate compliance with privacy laws and regulations, assuring citizens that their data will not be abused and automating the efficiency of operations (Nguyen, 2020).

13.6 IMPORTANCE OF QUALITY DATA FOR AI

13.6.1 A Guide to Data Quality in the Age of the Internet

As businesses compete in the ever-evolving digitized economy, organizations have to face the data quality challenge. Grounds of accurate and timely data aid organizations to take crucial decisions, drive operations, and stay ahead of the competition. We all know how data quality, explained through parameters such as accuracy, completeness, consistency, and timeliness, is the backbone of operational efficiency. Without quality assurance, poor quality decisions can cost businesses millions of dollars in lost revenue (Mishra et al., 2024b).

13.6.2 Dimensions of Data Quality: Accuracy, Completeness, and Timeliness

In addition, data integrity assures that the data being stored and processed mirrors real-world scenarios. It affects the trust in analytics and the decision-making process. Incorrect sales data, for example, a sales manager, can make poor decisions when managing the inventory, causing severe losses. Second, the completeness of a dataset, which means that all the data elements that should be there, is there. If a field is missing, such as a customer's address or phone number, customer service may be delayed, or communications may fail entirely. And proper timing is crucial, because old data usually lead to bad decisions. In domains such as healthcare, delayed data can pose dire repercussions, influencing operational efficacy alongside patient outcomes.

13.6.3 Financial Impact of Bad Data on Businesses

Data quality impacts every facet of your business. In fact, research suggests bad data costs U.S. businesses more than $3 trillion a year (Redman, 2016). These costs stem from inefficiencies, incorrect business insights, and wasted resources. Bad data results in failed marketing campaigns, inaccurate financial reports, and even lawsuits for failing to comply with data regulations, such as the GDPR. Moreover, when decisions are made from poor data that leads to unsatisfied customers, a company's reputation can take a hit. Organizations can eliminate the risk of poorly managed and inefficient data through investments in proper data management.

13.6.4 The use of AI in Data Cleansing and Validation

AI is proving to be exceedingly beneficial in solving these data quality problems. AI is transforming data cleansing, validation, and anomaly detection with its ability to automate complex processes.

13.6.5 AI Models for Automated Data Cleansing and Anomaly Detection

The next step is usually data cleansing, a straightforward process that is generally carried out by humans but has been assisted by AI. ML algorithms can sift through datasets to spot and fix errors, missing values, and duplicates significantly faster than human teams can. The AI models learn from patterns of historical data and anticipate and rectify mistakes before they affect business operations. One such use case is NLP, used to understand customer data and make sure it is standardized (e.g., no typos or incorrect formats).

AI helps detect anomalies, which is a core function in many fields, such as finance and telecommunications. In banking, for instance, anomaly detection is used to prevent fraud by detecting transactions that deviate from transaction norms. Anomaly detection tools, powered by AI, identify unusual trends using advanced statistical models and ML algorithms in real time, enabling firms to respond to potential problems immediately.

13.6.6 ML-Based Data Quality Monitoring Tools

ML facilitates the continuous monitoring of data quality at various levels of granularity. Using AI, data quality can be monitored and validated automatically and in real time, meaning anomalies are flagged and contained before they have a chance to filter through an organization's systems. Machine-learning-based methods of processing and analyzing large datasets, like Apache Spark's MLlib, are scalable and improve the quality of the data overall (Zaharia et al., 2016).

Some enterprises have deployed AI-enabled dashboards that provide real-time visibility into data quality measurements. These dashboards automatically alert you to disparities as and when they arise in data, reducing the manual effort involved in data monitoring and correction to a great extent. Over time, in addition, these tools learn from their environments, improving their abilities to detect data quality issues and providing recommendations when issues arise.

13.6.7 Data Quality Assurance Powered by AI

AI-driven quality assurance methods offer predictive insights about the potential problems of data quality and come with capabilities to improve the validation of datasets in real time.

13.6.8 Predictive Data Quality (PDQ) for AI-Powered Approaches

Predictive data quality (PDQ) leverages ML models to anticipate data quality problems before they occur. Through an examination of historical patterns in data errors, AI systems can anticipate the timing and location of future errors, enabling businesses to act preemptively. For example, AI can anticipate potential discrepancies in aligning financial data during quarterly reviews, allowing financial teams to focus on resolving them beforehand.

In environments where data is continuously ingested from multiple sources, AI models can also assist organizations in preserving the quality of their data. For applications that rely on binary analytics or involve external data sources such as Internet of Things (IoT) devices, predictive models may be utilized to uphold file patterns or data streams, ensuring that errors are identified and remedied in real time (Lee & Kang, 2020).

13.6.9 Tools and Technologies for Enhanced Real-Time Data Validation

AI-powered solutions are also implemented for real-time data validation. Tools, e.g., Google Cloud's Dataflow and Microsoft Azure's Data Factory, use ML to ensure accuracy and consistency in processing vast quantities of real-time data (Ghemawat et al., 2020). AI is also used by these platforms to recognize data inconsistencies while ingesting data so that organizations can mend them immediately.

Moreover, AI tools like IBM's Watson can validate data against relevant rules and patterns and provide immediate feedback on discrepancies. They save a ton of time looking for and fixing mistakes, giving companies validated data.

13.7 EXAMPLES OF AI-POWERED DATA QUALITY MANAGEMENT

The enabling of AI-driven data quality management is already generating value across industries by making their data more reliable and ultimately their operations more efficient.

13.7.1 Use Cases for Retail, Supply Chain, and Telecommunications

In retail, data quality solutions with AI capabilities have allowed companies like Amazon and Walmart to index their inventory data correctly. AI systems are constantly monitoring the levels of inventory, keeping predictive analysis on when stock should be restocked, thus minimizing errors in stock counts (Choi et al., 2018). This results in improved client experiences and streamlined stock management.

AI is already playing a role in the supply chain with companies using improved shipping data models for quality analytics on shipping times, costs, and inventory levels. For example, AI has been implemented by DHL to predict potential supply chain disruptions and enable real-time decision-making (Johnson, 2021).

Others, such as Verizon, are using AI to check their databases for quality. For instance, AI systems track network data for abnormal patterns in real time, updating customer data with correct information (Mishra & Darade, 2024b). It helps deliver better customer service and more accurate billing, improving the overall customer experience.

13.8 ARTIFICIAL INTELLIGENCE IN ENTERPRISE DATA MANAGEMENT

13.8.1 Sub-Components of Enterprise Data Management (EDM)

EDM refers to the managing process to ensure that data is accurate, accessible, secure, and usable across the organization. EDM itself covers many components, all of which are important to manage vast amounts of data in a modern business world.

Data integration, warehousing, architecture overview, and security
The main elements of EDM are as follows:

- Data Integration: Merging data from multiple sources—structured data, unstructured data—into one big picture. This enables businesses to have a complete and accurate view of their operations and customers, helping them make better decisions.
- Data Warehousing: Data warehouses are used to hold extensive historical data to be used for analysis. Warehousing solutions help organizations develop a data repository where data from different systems is collected and analyzed.
- Data Architecture: The architecture acts as the blueprint that defines how the data is collected, stored, and used. This involves not only the architecture of your databases, data lakes, and data warehouses but also all the infrastructure needed to support their scalable data management.
- Data Security: A key aspect of EDM is security, which is the protection of sensitive information against theft, breaches, or unauthorized access. Data is protected by security measures, access is restricted, and laws such as GDPR and CCPA are followed (GDPR).

EDM is a key factor in an enterprise to make decisions based on data and to keep operations moving smoothly. Poor data integration, architecture, and security management could lead to inconsistencies, vulnerabilities, and non-compliance with industry regulations for businesses.

13.8.2 The Data Domain, Underpinned by AI-Enabled Integration and Architecture

AI is altering enterprises' approaches to data management to a great degree. There are two areas in which AI is more pronounced in the context of EDM data integration and data architecture.

- How AI Makes Data Integration from Various Sources Easy

Different systems produce data with varied formats, quality, and velocity, making data integration typically complex. AI-powered solutions narrow down this burden by automating the data mapping, normalization, and validation processes that are often manual and time-intensive. The ML algorithms can take in the patterns in different data structures and automatically join datasets from different sources without losing the quality.

AI's ability to integrate real-time data. AI systems can constantly monitor traffic flows, enabling the detection of misconfigurations or anomalies and instantaneously repairing them. Tools like Talend and Informatica, which are based on AI, use ML algorithms to bypass a lot of these processes and thus, greatly minimize the time to prepare data to be analyzed.

- How AI Is Being Used to Build Efficient, Scalable Data Architectures

Thus, this is why we now can rapidly create more and more sophisticated data architecture systems. However, with the scenario of big data, the common data architectures dealt with the growing load poorly. AI-led architectures like Google

Cloud and Amazon Web Services (AWS) are self-optimization in nature, which self-adjust to the data workloads.

AI can forecast future data storage and processing needs, allowing organizations to allocate resources optimally. Additionally, it assists in developing architectures that are capable of accommodating both structured and unstructured data instantaneously, making them more versatile as the data landscape evolves.

13.8.3 Data Integration and Architecture Powered by AI

AI is playing a major role in how enterprises will manage their data. AI has a significant role to play in EDM with respect to two aspects: data integration and data architecture.

How AI Is Making It Easy to Integrate Data from Various Sources

To make it more complicated, the format, quality, and velocity of data produced in different systems differ wildly from each other, making data integration a complex task. AI-driven solutions make this process straightforward by automating the data mapping, normalization, and validation functions that are typically manual and time-consuming. ML algorithms identify the patterns in the data structure and can automatically merge data sets from different sources without any degradation in quality.

AI is also adept at integrating real-time data. AI systems work by constantly analyzing data flows, identifying discrepancies or anomalies, and adjusting accordingly. Tools such as Talend and Informatica that are powered by AI utilize ML algorithms to automate these processes and cut down the time to get data ready for analysis.

- Leveraging AI for Data Architecture Design and Scalability

Altogether, AI is innovating data architecture by allowing more flexible and scalable system design. The conventional data architectures can rarely keep pace with the growing load, given the explosion of big data. AI-powered architectures, for example, Google Cloud, and AWS, are self-optimizing, with analytics built into the infrastructure that automatically optimizes data workload.

AI helps you estimate the storage and processing requirements for the future so the organization can allocate resources in an intelligent manner. It further assists in developing architectures that can process structured and unstructured data in real time, enhancing their flexibility to the changing data environment.

13.8.4 Data Warehousing and Cloud Platforms in AI

AI has been playing a significant role as data warehousing is vital for modern enterprises, and cloud platforms are helping enhance these technologies and scale them.

- AI: Improving Data Storage, Retrieval, and Performing Real-Time Analytics

Data warehousing solutions powered by AI lead to more accurate and quicker data storage and retrieval. Traditional data warehouses use static and batch-processing methods; however, with the advent of AI, data can be processed dynamically and in real time. This permits the impartation of the data in real time, allowing enterprises to make faster, informed decisions based on data.

Query optimization in data warehouses is also being made more efficient by AI models. AI algorithms, when applied, can determine the most relevant data for future queries by analyzing past ones and therefore, making the retrieval of critical information less time-consuming (Chaudhuri et al., 2011).

Moreover, data analytics is already being simplified through the use of AI, like platforms such as IBM Watson and Microsoft Azure's Cognitive Services. They harness AI to detect trends, patterns, and insights that humans would struggle or take ages to uncover.

• Scalable EDM Powered by AI-Based Cloud Solutions
Cloud services like AWS, Google Cloud, and Microsoft Azure are increasingly embedding AI to streamline enterprise data. These cloud solutions, powered by AI, offer scalable EDM based on ML, NLP, deep learning, etc.

Example: AWS Glue is a fully managed ETL (extract, transform, load) service that uses AI to automate common data preparation tasks, helping data engineers automate how they move and manage data between systems. In the same vein, Google Cloud's BigQuery ML enables users to create and execute ML models directly in their data warehouse environment, with the promise of scalable, AI-driven analytics without complex infrastructure (Ghoting et al., 2019).

Cloud solutions like this one rely on AI, allowing businesses to scale their Data Architecture based on the continuously growing needs of the business without the burden of monitoring physical infrastructure.

13.8.5 Examples of AI-Powered Enterprise Data Management

Many organizations in various industries are already adopting AI to evolve their EDM. Here are some examples spanning several industries.

Examples of Companies Leveraging AI for Next-Gen EDM

- Retail: AI-driven EDM is exemplified by companies like Walmart. Utilizing AI and ML, the retail giant optimizes its supply chain to ensure accurate inventory data and real-time replenishment of products. By streamlining its data via its AI blockchain-based system, Walmart has reported an ability to forecast trends better, allowing for greater efficiency and minimal operational costs (Choi, 2018a).
- Healthcare: AI is utilized in health care for managing at-scale patient data that enhances the quality and accessibility of data for different use cases. Mayo Clinic, for example, has embedded AI-based capabilities into its data architecture to predict employee outcomes, and to ensure patient records are consistent across different systems (Mishra & Darade, 2024a). AI identifies anomalies in clinical trials to improve the quality of data to minimize trial errors.
- Finance: JPMorgan Chase has implemented AI to mitigate risk data more efficiently. To accomplish this, the company employs ML algorithms to amalgamate data from multiple financial systems and conduct real-time analytics

on the data, which has enhanced its capability to identify potentially fraudulent transactions and evaluate financial risk (Mishra et al., 2025c).

- Manufacturing: Robotics manufacturer GE (General Electric) has employed data management platforms driven by AI to track the performance of its machines in the industrial sector. Through predictive maintenance, by leveraging data from the IoT devices, for example, GE can anticipate failures of the Yale lock, leading to proactive scheduling of its repairs to minimize operational costs (Lee et al., 2015).

13.9 THE INFLUENCE OF AI ON THE DATA MANAGEMENT INNOVATION

13.9.1 AI and Data Excellence Driving Innovation

Data-driven enterprises have seen AI become a game-changer, unlocking the value of their data for creative solutions. Whereas data management used to be a manual and repetitive process, AI has changed the game by automating data collection, integration, and analytics. AI systems can leverage large amounts of increasingly complex structured and unstructured data in seconds to generate insights that would otherwise have been littered with hypotheses (Mishra et al., 2025a).

Not only is AI enhancing operational efficiency, but it is also fundamental for unlocking new business models and revenue streams. But with AI algorithms, businesses can mining their data to find innovating patterns, trends, and correlations. AI-empowered analytical platforms, for example, enable enterprises to cover a wider ground with personalized services and unlock new markets and segments by discovering opportunities that serve an unmet need. For instance, in the finance sector, AI helps to enhance trading strategies by examining historical data and market trends, which translates into more efficient decision-making and further income opportunities (Mishra et al., 2025b).

13.9.2 Business Decision-Making through AI and Predictive Analytics

One of the most valuable applications of AI in business decision-making is predictive analytics. It uses ML models to analyze historical information and make predictions, allowing predictive analytics to reach a new level. Inferences can be made by AI-driven predictive models; AI can help businesses to overcome risks and seize opportunities.

A glaring example of AI-powered predictive analytics is sales forecasting. By analyzing customer buying patterns, economic conditions, and seasonal trends, AI tools can forecast sales trends and help businesses plan inventory and marketing activities (Colson et al., 2015). For example, in customer relationship management (CRM), AI-based predictive analysis can predict customer behavior, including churn rates or potential upsell opportunities.

13.9.3 Implementation of AI in Product Development using Data

AI is also revolutionizing research and development (R&D) as it has accelerated the pace of innovation in product development. AI reduces the time, effort, and cost needed for R&D activities through automating data analysis and simulation processes. Because the AI algorithms can test multiples hypotheses quickly, measure how the products perform in different conditions, and optimize design parameters using the data from the simulations.

Moreover, insights gained from AI about consumer preferences and industry trends help organizations design their offerings to better align with customer requirements. In the automobile sector, for instance, AI analyzes drivers, and creates new feasible features and design of vehicles according to changing consumer lifestyle.

13.9.4 AI-Powered Innovation by Data Excellence: Case Studies

Many industries have opened using data management powered by AI for innovations. For pharmaceuticals, AI accelerates the drug discovery process through the analysis of millions of datasets involving molecular structures and clinical trial successes. AI-enabled platforms assist researchers in discovering viable compounds faster than conventional means, shortening the timeframe from discovery to testing (Mak & Pichika, 2019).

The same applies to fintech, where AI has been the catalyst behind innovations such as fraud detection and personalized banking options. Real-time transaction data analysis using AI algorithms can also prevent fraudulent activity and help improve the customer experience by providing customized financial products. AI is used in manufacturing to enhance supply chain management by predicting equipment breakdowns, minimizing downtime, and boosting efficiency in operations (Tao et al., 2018).

13.10 AI-POWERED DATA QUALITY: ETHICS AND DATA QUALITY CHALLENGES

13.10.1 Data Privacy and Security in AI-Based Systems

AI-enabled data management systems can open great opportunities for organizations that want to work with enormous amounts of data. But these opportunities raise serious ethical issues about how data is collected, used, and secured. The main ethical issue here revolves around the way AI handles personal data and to what extent it influences individual privacy. For this reason, robust AI systems, such as those used for predictive analytics, depend on extensive data supplies, including private personal data (Mishra & Darade, 2024a). The collection, storage, and processing of this data are pronel to misuse, unauthorized access, and data breaches.

However, this is a challenge as AI models are overall highly complex. AI algorithms function as "black boxes," so decision-making processes can be opaque, and ensuring compliance with data privacy regulations can be difficult. Moreover, the data used for AI are often large datasets that can be attacked and threaten sensitive personal data related to identity theft and fraud (Mishra & Darade, 2024b). However, to feel secure with such a service, you'll need data encryption, secure storage, and a tiny access control. Additionally, privacy laws, like the GDPR, are essential for making sure that AI models comply with global data privacy regulations.

13.10.2 Reducing AI Bias in Data-Based Decision-Making

Although prominent, AI algorithms are prone to biases that affect the results of decision-making. Historical data, societal stereotypes, or poorly designed models can introduce bias directly into AI systems, resulting in unfair and discriminatory outcomes (Johnson, 2021). For example, if biased training data are used to train AI models, it will have an impact on the models that were used to determine how a decision is made, which can be disproportionately disadvantageous for certain demographic groups. In data-driven decision-making, biased AI systems might skew hiring practices, lending decisions, or even criminal justice outcomes.

AI and Data Ethics Ensuring Fairness and Transparency in AI the integrity of AI and the organizational commitment to ethical practices closely ties to the quality of data used in a data science process. By using larger, more comprehensive datasets for ML, which include samples from all relevant subpopulations, it should be possible to lessen AI bias. Furthermore, AI systems must undergo periodic auditing to assess fairness, and businesses must take proactive steps to detect biases and discrepancies to minimize any potential inequity in their algorithmic outputs (Mehrabi et al., 2021). Developed by the European Commission, ethical AI frameworks and guidelines offer approaches for measuring fairness and transparency and will assist organizations in developing AI models that encourage equity and inclusiveness.

13.10.3 Compliance with Laws and Regulations

Also, the legal and regulatory landscape around AI is becoming more complex; companies will need to learn to navigate that. AI tools can help organizations meet compliance standards by ensuring that data usage parameters are monitored, regulatory breaches are identified, and data governance standards are adhered to (Mishra & Darade, 2024b). Where AI models are operating in spaces that are still emerging, legal challenges abound. For instance, AI systems that manage data (i.e., systems utilized in sectors like healthcare or finance) must adhere to strict regulations regarding data security, privacy, and transparency. Release 5 AI also poses a unique challenge when it comes to establishing EU policies regarding international data transfer (i.e., data transferred outside the EU). Such breaches can result in legal actions, penalties, or harm to reputation.

One of the key legal issues related to implementing AI-driven data management activities concerns data compliance with existing laws such as GDRP, CCPA, and other sector-specific regulations. This makes it challenging for organizations to prove how their AI models are compliant with these laws, especially in contexts where AI decision-making is a black box. In response to these challenges, businesses should create AI governance frameworks that specify who is accountable for decisions made by AI systems and ensure that legal teams are part of the development of the AI system at the beginning (Mishra et al., 2025c).

13.10.4 Striking a Balance between Innovation and the Ethics of Artificial Intelligence

While AI can lead to innovation and growth in data management technology, it should not compromise on ethics. Responsible AI usage will keep that trust with customers and stakeholders. Creating ethical AI policies that prioritize transparency, accountability, and fairness in AI systems can help organizations encourage innovation while also fostering responsible AI practices (Mishra et al., 2025c).

Ethics boards or committees can also be created by organizations to review AI projects and to make sure that ethical considerations are integrated into each stage of the AI lifecycle. Furthermore, companies need to encourage a culture of responsible usage of AI, a practice that helps organizations ensure their employees remain committed to values of ethics and innovation, rather than overly focusing on results and potentially causing societal harm in the process while ensuring that businesses can be successful in the long term. Balancing ethical AI practices with innovation will be imperative as AI continues to mature and develop into sustainable, socially responsible AI systems (Johnson, 2021).

13.11 WHAT'S NEXT—AI AND DATA EXCELLENCE

13.11.1 The Future of AI in Data Governance and Quality

This indicates how the future of AI data governance and quality is interconnected with other emerging technologies like, blockchain and quantum computing. Moreover, blockchain technology integrated with AI has the potential to immensely improve data governance by enabling transparent, tamper-proof records of data utilization, thereby raising levels of accountability and ensuring data integrity (Mishra & Darade, 2024a). As a decentralized record that eliminates intermediaries, the infrastructure provided by blockchain would pave the way for AI systems to validate input data across networks whilst ensuring compliance with data regulations (such as GDPR) and build mutual credibility (Johnson, 2021).

The advancements in quantum computing, on the other hand, is yet another paradigm shift technology that will change the game for AI, enabling AI to process and analyze vast datasets more efficiently by providing the computational power that is required. This can result in AI models that can process data more accurately and faster, which can enable more robust data quality management strategies (Mishra & Darade, 2024b). In the next ten years, AI and Quantum Computing will come together to facilitate real-time data validation, cleansing, and governance. The AI models are expected to grow into more autonomous so as to minimize the overhead in data management (Mishra & Darade, 2024b).

13.11.2 Convergence of AI and Advanced Data Analytics

The combination of AI and advanced data analytics is already transforming industries and will continue to expand. Real-time data is generated on such a massive scale by the IoT, that it will more and more be analyzed by AI for actionable insights (Davis, 2022). For instance, in healthcare, AI-based analytics will help to enhance patient outcomes by diagnosing diseases at an early stage, and in the retail industry, it will enable the firms to capitalize on real-time data checks and analytics, which will improve supply chain management and customer personalization (Anderson, 2023).

In addition to that, AI and big data analytics are becoming more and more interrelated. AI models need large datasets for learning, and advanced analytics systems rely on AI to search through tons of data for trends and patterns. AI analytics platforms that automate the detection of data anomalies, market trend forecasting, and predictive model generation for better business decision-making, will likely be the standard in the future (Mishra et al., 2025c). These integrated platforms will help unlock the true potential of data by accelerating the scalability of data-driven enterprises, enabling businesses to more effectively harness rich and complex datasets across various sectors.

13.11.3 How AI Is Reshaping the Data-Driven Enterprise

From transforming the way, we think about data-driven enterprises to empowering businesses and sectors in every aspect, we can expect AI to play a leading role moving forward. AI allows for fast data processing, which in turn provides real-time insights that can accelerate decision-making and enhance operational efficiency in sectors like finance, healthcare, and manufacturing (Mishra et al., 2025c). AI piloting data management will allow businesses to unlock new business models, revenue streams, and operational efficiency. AI-enabled predictive maintenance in manufacturing, for example, is already starting to hit mainstream as companies minimize downtime and/or operational costs (Mishra & Darade, 2024b).

By incorporating deep learning models into enterprise-level decision-making processes, AI will also transform data-driven strategies. The models will analyze both historical and real-time data to offer insights that can uncover opportunities for innovative products and business growth (Mishra et al., 2025c). The continuous infusion of AI into how enterprises operate in order to maximize competitiveness will by its very nature alter the way businesses leverage their data to scale.

13.12 CONCLUSION

In this chapter, we have discussed the transformation that AI can bring in the context of data management, governance, and quality, and how it can lead to the transformation of traditional systems moving from data management frameworks to a more automated, efficient, and scalable solution. The confluence of AI with other advanced technologies like big data, IoT, blockchain, quantum computing paves the way for new business models, accelerated decision-making, and revolutionized innovation in product development.

The AI has become an essential building block to fortify data governance frameworks that are necessarily responsive to intricate compliance obligations like GDPR and CCPA. This is where ML comes into play, as it helps automate processes such as tracking data lineage, anomaly detection, prediction, analytics and ensures organizations maintain robust governance structures while dramatically improving data quality and usability. Together, they serve as the foundation of innovation—where organizations have reliable data they can trust, while using this data to generate transformative results.

Fall on the frontiers of AI adoption in verticals: quality of data in healthcare, security in financial services, and governance at scale in telecommunication, founders are pulling the levers to supercharge effects. Moreover, AI-driven predictive analytics has transformed the decision-making process, allowing companies to anticipate customer needs, forecast market trends, and design data-driven strategies with unmatched accuracy. Pharma takes pipelines down to weeks instead of years, advancing product designs and revenue through AI-augmented data elevation.

Simultaneously, AI brings new ethical dilemmas like personal data privacy, security issues, and algorithmic prejudice. As these problems are brought to light, societies must adopt governance frameworks that prioritize fairness, transparency, and accountability to rectify them. While innovation will be critical for AI initiatives to succeed, organizations must balance that desire for innovation with an ethical responsibility to build trust with customers.

Going forward, AI capabilities to enable transformation, data governance, and quality remain enormous. Emerging technologies would continue to be a major catalyst to transform the landscape and harness the opportunities for operational excellence, growth, and innovation. Nonetheless, organizations should increasingly frame their approach to these issues in terms of the ethical and regulatory challenges they must overcome to reap the full benefits of these advances.

Some closing thoughts: the convergence of governance, quality, and AI represents a profound lever for innovation. As a result, organizations that recognize and harness this integration and embed AI in their data strategies will have a head start in realizing the full potential of their data assets. Leaning into sound governance principles and quality data will set companies on the path to sustainable innovation while helping them secure competitive advantage in the emerging data-driven economy.

REFERENCES

Abraham, R., Schneider, J., & Vom Brocke, J. (2019). Data Governance: A Conceptual Framework, Structured Review, and Research Agenda. *International Journal of Information Management*, 49, 424–438.

Anderson, T. (2023). *AI and Data Governance: Automation for Compliance*. Cambridge Press.

Armbrust, M., Fox, A., Griffith, R., et al. (2010). A View of Cloud Computing. *Communications of the ACM*, 53(4), 50–58.

Bose, I., & Leung, A. C. (2018). Predicting Financial Distress: An Empirical Study of Machine Learning Techniques. *International Journal of Forecasting*, 34(3), 635–646.

Brous, P., Janssen, M., & Herder, P. (2019). The Dual Effects of Data Governance on the Performance of Data Science. *Journal of Decision Systems*, 28(2), 99–119.

Cavoukian, A. (2012). Privacy by Design: Origins, Meaning, and Prospects. In *Privacy Protection: Governance & Technologies in Business Organizations: Aspects and Standards* (edited by George O. M. Yee), pp. 207–225. Springer.

Chaudhuri, S., Narasayya, V., & Ramakrishnan, R. (2011). "An Overview of Query Optimization in Relational Systems." *Proceedings of the VLDB Endowment*, 4(10), 1257–1267

Chen, J., & Lin, C. (Jan. 2014). Big Data Management Systems: Beyond SQL and NoSQL. *IEEE Transactions on Knowledge and Data Engineering*, 26(1), 44–53. doi: 10.1109/TKDE.2013.112

Chen, M., Mao, S., & Liu, Y. (2014). Big Data: A Survey. *Mobile Networks and Applications*, 19(2), 171–209.

Choi, T.-M., Wallace, S. W., & Wang, Y. (2018). Big Data Analytics in Operations Management. *Production and Operations Management*, 27(10), 1868–1883.

Codd, E. F. (1970). A Relational Model of Data for Large, Shared Data Banks. *Communications of the ACM*, 13(6), 377–387.

Colson, E., Choplain, M., & Carter, E. (2015). Demand Forecasting in the Age of Big Data: The Netflix Approach. *Interfaces*, 45(1), 16–32.

Davenport, T. H., & Ronanki, R. (2018). Artificial Intelligence for the Real World. *Harvard Business Review*, 96(1), 108–116.

Davis, R. (2022). *The Role of AI in Modern Data Management*. Springer.

Dean, J., & Ghemawat, S. (2004). MapReduce: Simplified Data Processing on Large Clusters. *OSDI*, 4, 137–150.

Fennelly, D., & Kaspersky, D. (2019). Regulatory Technology: Innovation in Compliance Monitoring. *Journal of Risk and Compliance*, 10(2), 107–125.

Garg, G., Versteeg, S., & Buyya, R. (July 2013). SLA-Based Resource Provisioning for Heterogeneous Workloads in a Cloud Datacenter. *IEEE Transactions on Cloud Computing*, 1(2), 129–141. doi: 10.1109/TCC.2013.13

Ghemawat, S., Gobioff, H., & Leung, S. (2020). *Google Cloud's Dataflow for Real-Time Data Processing*. Cloud Computing Innovations.

Ghoting, A., Krishnamurthy, A., & Das, G. (2019). ML in the Data Warehouse: Bridging the Gap Between Data Science and Data Engineering. *IEEE Data Engineering Bulletin*, 42(1), 12–23

Goodfellow, I., Bengio, Y., & Courville, A. (2016). *Deep Learning*. MIT Press.

Heinrich, B., Kaiser, M., & Klier, M. (2018). A Data Governance Maturity Model for Big Data: Theoretical and Empirical Insights. *Journal of Business & Information Systems Engineering*, 60(1), 15–33.

Johnson, A. (2021). *AI in Supply Chain Management: Improving Logistics through Data*. Logistics Insights.

Johnson, A. (2023). *Data Excellence and Business Strategy: Unlocking the Power of Data*. Oxford University Press.

Khatri, V., & Brown, C. V. (2010). Designing Data Governance. *Communications of the ACM*, 53(1), 148–152.

LeCun, Y., Bengio, Y., & Hinton, G. (2015). Deep Learning. *Nature*, 521(7553), 436–444.

Lee, J., Bagheri, B., & Kao, H.-A. (2015). A Cyber-Physical Systems Architecture for Industry 4.0-based Manufacturing Systems. *Manufacturing Letters*, 3, 18–23.

Lee, S., & Kang, H. (2020). *Predictive Data Quality Models for IoT: A Comprehensive Study*. IoT Innovations.

Mak, K.-K., & Pichika, M. R. (2019). Artificial Intelligence in Drug Development: Present Status and Future Prospects. *Drug Discovery Today*, 24(3), 773–780.

Mehrabi, N., Morstatter, F., Saxena, N., Lerman, K., & Galstyan, A. (2021). A survey on bias and fairness in machine learning. *ACM Computing Surveys*, 54(6), 1–35.

Mishra, V., & Darade, H. (2024a). "Harnessing Big Data: An Expert Review of Trends, Challenges, and Future Directions with a Focus on Data Quality," In *Proceedings of the Third International Conference on Artificial Intelligence and Data Science (AIDeS 2024)*. IEEE. https://doi.org/10.1109/AIDeS.2024.10957681

Mishra, V., & Darade, H. (2024b). "Advancements in Data Quality Management in the Big Data Era: A Comprehensive Review." In *Proceedings of the Third International Conference on Artificial Intelligence and Data Science (AIDeS 2024)*. IEEE.

Mishra, V., Banerjee, M., Rajendran, V., & Darade, H. (2025a). "Optimizing Travel Through AI: The Intersection of Data Quality and Customer Experience". *2025 International*

Conference on Artificial Intelligence and Data Engineering (AIDE), Nitte, India, 2025, pp. 509–513.

Mishra, V., Sachan, R. C., & Lakhani, R. (2025b). "Navigating the Evolution and Management of AI-Driven Data Platforms: An In-Depth Review & Expert Analysis." In C. Iwendi, Z. Boulouard, & N. Kryvinska (Eds.), *Proceedings of the 4th International Conference on Advances in Communication Technology and Computer Engineering (ICACTCE'24)* (Vol. 1, pp. 621–630). Springer. https://doi.org/10.1007/978-3-031-94620-2_54.

Mishra, V., Ujjwal, K., Vasanth, R., Monojit, B., & Harshal, D. (2025c). "Ethical frameworks for artificial intelligence: A comparative study." *Proceedings of the 2025 International Conference on Artificial Intelligence and Data Engineering (AIDE)*, pp. 502–508.

Nguyen, T. (2020). AI in Public Services: Transforming Governance and Service Delivery. *Journal of Public Administration and Policy*, 12(4), 200–218.

Otto, B. (2011). "A Morphology of the Organization of Data Governance," *Proceedings of the 19th European Conference on Information Systems (ECIS)*, Helsinki, Finland.

Prasad, S. (2020). *Predictive Analytics and AI in Enterprise Data Strategies*. MIT Press.

Psychoula, I., Gutmann, A., Mainali, P., Lee, S. H., Dunphy, P., & Petitcolas, F. A. P. (2021) Explainable Machine Learning for Fraud Detection. *Proceedings of the 15th Int. Conf. on Security and Cryptography* (SECRYPT 2021).

Reddy, S., & Aggarwal, A. (2020). Data Governance in AI: Challenges and Solutions. *International Journal of Data Management and Governance*, 8(3), 145–162.

Redman, T. C. (1996). *Data Quality: The Field Guide*. Digital Press.

Redman, T. C. (2016). Bad Data Costs the U.S. $3 Trillion Per Year. *Harvard Business Review*. https://hbr.org/2016/09/bad-data-costs-the-u-s-3-trillion-per-year

Schiff, D., Borenstein, J., Burton, E., & White, J. (2020). AI Ethics in Practice: Case Studies from the Field. *Journal of Responsible AI*, 1(2), 45–63.

Shin, D. (2020). The Effects of Explainability and Causability on Perception, Trust, and Acceptance: Implications for Explainable AI. *International Journal of Human-Computer Studies*, 140, 102429.

Tao, F., Qi, Q., Liu, A., & Kusiak, A. (2018). Data-driven smart manufacturing. *Journal of Manufacturing Systems*, 48, 157–169.

Topol, E. J. (2019). High-Performance Medicine: The Convergence of Human and Artificial Intelligence. *Nature Medicine*, 25(1), 44–56.

Vanberg, A. D., & Ünver, M. B. (2017). The Right to Data Portability in the GDPR and EU Competition Law: Odd Couple or Dynamic Duo? *European Journal of Law and Technology*, 8(1), 1–22.

Voigt, P., & Von dem Bussche, A. (2017). *The EU General Data Protection Regulation (GDPR): A Practical Guide*. Springer. doi: 10.1007/978-3-319-57959-7

Wang, R. Y., Strong, D. M., & Guarino, S. (2019). Beyond Accuracy: What Data Quality Means to Data Consumers. *Journal of Data Management & Governance*, 15(4), 245–262.

Watson, H. J. (2008). *Data Management: Databases and Organizations*. John Wiley & Sons.

Weber, R. H. (2020). Corporate Data Responsibility: Building a Trustworthy Framework for Data Governance. *International Data Privacy Law*, 10(1), 30–46. doi: 10.1093/idpl/ipz025

Zaharia, M., Xin, R. S., Wendell, P., Das, T., Armbrust, M., Dave, A., Meng, X., Rosen, J., Venkataraman, S., Franklin, M. J., Ghodsi, A., Gonzalez, J., Shenker, S., & Stoica, I. (2016). Apache Spark: A Unified Engine for Big Data Processing. *Communications of the ACM*, 59(11), 56–65.

Zhou, Y., Yang, W., & Wang, Z. (2019). Intelligent Data Management. *IEEE Access*, 7, 99885–99892.

14 Exploring the Frontiers of Generative AI

Applications, Hurdles, and the Synergy of AI and Human Interaction

Rachna Jain and Shikha Sharma

14.1 INTRODUCTION

Across disciplines and sectors, artificial intelligence (AI) has garnered a lot of attention (Hyder et al., 2019). According to Kaplan and Haenlein (2019, p. 15), AI is "a system's ability to correctly interpret external data, to learn from such data, and to use those learnings to achieve specific goals and tasks through flexible adaptation." AI has seen multiple phases of development and AI winters. AI showed promise in the first two decades (the 1950s and 1960s), with the development of successful programs like ELIZA (Weizenbaum, 1966) and General Problem Solver (Newell et al., 1959). Unfortunately, processing capacity constraints and a decline in funding for AI caused its development to stall.

AI has returned in recent years, as evidenced by the launch of ChatGPT in 2022 and AlphaGo in 2015. AI has gained international interest since the introduction of OpenAI's "Chat Generative Pre-Trained Transformer," or ChatGPT, application in late 2022. Since its launch on November 30, 2022, the keyword "ChatGPT" has drawn an increasing amount of search attention on Google. ChatGPT is a member of the generative pre-trained transformer (GPT) class of language models. Large language models (LLMs) that employ deep learning methods for lengthy training with massive volumes of data are within the GPT category (Cascella et al., 2023).

With ChatGPT's extensive knowledge base and tailored architecture for conversational use, it can generate responses that are as intelligent and human-like as possible. Generative AI is what makes ChatGPT possible. It is a kind of AI that can produce writing that is human-like and creative material (such as music and images), as well as combine data from many sources for analysis (Dasborough, 2023). By tricking users into thinking that ChatGPT was responding from humans rather than computers or AI, it was able to pass the Turing test (Woolf, 2022). It is anticipated that the introduction of this cutting-edge AI would completely transform society and the ways in which we work, learn, live, and communicate.

One extremely promising example of unsupervised machine learning is the generative AI model. Past generative models with limited generalization capacity include

DOI: 10.1201/9781003560463-17

Restricted Boltzmann Machines (Smolensky, 1986), Deep Belief Networks (Hinton et al., 2006), and Deep Boltzmann Machines (Salakhutdinov & Larochelle, 2010; Pan et al., 2019). Generative Adversarial Networks (GANs) were proposed as a unique procreative tool with remarkable ability for data production. Two competing neural networks, referred to as the discriminator and the generator, make up a GAN (Goodfellow et al., 2020). While the discriminator works to separate synthetic data from actual data, the generator attempts to provide data that is as realistic as possible.

Transformers are an alternate structure for neural networks that was demonstrated in the study "Attention Is All You Need." It is built on attention mechanisms that alternate between convolutions and recurrence (Vaswani et al., 2017). Self-attention allows the model to generate the output sequence while focusing on various aspects of the input sequence. These discoveries ushered in a new phase of generative AI research and development.

Multimodal items, such as text, music, images, videos, and even three-dimensional models, can be produced via generative AI. Several representative apps are DeepBrain for videos, Midjourney for photos, and ChatGPT for text. Text-to-image generating models (Qiao et al., 2022) and audio–visual correlation transformers (Wang et al., 2022) are two ways in which these models can be connected. A vast array of applications is made possible by the many types of AI-generated content (AIGC). AI has the capacity to produce textual content that can be difficult to distinguish from human-generated content, such as poems (Köbis & Mossink, 2021), political declarations (Bullock & Luengo-Oroz, 2019), and academic papers (Hu, 2023). Artworks (Gillotte, 2019), artificial faces (Whittaker et al., 2020), and magnetograms of the Sun (Liu et al., 2021) are a few examples of AI-generated images that span the humanities and sciences. However, there are possible legal, moral, and ethical problems that could arise from generative AI. These include copyright violations in AI-generated artworks (Gillotte, 2019), plagiarism and cheating in schools (University of Cambridge, n.d.), security and privacy of data (Siau & Wang, 2020), and the misuse of deepfakes and GANs for nefarious purposes (Whittaker et al., 2020) (Figure 14.1).

In many fields, including technology (explanation and information processing), business (decision-making and AI-enabled automation), education (personalized learning and intelligent tutors), healthcare (smart health and AI diagnosis), and the arts and humanities (human-centered design and cultural proximity), AI has brought forth both opportunities and challenges (Siau, 2018; Wang & Siau, 2019; Yin et al., 2022). The secret to overcoming obstacles and taking advantage of opportunities presented by generative AI is AI–human collaboration. Humans and generative AI can work together in a variety of ways. For instance, teachers should carefully review AI-generated resources before modifying them for the educational setting when using ChatGPT in science pedagogy (Cooper, 2023). Generative AI can help with brainstorming and the creation or improvement of ideas when it comes to problem-solving. Hydrologists tried using ChatGPT for simple problem-solving, but they weren't satisfied with the quality because of ChatGPT's inconsistent answers to both qualitative and quantitative queries (Halloran et al., 2023). Medical practice has the potential to benefit from ChatGPT (Cascella et al., 2023) and give public health recommendations (Biswas, 2023). According to Boßelmann et al. (2023),

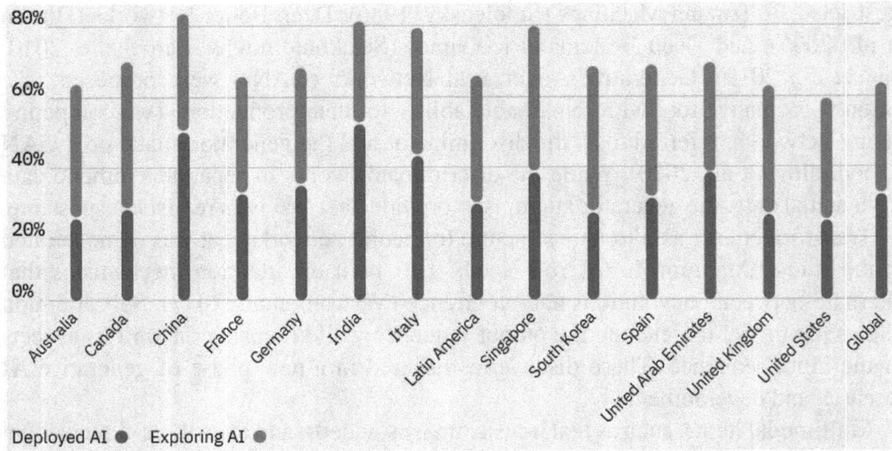

FIGURE 14.1 AI adoption globally. (https://newsroom.ibm.com/2022-05-19-Global-Data-from-IBM-Shows-Steady-AI-Adoption-as-Organizations-Look-to-Address-Skills-Shortages,-Automate-Processes-and-Encourage-Sustainable-Operations)

developments in AI can raise the general accuracy of epilepsy diagnosis and treatment. Additionally, new developments in the nursing field are anticipated to be brought about by ChatGPT, including a rise in the usage of robotics and digital technologies (Gunawan, 2023). On the other hand, there is a chance that misuse will occur, for example, when diagnosis and treatment suggestions are made without sufficient validation or supervision. Regulations and governance must therefore be established (Wan et al., 2022). The next sections cover some of the most important uses for generative AI, the difficulties it presents, and the significance of collaborating with generative AI from a human-centered perspective. Effective use of generative AI applications requires human-centered AI collaboration. In this work, we provide broad guidelines and recommendations to support collaborative AI that is human-centered.

14.2 METHODOLOGY

This study employs a qualitative and descriptive research design to explore the multifaceted impact of AI across four sectors: healthcare, business, education, and content creation. By examining the usage of AI, the associated challenges, and the interplay between AI systems and human interaction, the study aims to provide an in-depth understanding of the opportunities and limitations AI presents in these domains.

A qualitative approach was chosen to capture the nuanced experiences, perceptions, and practices surrounding AI implementation. This approach enables the exploration of subjective insights and contextual details that are often overlooked in quantitative studies. A descriptive design complements this by providing a comprehensive overview of AI's current applications and the challenges that emerge from its adoption in the four sectors under investigation. This methodology provides a robust framework for understanding the complex dynamics of AI usage and its interaction with human agency, offering valuable insights into how these sectors can navigate the challenges of integrating AI technologies.

14.3 GENERATIVE AI APPLICATIONS

ChatGPT, an example of generative AI, has potential uses in business, education, healthcare, and content creation, among other sectors. We go over several instances of generative AI applications in different sectors in this section.

14.3.1 Business

Businesses can benefit from generative AI, like ChatGPT, in a variety of areas, including operations, risk and legal, marketing and sales, HR, accounting and finance, and utility/employee optimization (Chui et al., 2022). It can be used as a chatbot to provide customer care, a virtual assistant to aid clients in completing activities, an accounting and human resource job performer, and a generator of marketing ideas or advertisements. Additionally, ChatGPT can serve as an internal or external partner for a range of business initiatives or campaigns. Therefore, ChatGPT has countless business applications that might be used.

But for corporations, generative AI – like ChatGPT – can be a double-sided coin. It can assist companies in being more efficient and in producing original content, such as ads or ad concepts. Profitability rises with lower costs and more marketing and sales. Conversely, ChatGPT can generate false or misleading information as well as hallucinations. Users are therefore unable to fully trust the data. Additionally, companies must make sure that none of their trade secrets or confidential information is disclosed to ChatGPT since it might be made public. Turning off the chat history would prevent the discussions from being utilized to train the ChatGPT models, which would be one way to protect confidential information.

Organizations and nations are enacting an increasing number of ChatGPT prohibitions and restrictions because of privacy and information security concerns (Stancati & Schechner, 2023; Ray, 2023). Two essential elements that can contribute to the success of ChatGPT applications for businesses are the ability to secure sensitive data and identify false or erroneous information. Using ChatGPT as the basis for the language model, AutoGPT creates a separate knowledge base from ChatGPT. It is critical for organizations to maximize the benefits and values that generative AI offers while limiting any losses or harm.

14.3.2 Education

The sector of education has seen significant upheaval and change because of ChatGPT. ChatGPT is a useful tool for teaching and learning activities. ChatGPT can help students with a range of tasks, including finding information online, providing answers to questions about subjects, and improving their writing in many languages. ChatGPT can help instructors with lesson planning, creating lesson materials (such as presentations, scripts, and quizzes), checking, and marking student work, and giving them feedback. Based on LLMs, ChatGPT can be used to tailor learning experiences, generate instructional content, and increase student engagement – all of which can enhance the effectiveness and efficiency of education delivery (Kasneci et al., 2023). ChatGPT can help with problem formulation, study design, data gathering and analysis, writing and composition evaluation and critique, and more in academic research (Susarla et al., 2023). Furthermore, ChatGPT is a helpful tool for self-learners who are autodidactic in open education since it offers customized guidance, assistance, and feedback (Firat, 2023).

Notwithstanding ChatGPT's benefits for education, there are worries that some students would use it to plagiarize essays or cheat on tests. According to the University of Cambridge (n.d.), it would be considered academic misconduct to include material created by ChatGPT and other AI systems. The University of Oxford views ChatGPT more positively and sees it as a helpful resource for instructors and students alike (University of Oxford, 2023). In response to the increasing use of generative AI, Turnitin, a well-known provider of plagiarism detection software, has added the capability of AI-generated text detection (Turnitin, n.d.). Similarly, a number of AI content detection services are provided, among others, by Originality.AI, Winston, AI Writing Check, CatchGPT, Content at Scale, Copyleaks, GPT Radar, GPTZero, and OpenAI's AI Text Classifier.

AI applications in education offer transformative benefits for marginalized groups, bridging gaps in access and quality. Tools powered by AI, such as adaptive learning platforms and virtual tutors, provide personalized education tailored to individual needs, helping students overcome barriers related to language, location, or learning pace. For underserved communities, AI-driven solutions enable access to high-quality educational resources, even in remote areas with limited infrastructure. Additionally, generative AI tools, like ChatGPT, assist in skill-building, language learning, and career readiness by offering cost-effective, accessible support. By fostering inclusivity and enabling equitable learning opportunities, AI has the potential to empower marginalized groups, fostering educational equity and social mobility (Salas-Pilco et al., 2022).

14.3.3 Healthcare

Another industry where ChatGPT and generative AI have a lot to offer is healthcare. As ChatGPT is rumored to have passed the US Medical Licensing Examinations, the world is beginning to get a better idea of how generative AI can change the healthcare sector (Kung et al., 2023). Regenerative AI driven by LLMs, like ChatGPT, has the potential to revolutionize the healthcare sector in several areas, including patient engagement, clinical diagnosis support, telemedicine, health education, health advising, and health promotion.

Extended reality opens new avenues for training workshops and pedagogy in medical arenas by providing seamless and safe experiences to the physicians in training, compared to in the traditional workspace, involving higher risks to skill improvement and advancements. Its applications enable better visual aids for precision and accuracy in the diagnosis and treatment of patients. Wearable patient monitoring devices can be utilized in Physical Therapy (PT) and Sensory Therapy for Speedy recovery with evident results improvements and support digital consultation with prescribed medicinal suggestions. Despite the benefits associated with the application of XR in the Healthcare Industry comes an Exorbitant cost structure (Dadario et al., 2021; Kelly et al., 2020).

It is still unclear if both medical professionals and patients would accept this new advancement in generative AI. Digital advancements like generative AI have found it difficult to break through the healthcare domain due to strict rules and high entry hurdles (Ozalp et al., 2022). There are ongoing worries about the moral application of AI, information accuracy, privacy, cybersecurity, and risk potential (Siau & Wang, 2020). In the highly regulated healthcare industry, where value generation is crucial, relying too much on AIGC might have disastrous implications, such as patient maltreatment. Furthermore, there is a risk that private and sensitive health data might be compromised or leaked when using generative AI in the healthcare sector.

14.3.4 Content Creation

The content industry is seeing a significant shift due to the application of generative AI in content creation. For instance, the marketing sector is utilizing generative AI to create individualized and synthetic ads for potential customers. Content based on artificial and automatic data production and manipulation is referred to as synthetic advertising (Arango et al., 2023). When humans and AI work together, they can create an unbelievable but plausible world using data and parameters from both sources, including deepfakes (Campbell et al., 2022). Another sector that generative AI has drastically altered is media. News production has made use of news robots like Quill and Xiaoming Bot. Therefore, using text and video, generative AI may create news stories that are more complicated (Wong et al., 2022). AI-driven innovations like ChatGPT, DALL-E 2, and Midjourney are also revolutionizing the art world. AI could promote collaborative content creation between humans and AI. Guo et al. (2023) present a paradigm for Parallel Art in which AI and humans work together to enhance artistic production.

The gaming industry has also seen changes because of generative AI. During ChatGPT's initial months of availability, users have created text-based games with it. According to Martinez-Arellano et al. (2016), AI has also been utilized to create character strategies, real-time 3D scenes, and game characters in fighting games without the need for human designer input.

Generative AI models boost generation efficiency primarily via automating the generation procedure, much as internal combustion engines, steam power, and electricity achieved during the Industrial Revolution improved the efficiency of the manufacturing of goods. By comparing the developments in content-generating automation to the industrial revolution, certain trends may be inferred. First, to increase productivity, workers are being asked to work together more and accept the use of generative AI. The demand for human content creators might decline, and people who lack the skills to work with AI could find themselves unemployed or even unherdable. Second, the revolution in content creation could lead to creative economic structures. Third, when algorithms grow in importance as nodes in the communication network, the structure of online social networks may become altered. The power to determine user perception and information exposure is increasingly delegated to AI algorithms, which could lead to a shift in communication power away from human users. As a result, businesses with the most sophisticated algorithms may have more sway and command over communication channels.

14.4 CHALLENGES WITH GENERATIVE AI

Society may face numerous difficulties because of generative AI. The chapter addresses these issues in this part from two major perspectives: the ethical perspective and the technological perspective.

14.4.1 Ethical Concerns

According to Fieser (n.d.), ethics is the systematization, defense, and advocacy of ideas about what constitutes right and wrong action. According to Siau and Wang (2020), ethical problems in the context of AI refer to the moral duties and obligations of an AI application and its designers. The main ethical problems and concerns related to generative AI are shown in Table 14.1. These difficulties include the growing digital divide, inappropriate or dangerous content, prejudice, over-reliance, abuse, security, and privacy issues.

TABLE 14.1
Ethical Challenges

References	Problems
Zhuo et al. (2023)	Unsuitable or unfit content – produced by generative AI could be violent, biased, or profane.
Zhuo et al. (2023)	Bias – Training data with limited population representation might lead to the creation of discriminatory norms.
Weidinger et al. (2021)	Monolingual (or non-multilingual) bias may result from training data in one specific language, or in a small number of languages.
Dwivedi et al. (2023); Chan (2022)	Cultural sensitivity – Decisions made by generative AI on employment are biased.
Iskender (2023); Van Dis et al. (2023)	Over-reliance – Without carefully verifying or fact-checking, users accept responses provided by generative AI.
Cotton et al. (2023); Thorp (2023); Susnjak (2022)	Misuse – Plagiarism for assignments and essays using texts generated by AI. One can utilize generative AI to cheat on tests or homework.
Siau and Wang (2020); Fang et al. (2017)	Privacy and security – Generative AI could reveal sensitive or confidential data.
Bozkurt and Sharma (2023); Dwivedi et al. (2023)	Digital divide – First-order digital divide for those lacking access to systems with generative AI. Second-order digital divide: generative AI may be more acceptable to certain persons and cultural groups than to others.

14.4.1.1 Unsuitable or Unfit Content

Generative AI-generated unsuitable or unfit content includes a wide range of things, including profanity, pornography, prejudiced content, and violent material. Despite the content regulation that OpenAI has put in place for ChatGPT, inappropriate or hazardous content may still emerge because of algorithmic limitations or jailbreaking, which is the process of removing imposed limits. Toxicology is the term used to describe language models' capacity to comprehend or produce offensive or damaging information (Zhuo et al., 2023). Toxicity has the potential to undermine communal cohesion and cause harm to society. It is crucial to ensure that any objectionable or hazardous content is eliminated when revealed to be included in the training set. In the same vein, no explicit, sexual, or erotic material should be included in the training data (Zhuo et al., 2023). It is necessary to have rules, governance, and policies in place to ensure that users do not have access to any undesirable content.

14.4.1.2 Bias

The term "bias" in the context of AI refers to the possibility that recommendations or replies produced by AI may unfairly favor or discriminate against a particular individual or group (Ntoutsi et al., 2020). Language models can produce material that contains biases of various kinds, which may result from the training data. Exclusionary norms, for instance, arise when a small portion of the population is represented in the training data (Zhuo et al., 2023). Like this, when training data is only available in one language, multilingualism suffers from monolingual bias (Weidinger et al., 2021). Due to ChatGPT's global operations, cultural sensitivity to various locales is essential to preventing bias (Dwivedi et al., 2023). Biases and opacity may occur when AI is utilized to support decision-making at various phases of employment (Chan, 2022). Generative AI recommendations often include prejudices against specific racial groups, genders, sexual orientations, and occupations. Therefore, to guarantee justice and prevent biases, the representativeness, completeness, and diversity of the training data are crucial (Gonzalez, 2023). Incorporating synthetic data into training can improve the dataset's diversity and mitigate sample-selection biases caused by class imbalances (Chen et al., 2021). A wide spectrum of consumers and industry experts should test and evaluate applications of generative AI.

14.4.1.3 Over-Dependence

Users may grow unduly reliant on ChatGPT and begin to accept the replies that it offers because of its seeming power and simplicity. As opposed to traditional search engines, which present users with a selection of information sources, ChatGPT creates customized answers for each request. While using ChatGPT might increase productivity by saving time and effort, users run the risk of developing a habit of accepting responses without question or justification. In addition to creating human automation bias owing to habitually accepting generative AI recommendations, an over-reliance on generative AI technology can hinder skills like creativity, critical thinking, and problem-solving (Iskender, 2023; Van Dis et al., 2023). AI literacy is therefore essential for every user. It is advised that users verify the answers produced by generative AI before adopting them rather than accepting them at face value.

14.4.1.4 Misuse

Any intentional use of generative AI that might have unfavorable, immoral, or improper effects is considered a misuse (Brundage et al., 2020). Education is a well-known area that is susceptible to abuse. In the age of ChatGPT, Cotton et al. (2023) have expressed worries about academic integrity. ChatGPT is a sophisticated plagiarism detection technology that finds patterns in big data sets to produce original material (Gefen & Arinze, 2023). Because ChatGPT and other generative AI systems can produce excellent responses in a matter of seconds, unmotivated students might not put in the time and effort necessary to complete their projects and essays. Because of this, evaluating the originality of student work in the age of generative AI may be challenging. Text created using ChatGPT is not appropriate and is considered plagiarism (Thorp, 2023). Exam cheating is another instance of misuse. Strict proctoring procedures will need to be implemented, and AIGC detectors like Turnitin may be employed to address possible misuse in education (Susnjak, 2022). The difficulties lie beyond content detection and examination proctoring, though, as it may be difficult to distinguish between proper and improper use of ChatGPT. Scholars have provided recommendations and ideas for the appropriate usage of ChatGPT in academic activities (Susarla et al., 2023).

14.4.1.5 Confidentiality and Safety

Data safety and confidentiality are major problems for ChatGPT and other generative AI systems. Sensitive personal information that owners do not want to share with others is related to privacy (Fang et al., 2017). Huge amounts of private and personal data were used to train ChatGPT throughout its development phase, endangering privacy (Siau & Wang, 2020). ChatGPT is becoming more and more integrated into people's daily lives, providing them with more convenience at the expense of gathering a massive quantity of confidential data about them. The prospect of confidential information being accidentally or intentionally made public raises concerns and hazards (Porter, 2023). Concerns about security and confidentiality of data impact not just individual users but also big businesses and governmental institutions. Businesses using AI, especially the biggest tech companies, should make the required efforts to educate users about ethical concerns connected to security and privacy, such as trade secret breaches and the "dos and don'ts" of sharing sensitive data with generative AI. In the meantime, regulations and policies should be in place to safeguard and maintain the confidentiality of information.

14.4.1.6 Digital Divide

The difference in access to computers and the Internet between those who do and do not (Van Dijk, 2006) is a common definition of the digital divide. A second-level digital divide, defined as the difference in Internet usage and skills between various groups and cultures, is being discussed as a potential worry as the Internet becomes more and more commonplace (Scheerder et al., 2017). Being a new technology, generative AI has the potential to increase society's already-existing digital divide. The relationship between people and technology has become more complex due to the "invisible" AI that underlies AI-enabled technologies (Carter et al., 2020). The first-level digital divide may expand between people who live in areas restricted by generative AI suppliers or websites, or between people who do not have access to gadgets or the Internet (Bozkurt & Sharma, 2023). Language and cultural hurdles may arise for people from marginalized or minority cultures if generative AI models do not thoroughly learn about or incorporate their traditions. Moreover, the second-level digital gap can appear or get wider for people who find it difficult to use the generative AI tool, including some older people (Dwivedi et al., 2023). More accessible AI and training in AI literacy would be helpful in addressing the digital divide.

14.4.2 Technology Concerns

Technological challenges are those that stem from the restrictions or limitations imposed by generative AI. For instance, one of the biggest obstacles to the development of generative AI models is the quality of training data. The limits of the algorithms also present issues with the output's legitimacy and explainability. The technological difficulties and problems related to generative AI are shown in Table 14.2. Authenticity, explainability, timely engineering, and the quality of training data are some of these challenges.

14.4.2.1 Quality of Training Data

Another difficulty generative AI faces is the quality of training data. The caliber of training data has a major impact on the caliber of generative AI models (Dwivedi et al., 2023; Su & Yang, 2023). The model's output might contain any factual mistakes, biased sources of information, or imbalances in the training set of data. Large volumes of training data are frequently needed for generative AI models, including ChatGPT and Stable Diffusion, which are text-to-image models (Gozalo-Brizuela &

TABLE 14.2
Technological Concern

References	Problems
Gozalo-Brizuela and Garrido-Merchan (2023); Dwivedi et al. (2023); Su and Yang (2023)	Quality of training data – Challenging to get enough generative AI training data to guarantee the dataset's quality.
Dwivedi et al. (2023)	Explainability – Complicated to decipher and comprehend generative AI outputs.
Rudin (2019)	Errors in generative AI outputs are difficult to find
Burrell (2016)	Users' confidence in generative AI is declining or nonexistent.
Rieder and Simon (2017)	Bias Difficult for regulators to determine if generative AI is biased or unfair.
Gragnaniello et al. (2022)	Authenticity – When content (such as pictures and movies) is altered, there are questions about its validity.
Liu and Chilton (2022)	Prompt Engineering – Digital literacy requires prompt engineering to fully utilize generative AI.

Garrido-Merchan, 2023). Obtaining not just good training datasets but also comprehensive and well-balanced training datasets is crucial. The quality dataset problem must be addressed by data purification for the training datasets; however, this is too costly given the volume of data. To eliminate sample-selection biases in the datasets and guarantee dataset diversity, synthetic training data may be employed (Chen et al., 2021).

14.4.2.2 Explainability

The inability of AI algorithms to describe their model, or the absence of information about how the algorithm generates its conclusions, is a common source of worry (Deeks, 2019). According to Dwivedi et al. (2023), generative AI models lack transparency into the reasoning behind the model's output. The lack of openness brings up several problems. First, consumers may find it challenging to decipher and comprehend the results (Dwivedi et al., 2023). Users would also have a hard time finding any possible errors in the output (Rudin, 2019). Furthermore, users may find it difficult to trust the system and its responses or recommendations if the interpretation and evaluation of the result are not accessible (Burrell, 2016). Furthermore, it would be challenging for the regulatory body to determine whether the generative AI system is potentially biased or unfair from the standpoint of law and regulations (Rieder & Simon, 2017).

14.4.2.3 Authenticity

It gets more difficult to determine a piece of work's legitimacy as generative AI advances. Images that appear to depict individuals or events in the actual world could be created by DeepFake AI. Large-scale image and video manipulation could result from generative AI's capability, exacerbating the issue of false information spreading on social media platforms (Gragnaniello et al., 2022). An algorithm in the arts might directly result in an exquisite artwork or piece of music. Since algorithms often provide standardized and repetitive outputs, some have argued that artwork produced by AI lacks authenticity (McCormack et al., 2019).

14.4.2.4 Prompt Engineering

The capacity to work with AI successfully and efficiently has become one of the most critical media literacies with the widespread use of generative AI. Therefore, it is essential that users of generative AI understand and put into practice the concepts of prompt engineering. Prompt engineering is the systematic process of deliberately constructing prompts or inputs for generative AI models in order to elicit valuable outputs. Because human languages are ambiguous, using prompts to communicate with robots might result in mistakes or miscommunications. As a result, quick quality matters. Debugging the prompts and enhancing the capacity to interact with generative AI present further challenges (Liu & Chilton, 2022). As a result, prompt engineering training is required, particularly for individuals who interface with generative AI on a regular basis.

14.5 FINDINGS ON HUMAN-CENTERED AI COLLABORATION

The design of AI with the understanding that it is a component of a broader system with human stakeholders is the focus of human-centered AI (Riedl, 2019). In addition to understanding humans from a sociocultural standpoint, Human-Centered AI ought to be able to assist individuals in understanding themselves (Riedl, 2019). A large part of the reason previous waves of AI failed was that they could not meet human desires to the necessary extent. The human aspects of working with AI are being addressed by recent advances in AI development, which also emphasize AI transparency and explainability (i.e., addressing AI's interpretability and comprehensibility), AI ethics and governance, and digital transformation through AI literacy and intelligence augmentation (Xu, 2019).

14.5.1 Healthcare and Education – Sensitivity and Human Needs

Neglecting socio-technical concerns and human needs and values is the root cause of many of the obstacles faced by generative AI. As a result, we propose the Human-Centered AI partnership to direct the development and implementation of generative AI. First, Human-Centered AI highlights that human values and needs should inform the design of intelligent systems. Second, "Human-in-the-loop" can mitigate some of the concerns over AI, such as threats to privacy and security, robot-driven deployment, and singularity (i.e., AI advancing beyond human control) (Shneiderman, 2020a; Xu et al., 2023). The creation of generative AI algorithms and products might encounter minimal ethical and legal difficulties if societal ethics and rules are incorporated. Reduced bias, job creation, augmenting rare data, maintaining human-level precision, incorporating subject-level experts, ensuring consistency and accuracy, ease of work, increased efficiency, accountability and transparency, and increased safety are all benefits of having "human-in-the-loop" technology (Monarch, 2021). The following concerns need to be taken into consideration to accomplish "human-centered AI collaboration."

Human needs are a fundamental part of Human-Centered AI (Auernhammer, 2020). One issue that might arise from a misalignment with human desires or needs is the digital gap. Due to ChatGPT's immense global appeal, there is a strong desire to use it in both personal and professional settings. Now, ChatGPT's digital divide encompasses both usability and accessibility. Efficiency, security, sustainability, and creativity are additional human demands and expectations for generative AI in addition to these two problems. We contend that to enhance human–AI cooperation, human needs should be considered while designing generative AI in the future. OpenAI, on the one hand, restricts access and forbids users from utilizing ChatGPT in some places. However, some users express dissatisfaction about finding it difficult to employ the appropriate suggestions to successfully meet their objectives.

14.5.2 Content Creation – Transparency and Explainability

Understanding the logic behind AI systems' output is referred to as AI transparency and explainability (Flyverbom, 2016). Improved algorithm transparency will enable consumers to comprehend the normally concealed "black boxes" of AI systems. In addition to allowing users to make educated decisions rather than relying solely on the system, AI transparency and explainability can improve the interoperability and comprehensibility of the AI system (Rader et al., 2018). The fundamental guidelines for creating an explainable system in the field of human-centered explainable AI are the things users need to know about the system (Sun et al., 2022). This principle states that the following are essential components of a transparent and understandable generative AI system: (1) an approachable overview of the model's general architecture and mechanism; (2) an explanation of the model's capabilities (such as functions and areas of expertise); (3) an explanation of the model's limitations (such as uncertainties and risks); (4) post-hoc explanations of the model's decision-making process; (5) contextual explanations applicable to particular contexts; and (6) additional information to meet users' needs that arise in a usage context (Sun et al., 2022).

14.5.3 Business – Ethics and Governance

Collaboration with AI that is human-centered should respect moral principles and human values. Diversity, privacy, and hazardous content are only a few of the issues that should be covered by ethical considerations. As was previously said, Human-Centered AI systems must comprehend people from a sociocultural standpoint (Riedl, 2019). As ChatGPT and other generative AI work in many sociocultural environments around the globe, more complex ethical considerations will need to be made. By abstaining from discrimination and showing respect for individuals of diverse racial, ethnic, cultural, and gender identities, a Human-Centered AI can be just and responsible. It should not contain any information that could be harmful to others or offensive, aggressive, or sensual content. Another crucial ethical issue in human-centered AI collaboration is privacy. These systems are gathering more and more organizational and personal data as generative AI helps us in more areas of our lives and work. To improve the credibility of generative AI and promote human–AI cooperation, data security and confidentiality should be guaranteed.

Policies and regulations are necessary to utilize emerging technology and accomplish targeted outcomes (Qadir et al., 2022). Some of the issues can be resolved by human-centered regulations and policies using a two-step procedure. To be effective, rules and policies must first be designed with people in mind (Qadir et al., 2022). The difficulty with copyright issues is one such. Any information on the Internet can be used by generative AI to produce a response or answer. Information pertaining to copyright or other intellectual property rights may have been utilized. Therefore, consideration of the interests and legal rights of human stakeholders will be necessary while formulating regulations and policies. Discrimination can also be effectively combated via a human-centered approach by removing offensive content. OpenAI has implemented a content policy to steer clear of discriminatory language in its generated responses. The effective integration of human-centered laws and policies into AI algorithms is the second stage. Despite OpenAI creating a content policy for ChatGPT, one counterexample is that dangerous content still occasionally appears due to jailbreaking or algorithmic errors. AI algorithms must thus be developed and enhanced to avoid these problems.

14.5.4 Education – Transforming Learning

The integration of ChatGPT into the education sector has brought about significant transformations in teaching, learning, and research practices. As a versatile tool,

ChatGPT enhances various educational activities. For students, it facilitates tasks such as finding online information, addressing subject-specific queries, and improving writing skills in multiple languages. For educators, ChatGPT aids in lesson planning, the creation of instructional materials like presentations, scripts, and quizzes, as well as evaluating and providing feedback on student work. Leveraging LLMs, ChatGPT customizes learning experiences, generates educational content, and fosters student engagement, contributing to the overall efficiency and effectiveness of education delivery (Kasneci et al., 2023).

In academic research, ChatGPT demonstrates its utility in diverse activities, including problem formulation, study design, data collection, analysis, and manuscript critique. Additionally, it supports self-learners by providing personalized guidance, assistance, and feedback, making it a valuable resource for autodidactic individuals in open education (Firat, 2023).

14.5.5 Society – AI Proficiency and Intellect Enhancement

> The goal of AI, according to the Human Centered AI approach, is to compliment people, not to replace them. Collaboration is the cornerstone to intelligence augmentation (IA), and user AI literacy is critical (Fugener et al., 2021). A collection of skills that enable people to understand and assess AI technologies, interact, and communicate with AI, and use AI efficiently and morally is known as AI literacy (Long & Magerko, 2020; Ng et al., 2021). Users who possess AI literacy are more equipped to utilize and operate together with AI to digitally enhance and alter work.
>
> (Fast & Horvitz, 2017)

Intelligence augmentation (IA) is the focus in the application of AI and has transformed the conflict between humans and AI into a symbiotic one, despite concerns that AI may lead to singularity (Zhou et al., 2021, 2023). A more comprehensible and controllable future for AI development may result from such a change (Shneiderman, 2020b). Three interrelated components make up the Human-Centered AI framework proposed by Xu et al. (2023): technologies, human factors, and ethics. The framework includes enhancing human skills as one of its goals. Four design principles for information architecture (IA) were presented by Zhou et al. (2023). These are human-centeredness, interpretability, simplicity, and ethics. A human-centered approach to AI, which considers ethical issues and human elements in human–machine systems, was emphasized by Paul et al. in 2022. AI is based on the complexities of human–machine interaction. Human creativity and problem-solving skills can be enhanced through Human-Centered AI collaboration.

14.8 DISCUSSION AND IMPLICATIONS

Generative AI, a rapidly evolving subset of AI, has demonstrated transformative potential across industries, including content creation, healthcare, and education. However, its rise also brings challenges, particularly concerning biases embedded within models and risks to user privacy. These issues, if left unaddressed, could perpetuate discrimination, undermine trust, and result in the misuse of sensitive data. This chapter identifies and suggests practical policy measures to mitigate these challenges, drawing insights from real-world regulatory frameworks and ethical guidelines. By addressing biases and ensuring robust privacy protections, the synergy between AI and human interaction can be harnessed for equitable and responsible technological advancement.

To mitigate biases in generative AI, governments and organizations must adopt proactive strategies. Regular bias audits can help identify and correct discriminatory patterns in AI models, ensuring fair outcomes. Additionally, using diverse and representative training datasets minimizes the risk of biased outputs. Policies promoting explainable AI require developers to disclose decision-making processes, enabling better understanding and accountability. Inclusive design practices, involving input from marginalized and underrepresented groups, further enhance equity in AI systems. These principles are reflected in frameworks like the proposed EU AI Act, which mandates risk assessments for AI models, and Singapore's Model AI Governance Framework, which emphasizes transparency and fairness.

Privacy protection is another critical area requiring policy interventions. Data anonymization is essential to prevent the re-identification of individuals in training datasets. Strict user consent protocols ensure ethical data usage, while differential privacy techniques add protective noise to datasets without compromising their utility. Access control mechanisms restrict sensitive data access to authorized personnel only. These measures align with regulations such as the EU's General Data Protection Regulation (GDPR), which enforces data minimization and the right to be forgotten, and the California Consumer Privacy Act (CCPA), which provides transparency and user control over data. The OECD AI Ethics Guidelines further emphasize robust data governance for safeguarding privacy.

Transparency and accountability are key to fostering trust in generative AI systems. Clear documentation of training data, algorithms, and testing methodologies ensures traceability and informed decision-making. Audit trails and third-party audits enable oversight and compliance with ethical standards. Periodic transparency reports can highlight performance metrics, bias incidents, and corrective actions taken, encouraging public trust. For instance, New York City's AI Hiring Law mandates bias audits and public disclosure, and Canada's Directive on Automated Decision-Making establishes stringent transparency requirements for government-deployed AI systems.

Ethical AI practices form the backbone of responsible generative AI deployment. Incorporating human oversight in critical decision-making processes, such as healthcare and criminal justice, ensures that AI outputs are carefully evaluated before implementation. Establishing ethics committees to review the societal impacts of AI systems further strengthens this approach. Adaptive regulations, which evolve alongside technological advancements, can help address emerging risks effectively. Guidelines like UNESCO's AI Ethics Recommendations advocate for human-centric AI development, while Australia's AI Ethics Framework emphasizes alignment with human rights and social values.

14.7 CONCLUSIONS

Among the most revolutionary technological developments in recent years are generative AI models, like ChatGPT, Midjourney, and DeepBrain (Dwivedi et al., 2023). Generative AI models are thought to be the next development in artificial general intelligence, capable of creating fresh text, image, and video content (Luo et al., 2023). The fields of business, education, healthcare, and content production can all benefit greatly from the use of generative AI. But there are drawbacks to generative AI as well. These fall into four categories: technology, ethics, laws and policies, and economics. Numerous of these issues stem from the lack of Human-Centered AI. Human-centeredness is a prerequisite for the development of generative AI; this means that it must consider human needs and empathy, transparency and comprehensibility, ethics and governance, and transformation via perceptive reasoning and AI literacy. AI will always exist. The field of generative AI is rapidly progressing, and its potential to upend entire industries and businesses will only grow. As generative AI is affecting our lives and work so profoundly, collaborating and working with it will soon become the norm, if it has not already. Education must alter to provide students with the hard and soft skill sets they will need to collaborate and engage with generative AI in both academic and professional situations. As AI develops and changes both our workplace and our lives, it will become increasingly important to learn new things constantly and adapt in order to upskill, reskill, and retool the workforce. We are living in an exciting and demanding period where it is imperative that we prepare for the generative AI future.

REFERENCES

Arango, L., Singaraju, S. P., & Niininen, O. (2023). Consumer responses to AI-Generated charitable giving ads. *Journal of Advertising*, 1–18. https://doi.org/10.1080/00913367.2023.2183285

Auernhammer, J. (2020). Human-centered AI: The role of human-centered design research in the development of AI. In S. Boess, M. Cheung, & R. Cain (Eds.), *Synergy – DRS International Conference 2020*, online (pp. 1315–1333). https://doi.org/10.21606/drs.2020.282

Biswas, S. S. (2023). Role of ChatGPT in public health. *Annals of Biomedical Engineering*, *51*(5), 868–869. https://doi.org/10.1007/s10439-023-03172-7

Boßelmann, C. M., Leu, C., & Lal, D. (2023). Are AI language models such as ChatGPT ready to improve the care of individuals with epilepsy? *Epilepsia*, *64*(5), 1195–1199. https://doi.org/10.1111/epi.17570

Bozkurt, A., & Sharma, R. C. (2023). Challenging the status quo and exploring the new boundaries in the age of algorithms: Reimagining the role of generative AI in distance education and online learning. *Asian Journal of Distance Education*, *18*(1), 1–8. https://doi.org/10.5281/zenodo.7755273

Brundage, M., Avin, S., Wang, J., Belfield, H., Krueger, G., Hadfield, G., Khlaaf, H., Yang, J., Toner,H., Fong, R., Maharaj, T., Koh, P., Hooker, S., Leung, J., Trask, A., Bluemke, E., Lebensold, J., O'Keefe, C., Koren, M., …, & Anderljung, M. (2020). Toward trustworthy AI development: Mechanisms for supporting verifiable claims. https://doi.org/10.48550/arXiv.2004.07213

Bullock, J., & Luengo-Oroz, M. (2019). Automated speech generation from UN general assembly statements: Mapping risks in AI generated texts. https://doi.org/10.48550/arXiv.1906.01946

Burrell, J. (2016). How the machine 'thinks': Understanding opacity in machine learning algorithms. *Big Data & Society*, *3*(1), 1–12. https://doi.org/10.1177/2053951715622512

Campbell, C., Plangger, K., Sands, S., & Kietzmann, J. (2022). Preparing for an era of deepfakes and AI-generated ads: A framework for understanding responses to manipulated advertising. *Journal of Advertising*, *51*(1), 22–38. https://doi.org/10.1080/00913367.2021.1909515

Carter, L., Liu, D., & Cantrell, C. (2020). Exploring the intersection of the digital divide and artificial intelligence: A hermeneutic literature review. *AIS Transactions on Human-Computer Interaction*, *12*(4), 253–275. https://doi.org/10.17705/1thci.00138

Cascella, M., Montomoli, J., Bellini, V., & Bignami, E. (2023). Evaluating the feasibility of ChatGPT in healthcare: An analysis of multiple clinical and research scenarios. *Journal of Medical Systems*, *47*(1), 1–5. https://doi.org/10.1007/s10916-023-01925-4

Chan, G. K. (2022). AI employment decision-making: Integrating the equal opportunity merit principle and explainable AI. *AI & Society*, 1–12. https://doi.org/10.1007/s00146-022-01532-w

Chen, R. J., Lu, M. Y., Chen, T. Y., Williamson, D. F. K., & Mahmood, F. (2021). Synthetic data in machine learning for medicine and healthcare. *Nature Biomedical Engineering*, *5*(6), 493–497. https://doi.org/10.1038/s41551-021-00751-8

Chui, M., Roberts, R., & Yee, L. (2022, December 20). *Generative AI is here: How tools like ChatGPT could change your business*. Quantum Black AI by McKinsey. www. mckinsey.com/capabilities/quantumblack/our-insights/generative-ai-is-here-how-tools-like-chatgpt-could-change-your-business#/

Cooper, G. (2023). Examining science education in ChatGPT: An exploratory study of generative artificial intelligence. *Journal of Science Education and Technology*, *32*(3), 444–452. https://doi.org/10.1007/s10956-023-10039-y

Cotton, D. R., Cotton, P. A., & Shipway, J. R. (2023). Chatting and cheating: Ensuring academic integrity in the era of ChatGPT. *Innovations in Education and Teaching International*, 1–12. https://doi.org/10.1080/14703297.2023.2190148

Dadario, N. B., Quinoa, T., Khatri, D., Boockvar, J., Langer, D., & D'Amico, R. S. (2021). Examining the benefits of extended reality in neurosurgery: A systematic review. *Journal of Clinical Neuroscience*, *94*, 41–53.

Dasborough, M. T. (2023). Awe-inspiring advancements in AI: The impact of ChatGPT on the field of organizational behavior. *Journal of Organizational Behavior*, *44*(2), 177–179. https:// doi.org/10.1002/job.2695

Deeks, A. (2019). The judicial demand for explainable artificial intelligence. *Columbia Law Review*, *119*(7), 1829–1850. www.jstor.org/stable/26810851

Dwivedi, Y. K., Kshetri, N., Hughes, L., Slade, E. L., Jeyaraj, A., Kar, A. K., Baabdullah, A. M., Koohang, A., Raghavan, V., Ahuja, M., Albanna, H., Albashrawi, M. A., Al-Busaidi, A. S.,Balakrishnan, J., Barlette, Y., Basu, S., Bose, I., Brooks, L., Buhalis, D., …, & Wright, R. (2023). Opinion paper: "So what if ChatGPT wrote it?" Multidisciplinary perspectives on opportunities, challenges and implications of generative conversational AI for research, practice and policy. *International Journal of Information Management*, *71*, 1–63. https://doi.org/10.1016/j.ijinfomgt.2023.102642

Fang, W., Wen, X. Z., Zheng, Y., & Zhou, M. (2017). A survey of big data security and privacy preserving. *IETE Technical Review*, *34*(5), 544–560. https://doi.org/10.1080/02564602.2016.1215269

Fast, E., & Horvitz, E. (2017, February). Long-term trends in the public perception of artificial intelligence. In *Proceedings of the AAAI Conference on Artificial Intelligence*, San Francisco, California, USA, 31(1), 963–969.

Fieser, J. (n.d.). Ethics. *Internet Encyclopedia of Philosophy*. Retrieved May 17, 2023, https://web.archive.org/web/20180119084940/http://www.iep.utm.edu/ethics

Firat, M. (2023). How ChatGPT can transform autodidactic experiences and open education? *OSF Preprints*, 1–5. https://doi.org/10.31219/osf.io/9ge8m

Flyverbom, M. (2016). Transparency: Mediation and the management of visibilities. *International Journal of Communication*, *10*(1), 110–122. http://ijoc.org/index.php/ijoc/article/download/4490/1531

Fugener, A., Grahl, J., Gupta, A., & Ketter, W. (2021). Will humans-in-the-loop become Borgs? Merits and pitfalls of working with AI. *MIS Quarterly*, *45*(3b), 1527–1556. https://doi.org/10.25300/MISQ/2021/16553

Gefen, D., & Arinze, O. (2023). ChatGPT and usurping academic authority. *Journal of Information Technology Case & Application Research*, *25*(1), 3–9. https://doi.org/10.1080/15228053.2023.2186629

Gillotte, J. L. (2019). Copyright infringement in AI-generated artworks. *UC Davis Law Review*, *53*, 2655.https://ssrn.com/abstract=3657423

Gonzalez, W. (2023, April 25). How businesses can help reduce bias in AI. *Forbes*. Available at www.forbes.com/sites/forbesbusinesscouncil/2023/04/25/how-businesses-can-help-reduce-bias-in-ai/

Goodfellow, I., Pouget-Abadie, J., Mirza, M., Xu, B., Warde Farley, D., Ozair, S., Courville, A., & Bengio, Y. (2020). Generative adversarial networks. *Communications of the ACM*, *63*(11), 139–144. https://doi.org/10.1145/3422622

Gozalo-Brizuela, R., & Garrido-Merchan, E. C. (2023). ChatGPT is not all you need. A state-of-the-art review of large generative AI models. https://doi.org/10.48550/arXiv.2301.04655

Gragnaniello, D., Marra, F., & Verdoliva, L. (2022). Detection of AI-Generated synthetic faces. In *Handbook of digital face manipulation and detection: From DeepFakes to morphing attacks* (pp. 191–212). Springer International Publishing.

Gunawan, J. (2023). Exploring the future of nursing: Insights from the ChatGPT model. *Belitung Nursing Journal*, *9*(1), 1–5. https://doi.org/10.33546/bnj.2551

Guo, C., Lu, Y., Dou, Y., & Wang, F. Y. (2023). Can ChatGPT boost artistic creation: The need of imaginative intelligence for parallel art. *IEEE/CAA Journal of Automatica Sinica*, *10*(4), 835–838. https://doi.org/10.1109/JAS.2023.123555

Halloran, L. J., Mhanna, S., & Brunner, P. (2023). AI tools such as ChatGPT will disrupt hydrology, too. *Hydrological Processes*, *37*(3), 1–3. https://doi.org/10.1002/hyp.14843

Hinton, G. E., Osindero, S., & Teh, Y. W. (2006). A fast learning algorithm for deep belief nets. *Neural Computation*, *18*(7), 1527–1554. https://doi.org/10.1162/neco.2006.18.7.1527

Hu, G. (2023). Challenges for enforcing editorial policies on AI-generated papers. *Accountability in Research*, 1–4. https://doi.org/10.1080/08989621.2023.2184262

Hyder, Z., Siau, K., & Nah, F. F. H. (2019). Artificial intelligence, machine learning, and autonomous technologies in mining industry. *Journal of Database Management*, *30*(2), 67–79. https://doi.org/10.4018/JDM.2019040104

Iskender, A. (2023). Holy or unholy? Interview with open AI's ChatGPT. *European Journal of Tourism Research*, *34*(3414), 1–11. https://doi.org/10.54055/ejtr.v34i.3169

Kaplan, A., & Haenlein, M. (2019). Siri, Siri, in my hand: Who's the fairest in the land? On the interpretations, illustrations, and implications of artificial intelligence. *Business Horizons*, *62*(1), 15–25. https://doi.org/10.1016/j.bushor.2018.08.004

Kasneci, E., Seßler, K., Küchemann, S., Bannert, M., Dementieva, D., Fischer, F., Gasser, U., Groh, G., Günnemann, S., Hüllermeier, E., Krusche, S., Kutyniok, G., Michaeli, T., Nerdel, C., Pfeffer, J., Poquet, O., Sailer, M., Schmidt, A., Seidel, T., …, & Kasneci, G. (2023). ChatGPT for good? On opportunities and challenges of large language models for education. *Learning and Individual Differences*, *103*, 1–9. https://doi.org/10.1016/j.lindif.2023.102274

Kelly, A. H., Lezaun, J., Löwy, I., Matta, G. C., de Oliveira Nogueira, C., & Rabello, E. T. (2020). Uncertainty in times of medical emergency: Knowledge gaps and structural ignorance during the Brazilian Zika crisis. *Social Science & Medicine*, *246*, 112787.

Köbis, N., & Mossink, L. D. (2021). Artificial intelligence versus Maya Angelou: Experimental evidence that people cannot differentiate AI-generated from human-written poetry. *Computers in Human Behavior*, *114*, 1–13. https://doi.org/10.1016/j.chb.2020.106553

Kung, T. H., Cheatham, M., Medenilla, A., Sillos, C., De Leon, L., Elepaño, C., Madriaga, M., Aggabao, R., Diaz-Candido, G., Maningo, J., & Tseng, V. (2023). Performance of ChatGPT on USMLE: Potential for AI-assisted medical education using large language models. *PLOS Digital Health*, *2*(2), 1–12. https://doi.org/10.1371/journal.pdig.0000198

Liu, J., Wang, Y., Huang, X., Korsós, M. B., Jiang, Y., Wang, Y., & Erdélyi, R. (2021). Reliability of AI-generated magnetograms from only EUV images. *Nature Astronomy*, *5*(2), 108–110. https://doi.org/10.1038/s41550-021-01310-6

Liu, V., & Chilton, L. B. (2022, April). Design guidelines for prompt engineering text-to-image generative models. In *Proceedings of the 2022 CHI Conference on Human Factors in Computing Systems*, New Orleans, Louisiana, USA (pp. 1–23).

Long, D., & Magerko, B. (2020, April). What is AI literacy? Competencies and design considerations. In *Proceedings of the 2020 CHI Conference on Human Factors in Computing Systems*, Honolulu, Hawaii, USA (pp. 1–16).

Luo, G., Zhou, Y., Ren, T., Chen, S., Sun, X., & Ji, R. (2023). Cheap and quick: Efficient vision-language instruction tuning for large language models. *Advances in Neural Information Processing Systems*, *36*, 29615–29627.

Martinez-Arellano, G., Cant, R., & Woods, D. (2016). Creating AI characters for fighting games using genetic programming. *IEEE Transactions on Computational Intelligence and AI in Games*, *9*(4), 423–434. https://doi.org/10.1109/TCIAIG.2016.2642158

McCormack, J., Gifford, T., & Hutchings, P. (2019). Autonomy, authenticity, authorship and intention in computer generated art. In *8th International Conference on Computational Intelligence in Music, Sound, Art and Design*, Leipzig, Germany (pp. 35–50). Springer International Publishing.

Monarch, R. (2021). *Human-in-the-loop machine learning*. Manning.

Murray, M. D. (2023). Generative and AI authored artworks and copyright law.

Newell, A., Shaw, J. C., & Simon, H. A. (1959). Report on a general problem-solving program. In *Proceedings of the International Conference on Information Processing,* Paris, France (pp. 256–264).

Ng, D. T. K., Leung, J. K. L., Chu, S. K. W., & Qiao, M. S. (2021). Conceptualizing AI literacy: An exploratory review. *Computers and Education: Artificial Intelligence*, *2*, 1–11. https://doi. org/10.1016/j.caeai.2021.100041

Ntoutsi, E., Fafalios, P., Gadiraju, U., Iosifidis, V., Nejdl, W., Vidal, M. E., Ruggieri, S., Turini, F.,Papadopoulos, S., Krasanakis, E., Kompatsiaris, I., Kinder-Kurlanda, K., Wagner, C., Karimi,F., Fernandez, M., Alani, H., Berendt, B., Kruegel, T., …, & Staab, S. (2020). Bias in data-driven artificial intelligence systems—An introductory survey. *Wiley Interdisciplinary Reviews: Data Mining and Knowledge Discovery*, *10*(3), 1–14. https://doi.org/10.1002/widm.1356

Ozalp, H., Ozcan, P., Dinckol, D., Zachariadis, M., & Gawer, A. (2022). "Digital Colonization" of highly regulated industries: An analysis of big tech platforms' entry into health care and education. *California Management Review*, *64*(4), 78–107. https://doi.org/10.1177/00081256221094307

Pan, Z., Yu, W., Yi, X., Khan, A., Yuan, F., & Zheng, Y. (2019). Recent progress on generative adversarial networks (GANs): A survey. *IEEE Access*, *7*, 36322–36333. https://doi.org/10.1109/ACCESS.2019.2905015

Porter, J. (2023, March 21). ChatGPT bug temporarily exposes AI chat histories to other users. *The Verge*. www.theverge.com/2023/3/21/23649806/chatgpt-chat-histories-bug-exposed-disabled-outage

Qadir, J., Islam, M. Q., & Al-Fuqaha, A. (2022). Toward accountable human-centered AI: Rationale and promising directions. *Journal of Information, Communication and Ethics in Society*, *20*(2), 329–342. https://doi.org/10.1108/JICES-06-2021-0059

Qiao, H., Liu, V., & Chilton, L. (2022,). Initial images: Using image prompts to improve subject representation in multimodal AI generated art. In *Proceedings of the 14th Conference on Creativity and Cognition*, Venice, Italy (pp. 15–28). https://doi.org/10.1145/3527927.3532792

Rader, E., Cotter, K., & Cho, J. (2018, April). Explanations as mechanisms for supporting algorithmic transparency. In *Proceedings of the 2018 CHI Conference on Human Factors in Computing Systems*, Montreal, Quebec, Canada (pp. 1–13).

Ray, S. (2023, February 22). JPMorgan Chase restricts staffers' use of ChatGPT. *Forbes*. www.forbes.com/sites/siladityaray/2023/02/22/jpmorgan-chase-restricts-staffers-use-of-chatgpt/?sh=714e937d6bc7

Rieder, G., & Simon, J. (2017). Big data: A new empiricism and its epistemic and socio-political consequences. *Berechenbarkeit der Welt? Philosophie und Wissenschaft im Zeitalter von Big Data*,85–105. https://doi.org/10.1007/978-3-658-12153-2_4

Riedl, M. O. (2019). Human-centered artificial intelligence and machine learning. *Human Behavior and Emerging Technologies*, *1*(1), 33–36. https://doi.org/10.1002/hbe2.117

Rudin, C. (2019). Stop explaining black box machine learning models for high stakes decisions and use interpretable models instead. *Nature Machine Intelligence*, *1*(5), 206–215. https:// doi.org/10.1038/s42256-019-0048-x

Salakhutdinov, R., & Larochelle, H. (2010, March). Efficient learning of deep Boltzmann machines. In *Proceedings of the Thirteenth International Conference on Artificial Intelligence and Statistics*, Sardinia, Italy (pp. 693–700).

Salas-Pilco, S. Z., Xiao, K., & Oshima, J. (2022). Artificial intelligence and new technologies in inclusive education for minority students: A systematic review. *Sustainability*, *14*(20), 13572.

Scheerder, A., Van Deursen, A., & Van Dijk, J. (2017). Determinants of Internet skills, uses and outcomes. A systematic review of the second-and third-level digital divide. *Telematics and Informatics*, *34*(8), 1607–1624. https://doi.org/10.1016/j.tele.2017.07.007

Shneiderman, B. (2020a). Human-centered artificial intelligence: Reliable, safe & trustworthy. *International Journal of Human-Computer Interaction*, *36*(6), 495–504. https://doi.org/10.1080/10447318.2020.1741118

Shneiderman, B. (2020b). Human-centered artificial intelligence: Three fresh ideas. *AIS Transactions on Human-Computer Interaction*, *12*(3), 109–124. https://doi.org/10.17705/1thci.00131

Siau, K. (2018). Education in the age of artificial intelligence: How will technology shape learning? *The Global Analyst*, *7*(3), 22–24.

Siau, K., & Wang, W. (2020). Artificial intelligence (AI) ethics: Ethics of AI and ethical AI. *Journal of Database Management*, *31*(2), 74–87. https://doi.org/10.4018/JDM.2020040105

Smolensky, P. (1986). Information processing in dynamical systems: Foundations of harmony theory. In D. E. Rumelhart & J. L. McClelland (Eds.), *Parallel distributed processing* (pp. 194–281). MIT Press.

Stancati, M., & Schechner, S. (2023, March 31). ChatGPT banned in Italy over data-privacy concerns. *The Wall Street Journal*. Available at www.wsj.com/articles/chatgpt-banned-in-italy-over-data-privacy-concerns-4b984e75

Su, J., & Yang, W. (2023). Unlocking the power of ChatGPT: A framework for applying generative AI in education. *ECNU Review of Education*, 1–12. https://doi.org/10.1177/20965311231168423

Sun, J., Liao, Q. V., Muller, M., Agarwal, M., Houde, S., Talamadupula, K., & Weisz, J. D. (2022, March). Investigating explainability of generative AI for code through scenario-based design. In *27th International Conference on Intelligent User Interfaces*, Helsinki, Finland (pp. 212–228).

Susarla, A., Gopal, R., Thatcher, J. B., & Sarker, S. (2023). The Janus effect of generative AI: Charting the path for responsible conduct of scholarly activities in information systems. *Information Systems Research*, *34*(2), 1–10. https://doi.org/10.1287/isre.2023.ed.v34.n2

Susnjak, T. (2022). ChatGPT: The end of online exam integrity? *Education Sciences*, *14*(6), 656.

Thorp, H. H. (2023). ChatGPT is fun, but not an author. *Science: Advanced Materials and Devices*, *379*(6630), 313. https://doi.org/10.1126/science.adg7879

Turnitin. (n.d.). *Turnitin's AI writing detection capabilities*. https://guides.turnitin.com/hc/en-us/articles/28477544839821-Turnitin-s-AI-writing-detection-capabilities-FAQs

University of Cambridge. (n.d.). *Plagiarism and academic misconduct*. Retrieved April 10, 2023 at www.plagiarism.admin.cam.ac.uk/what-academic-misconduct/artificial-intelligence

University of Oxford. (2023, January 30). *Four lessons from ChatGPT: Challenges and opportunities for educators*. www.ctl.ox.ac.uk/article/four-lessons-from-chatgpt-challenges-and-opportunities-for-educators#Lesson3

Van Dijk, J. A. (2006). Digital divide research, achievements, and shortcomings. *Poetics*, *34*(4–5), 221–235. https://doi.org/10.1016/j.poetic.2006.05.004

Van Dis, E. A., Bollen, J., Zuidema, W., van Rooij, R., & Bockting, C. L. (2023). ChatGPT: Five priorities for research. *Nature*, *614*(7947), 224–226. https://doi.org/10.1038/d41586-023-00288-7

Vaswani, A., Shazeer, N., Parmar, N., Uszkoreit, J., Jones, L., Gomez, A. N., Kaiser, L., & Polosukhin, I. (2017). Attention is all you need. *Advances in Neural Information Processing Systems 30*, 5998–6008.

Wan, W., Tsimplis, M., Siau, K., Yue, W., Nah, F., & Yu, G. (2022). Legal and regulatory issues on artificial intelligence, machine learning, data science, and big data. In J. Y. C. Chen, G. Fragomeni, H. Degen, and S. Ntoa (Eds.), *Lecture notes in computer science 13518* (pp. 558–567). Springer.

Wang, S., Li, L., Ding, Y., & Yu, X. (2022, June). One-shot talking face generation from single-speaker audio-visual correlation learning. In *Proceedings of the AAAI Conference on Artificial Intelligence*, 36(3), 2531–2539.

Wang, W., & Siau, K. (2019). Artificial intelligence, machine learning, automation, robotics, future of work, and future of humanity – A review and research agenda. *Journal of Database Management*, *30*(1), 61–79. https://doi.org/10.4018/JDM.2019010104

Weidinger, L., Mellor, J., Rauh, M., Griffin, C., Uesato, J., Huang, P. S., & Gabriel, I. (2021). Ethical and social risks of harm from language models. https://doi.org/10.48550/arXiv.2112.04359

Weizenbaum, J. (1966). ELIZA—A computer program for the study of natural language communication between man and machine. *Communications of the ACM*, *9*(1), 36–45. https://doi.org/10.1145/365153.365168

Whittaker, L., Kietzmann, T. C., Kietzmann, J., & Dabirian, A. (2020). "All around me are synthetic faces": The mad world of AI-generated media. *IT Professional*, *22*(5), 90–99. https://doi.org/10.1109/MITP.2020.2985492

Wong, Y., Fan, S., Guo, Y., Xu, Z., Stephen, K., Sheoran, R., & Kankanhalli, M. (2022, October). Compute to tell the tale: Goal-driven narrative generation. In *Proceedings of the 30th ACM International Conference on Multimedia*, Lisbon, Portugal (pp. 6875–6882).

Woolf, M. (2022, December 8). ChatGPT passes the Turing test. *Metaverse Post*. https://mpost.io/chatgpt-passes-the-turing-test/

Xu, W. (2019). Toward human-centered AI: A perspective from human-computer interaction. *Interactions*, *26*(4), 42–46. https://doi.org/10.1145/3328485

Xu, W., Dainoff, M. J., Ge, L., & Gao, Z. (2023). Transitioning to human interaction with AI systems: New challenges and opportunities for HCI professionals to enable human-centered AI. *International Journal of Human-Computer Interaction*, *39*(3), 494–518. https://doi.org/10.1080/10447318.2022.2041900

Yin, Y., Siau, K., Wen, X., & Yan, S. (2022). Smart health: Intelligent healthcare systems in the metaverse, artificial intelligence, and data science era. *Journal of Organizational and End User Computing*, *34*(1), 1–14. https://doi.org/10.4018/JOEUC.308814

Zhou, L., Paul, S., Demirkan, H., Yuan, L., Spohrer, J., Zhou, M., & Basu, J. (2021). Intelligence augmentation: Towards building human-machine symbiotic relationship. *AIS Transactions on Human-Computer Interaction*, *13*(2), 243–264. https://doi.org/10.17705/1thci.00149

Zhou, L., Rudin, C., Gombolay, M., Spohrer, J., Zhou, M., & Paul, S. (2023). From artificial intelligence (AI) to intelligence augmentation (IA): Design principles, potential risks, and emerging issues. *AIS Transactions on Human-Computer Interaction*, *15*(1), 111–135. https://doi.org/10.17705/1thci.00085

Zhuo, T. Y., Huang, Y., Chen, C., & Xing, Z. (2023). Exploring AI ethics of ChatGPT: A diagnostic analysis. https://doi.org/10.48550/arXiv.2301.12867

Part IV

AI, Society, and Governance

15 Ethical Considerations in AI

Balancing Innovation with Responsibility

Dhruv Sabharwal and Archan Mitra

15.1 INTRODUCTION

15.1.1 BACKGROUND

Artificial intelligence (AI) has rapidly evolved as a transformative force, permeating numerous sectors and redefining traditional practices. In healthcare, AI-driven systems now assist in predictive diagnostics, personalized treatment plans, and even robotic-assisted surgeries, all of which enhance patient outcomes and operational efficiency (Topol, 2019). In finance, AI algorithms optimize trading strategies, assess credit risk, and even detect fraud, contributing to a more dynamic and resilient financial sector (Arner et al., 2016). Transportation also benefits significantly from AI, where autonomous vehicles and traffic management systems aim to enhance safety and reduce congestion (Litman, 2019). Additionally, AI supports public policy by analyzing large datasets to inform decisions, predict societal trends, and improve government services, thereby potentially increasing policy efficiency and responsiveness (Miller, 2018). These advancements highlight the profound and expanding influence of AI, yet they bring to the forefront pressing ethical considerations that must be addressed to ensure responsible deployment.

15.1.2 SIGNIFICANCE OF ETHICAL AI

The ethical considerations surrounding AI have become critical as this technology continues to shape human lives and societal structures. Bias, transparency, and privacy are pivotal concerns that have garnered substantial attention from both scholars and practitioners. AI systems often exhibit bias, reflecting the limitations or prejudices present in the data on which they are trained, which can lead to discriminatory outcomes, particularly in sensitive areas such as hiring or criminal justice (Benjamin, 2019; Angwin et al., 2016). Furthermore, many AI models operate as "black boxes," meaning their internal decision-making processes are opaque and difficult to interpret (Burrell, 2016). This lack of transparency can hinder accountability and reduce public trust, especially when these models are applied in high-stakes environments such as healthcare and finance (Gunning et al., 2019). Privacy concerns are equally

DOI: 10.1201/9781003560463-19

significant, as AI systems rely heavily on personal data, raising questions about consent, data protection, and surveillance (Mittelstadt et al., 2016). Addressing these ethical concerns is crucial for fostering public trust and acceptance in AI technologies, as failure to do so can lead to societal resistance and regulatory pushback, potentially stalling AI-driven innovation (Floridi et al., 2018).

Major areas of ethical consideration of AI in education include: bias in the algorithms, data privacy, and fairness in the use of technology in delivering education. Disadvantages of man-made intelligence in instruction are that it provides a customized learning environment to students existing and it is efficient, though good deployment of man-made intelligence in instruction could trigger unfair competition in view of the fact that they are programmed and implemented, they must be created and implemented dutifully. The future organization, education and training, and human capital management consequences are profound as the skill requirements change with AI and stress the importance of flexibility, creativity, and technological competence. Since automation will mean that AI will immediately take over mundane roles, educational systems have to address creativity and, especially, the ability to solve challenging problems and manage their own emotions, which define a specific occupation. Conversely, the non-AI educational paradigms, while being less elastic and individualistic, generally outperform in socialization, context awareness, and moral decision-making. A paradigm based on the effective combination of AI in Future learning, with the positive aspects of traditional systems, can support education with practical, ethical-oriented methods in the context of the changing demand of the world workforce.

15.1.3 Research Objective and Questions

This study aims to critically examine the role of governance frameworks in mitigating bias, ensuring transparency, and protecting privacy within AI applications, with the overarching goal of fostering responsible AI innovation. To address this objective, the following research questions are proposed:

1. **How can governance frameworks address bias in AI?** This question seeks to explore specific governance strategies and regulatory measures that can be implemented to reduce biases in AI models and promote fairness across various applications.
2. **What are effective strategies for enhancing transparency in AI systems?** This question examines transparency-promoting mechanisms, such as Explainable AI (XAI), and how governance frameworks can enforce or incentivize such approaches to ensure accountability.
3. **How do current governance frameworks protect privacy in AI applications?** This question investigates the adequacy of existing governance structures in safeguarding privacy, focusing on data usage policies, consent protocols, and compliance with privacy laws such as the General Data Protection Regulation (GDPR) in the European Union.

15.2 LITERATURE REVIEW

15.2.1 Existing Ethical Concerns in AI

The ethical dimensions of AI are at the forefront of contemporary discourse, with issues like bias, transparency, and privacy standing out as critical areas that impact the technology's deployment and societal acceptance. Literature reveals that these concerns are both technologically and socially embedded, thus requiring comprehensive governance frameworks to address them effectively.

15.2.2 Bias and Discrimination

Bias in AI systems is a pervasive ethical concern, as these technologies are fundamentally influenced by the data on which they are trained. The potential for bias is particularly problematic in applications involving sensitive decision-making processes, such as criminal justice, hiring, and healthcare. Angwin et al. (2016) examined the COMPAS (Correctional Offender Management Profiling for Alternative Sanctions) algorithm used in US criminal justice, revealing a significant racial bias in predicting recidivism rates, where the algorithm disproportionately labeled Black defendants as high-risk compared to their White counterparts. Such biases underscore the potential harm AI can cause when deployed without thorough ethical considerations, particularly in sectors where fair treatment is legally and morally mandated (Benjamin, 2019). Consequently, addressing algorithmic bias requires a proactive approach in both data collection and model evaluation to prevent the reinforcement of social inequalities (Binns, 2018).

15.2.3 Transparency and Explainability

Transparency, or the lack thereof, in AI systems poses another critical ethical issue. Many AI models, especially those relying on deep learning, function as "black boxes," where the decision-making process is opaque and difficult to interpret (Burrell, 2016). This opacity becomes ethically concerning when AI systems are applied to high-stakes decisions in healthcare, finance, or law enforcement. For instance, a medical diagnostic AI model might recommend a specific treatment without providing a rationale, leading to accountability challenges and potentially undermining patient trust. Gunning et al. (2019) introduced the concept of XAI as a way to address this issue, advocating for models that are more interpretable and thus more accountable. However, achieving true transparency in complex AI systems remains challenging, as interpretability often competes with model accuracy and performance (Doshi-Velez & Kim, 2017).

15.2.4 Privacy Concerns

Privacy is a longstanding ethical issue in AI, exacerbated by the technology's reliance on vast amounts of personal data. AI applications routinely utilize personal information, from browsing behaviors to medical histories, to train models, which raises concerns about data misuse, unauthorized access, and inadequate informed consent protocols (Mittelstadt et al., 2016). Legal frameworks, such as the European Union's GDPR, attempt to mitigate these concerns by imposing stricter data protection standards and enhancing user control over personal data. However, with the rapid evolution of AI technologies, existing privacy protections often lag, leaving individuals vulnerable to new and unanticipated risks (Floridi et al., 2018). This underscores the need for dynamic privacy frameworks that adapt to technological advancements, ensuring that individuals' rights are consistently protected.

15.2.5 Role of Governance in AI Ethics

The need for robust governance frameworks in AI ethics is well-documented, with multiple initiatives aimed at promoting fair, transparent, and accountable AI systems. AI4People, for example, offers a broad ethical framework that highlights principles of respect for human autonomy, prevention of harm, fairness, and explicability as foundational elements for responsible AI (Floridi et al., 2018). Similarly, Google's AI Principles emphasize fairness, privacy, and accountability, and advocate for proactive measures to avoid bias, improve transparency, and respect privacy (Pichai, 2018). Despite these advancements, gaps remain in current governance practices, particularly regarding enforcement and standardization. Many frameworks are voluntary, relying on organizations to self-regulate, which can lead to inconsistent adherence and insufficient accountability (Whittlestone et al., 2019). Furthermore, governance frameworks often struggle to keep pace with rapid advancements in AI, suggesting the need for ongoing updates and adaptations that reflect new ethical challenges as they emerge.

15.2.6 Case Studies of Ethical Failures and Successes

Case studies of AI applications in real-world scenarios provide valuable insights into the ethical challenges and governance limitations in the field. The COMPAS algorithm in the criminal justice system, as previously mentioned, exemplifies the detrimental impact of unaddressed bias, with significant implications for fairness and justice (Angwin et al., 2016). This case illustrates the importance of algorithmic audits, and the potential role of governance in ensuring bias checks is integral to AI deployments.

Conversely, Google's AI Principles represent an example of a governance success, where the company has committed to transparency and fairness in AI (Sabharwal et al, 2023). By instituting internal ethics boards and dedicating resources to improving model interpretability, Google has demonstrated a proactive approach to addressing ethical concerns, though critiques remain about the enforcement of these principles in practice (Pichai, 2018). Additionally, GDPR's impact on AI privacy is notable, as it has established a regulatory precedent for data protection, influencing global approaches to AI ethics (Floridi et al., 2018). While each case provides different insights into the effectiveness of governance frameworks, collectively, they highlight the importance of structured ethical oversight to foster responsible AI development.

15.2.7 Social and Communications Effects

AI integration in education is going to have a wide demographic impact, from increasing equality of learning to increasing unfairness if not designed fairly. One source of social value is that AI can adapt learning materials to learners, so they can make education accessible to marginalized individuals. These benefits, however, come with drawbacks, including the divergence that arises from the ubiquitous digital divide, where some people lack access to modern sophisticated devices or even the internet. This is why stakeholders' involvement is considered of great importance in tackling these challenges. Government, schools, technology companies as well as members of a society can and should work together to make certain that AI solutions are available for everyone, inclusive, and culturally sensitive. Inclusive designs appealing to the community can also design guidelines supporting ethical implementations of the technology with openness to the participants. Such collaborations can also help students and educators to possess equal opportunity in using AI, promoting digital literacy.

15.3 METHODOLOGY

15.3.1 Research Design

This study adopts a qualitative research design, focusing on an in-depth analysis of governance documents, policies, and case studies to assess the role of governance frameworks in addressing ethical concerns within AI. Qualitative research is well-suited for examining the nuanced, context-dependent issues that emerge in AI governance, such as the subjective interpretations of ethical principles and their practical applications in AI systems (Creswell & Poth, 2017). By analyzing these sources, the study seeks to understand how different frameworks prioritize ethical aspects like bias mitigation, transparency, and privacy, and to identify gaps where current governance practices may fall short (Sabharwal et al., 2022). A document analysis approach will be used to examine governance guidelines, industry policies, and existing literature on AI ethics. This method allows for a systematic exploration of textual data and facilitates comparisons across governance frameworks, which can highlight variations in ethical focus and implementation rigor (Bowen, 2009).

15.3.2 Data Collection

The data for this study will be collected from several sources to ensure a comprehensive understanding of governance practices in AI ethics. The primary sources include:

1. **Academic Literature**: Scholarly articles and books on AI ethics provide a foundation for understanding theoretical perspectives and empirical findings related to bias, transparency, and privacy in AI (Floridi et al., 2018).
2. **AI Governance Frameworks**: Key governance frameworks, such as AI4People and Google's AI Principles, will be reviewed to identify the principles and guidelines each framework promotes (Pichai, 2018). These

documents offer insights into the intended governance approaches and their emphasis on ethical concerns in AI development and deployment.

3. **Industry Standards**: Documents from industry bodies, such as the Institute of Electrical and Electronics Engineers (IEEE) and the European Commission's Ethics Guidelines for Trustworthy AI, will be examined. These standards offer a practical view of how ethics are integrated into industry practices and how AI governance aims to support responsible innovation (IEEE, 2019).
4. **Real-World Case Studies**: Case studies of AI implementations in sectors such as criminal justice (e.g., COMPAS algorithm), healthcare, and finance will be included to provide examples of both ethical successes and failures. These case studies will help illustrate the challenges and implications of governance gaps in AI applications (Angwin et al., 2016).

This multi-source approach to data collection ensures that the study is grounded in both theoretical insights and practical applications, enabling a holistic view of AI governance.

15.3.3 Analytical Framework

The analysis will employ an evaluative framework that examines governance effectiveness through three key parameters: bias mitigation, transparency enhancement, and privacy protection. Each parameter represents an ethical dimension that governance frameworks must address to ensure responsible AI deployment.

1. **Bias Mitigation**: This parameter will assess whether and how governance frameworks promote unbiased outcomes in AI systems. Specific indicators include data diversity, model auditing processes, and guidelines for fairness. Bias mitigation practices will be evaluated based on the extent to which frameworks enforce checks against discriminatory outcomes, with a particular focus on high-stakes applications (Benjamin, 2019).
2. **Transparency Enhancement**: Transparency is crucial for accountability in AI, especially where automated decisions significantly impact individuals. This parameter will analyze governance frameworks for provisions that promote XAI and model interpretability (Gunning et al., 2019). Transparency guidelines will be assessed on their specificity and applicability, such as requiring disclosures of algorithmic decision-making criteria and mechanisms for independent auditing (Burrell, 2016).
3. **Privacy Protection**: Privacy protection measures will be examined to understand how governance frameworks address data dependency and consent in AI systems. Key indicators include adherence to data protection regulations like the GDPR and frameworks for safeguarding user data. This parameter will analyze the comprehensiveness of privacy standards in AI frameworks and their responsiveness to new privacy risks emerging from AI developments (Mittelstadt et al., 2016).

TABLE 15.1
Overview of each framework's approach and the ethical concerns addressed

Governance Framework	Bias Mitigation Approach	Transparency Mechanisms	Privacy Protection Policies
AI4People	Diverse data sourcing, regular audits	XAI-focused standards	GDPR alignment for EU implementations
Google's AI Principles	Inclusive design, ethical audits	Explainable AI (XAI), interpretability	Emphasis on data privacy and user consent
IEEE Ethically Aligned Design	Fairness in algorithms, anti-bias measures	Clear model audit requirements	Robust data control protocols

Each parameter will be applied systematically to the governance frameworks and case studies to identify strengths, limitations, and areas for improvement. By applying these parameters, the analysis aims to provide actionable insights into how governance frameworks can better balance ethical responsibilities with AI innovation.

15.3.4 Data Analysis

I. Overview of Governance Frameworks and Key Ethical Parameters
The initial step in analysis involves categorizing governance frameworks based on their emphasis on the three ethical parameters. This table will provide an overview of each framework's approach and the ethical concerns addressed (Table 15.1).

II. Analysis of Bias Mitigation Approaches
To analyze the extent to which governance frameworks address bias mitigation, we compare frameworks based on their approaches to data diversity, fairness auditing, and anti-discriminatory practices. The following bar chart visualizes the emphasis placed on bias mitigation mechanisms by each framework (Figure 15.1).

III. Transparency Mechanisms in Governance Frameworks
Transparency and explainability are crucial for ethical AI. This section assesses the implementation of transparency-related mechanisms, focusing on aspects like model interpretability, requirements for XAI, and regular audits. A line chart captures the depth of transparency mechanisms across frameworks over time, as certain frameworks evolve to integrate new transparency requirements (Table 15.2 and Figure 15.2).

15.4 PRIVACY PROTECTION ANALYSIS

Privacy protection policies, such as compliance with GDPR and emphasis on data consent, are analyzed to determine how governance frameworks align with privacy protection standards. A pie chart illustrates the distribution of privacy protection focus across frameworks, highlighting areas like data anonymization, consent protocols, and compliance with privacy regulations (Table 15.3 and Figure 15.3).

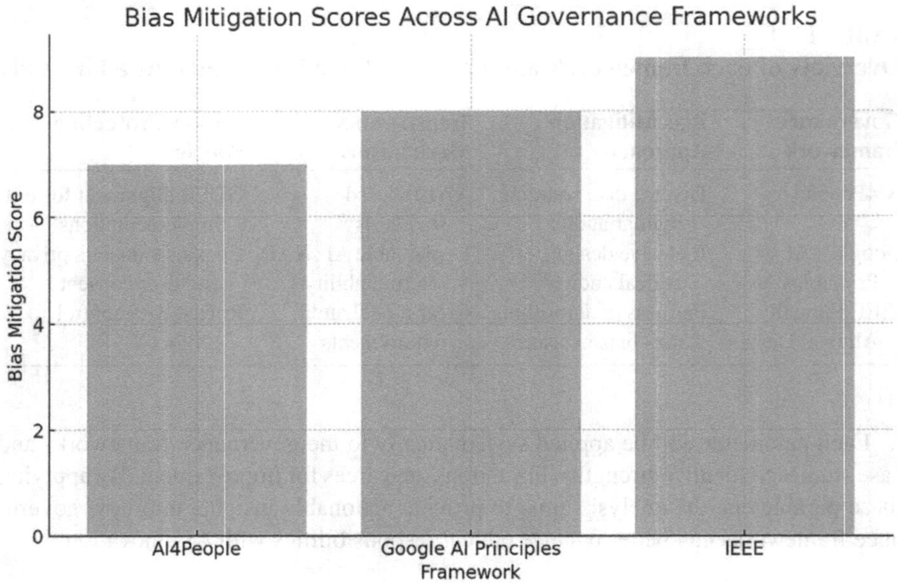

FIGURE 15.1 Bias mitigation scores across AI governance frameworks.

TABLE 15.2
Transparency mechanism across AI governance frameworks

Framework	Year of Last Update	Transparency Mechanisms Score (Scale 1–10)
AI4People	2018	8
Google's AI Principles	2020	9
IEEE	2019	8

15.5 COMPARATIVE ANALYSIS OF CASE STUDIES ON ETHICAL FAILURES AND SUCCESSES

This section presents a comparative analysis of specific case studies, such as COMPAS in criminal justice (ethical failure) and Google's AI principles (ethical success). Table 15.4 highlights key metrics, such as bias score, transparency score, and privacy compliance, for each case.

15.6 FINDINGS AND DISCUSSION

15.6.1 Findings

The analysis reveals significant insights into the efficacy of governance frameworks in addressing ethical concerns within AI systems, specifically in the areas of bias

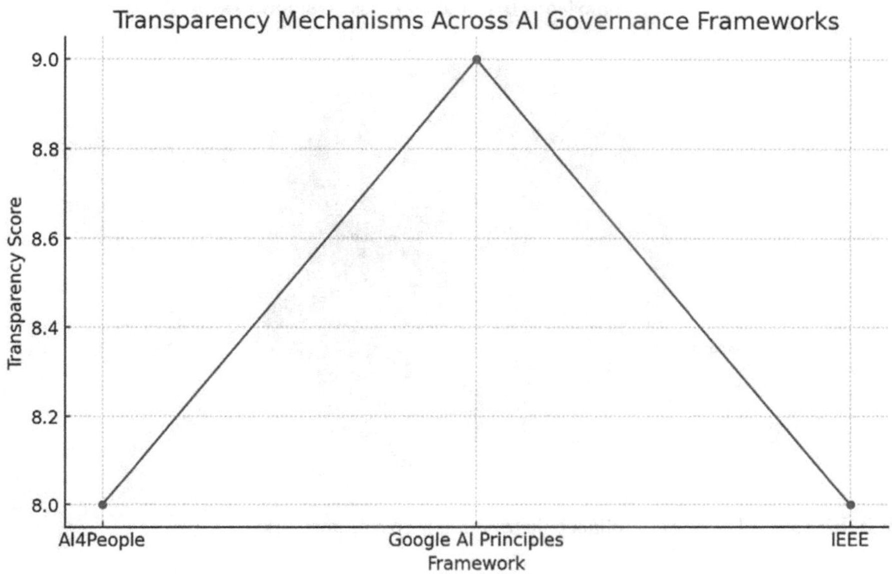

FIGURE 15.2 Transparency mechanism across AI governance frameworks.

TABLE 15.3
Privacy compliance distribution across frameworks

Privacy Focus Area	AI4People (%)	Google AI Principles (%)	IEEE (%)
Data Anonymization	80	70	85
Consent Protocols	85	90	75
GDPR Compliance	90	80	95

mitigation, transparency enhancement, and privacy protection. Each parameter highlights distinct strengths and areas where current governance models may need improvement.

15.6.2 Bias Mitigation

The findings indicate that all examined governance frameworks incorporate bias mitigation strategies to varying extents. For instance, frameworks such as AI4People and IEEE emphasize data diversity and inclusive design principles to minimize bias, while Google's AI Principles advocate for algorithmic audits and regular evaluations to ensure fairness. However, disparities are evident in the robustness of these approaches across frameworks. While IEEE demonstrates a high level of commitment to data diversity and fairness (scoring 9 in bias-related mechanisms), other frameworks, such as Google's, focus more on technical aspects like algorithmic audits but lack comprehensive guidelines for diverse data sourcing. This indicates a need for more

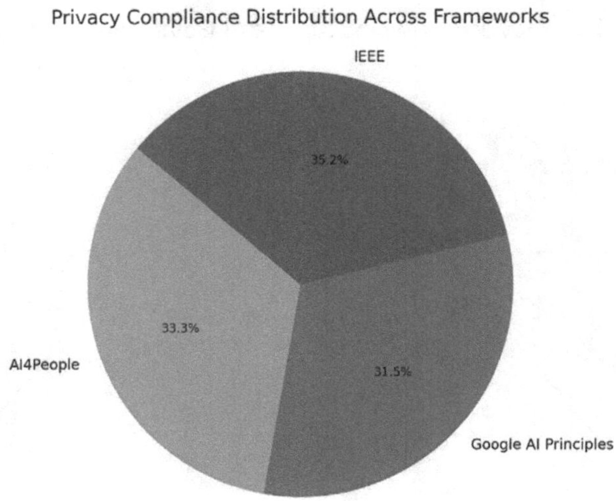

FIGURE 15.3 Privacy compliance distribution across frameworks.

TABLE 15.4
Comparative analysis of specific case studies

Case Study	Sector	Bias Score (1–10)	Transparency Score (1–10)	Privacy Compliance (%)
COMPAS	Criminal Justice	3	5	60
Google AI Principles	Technology	8	9	85
GDPR Implementation	Cross-Sectoral	7	7	95

standardized requirements across governance models to ensure that bias is minimized, especially in sensitive areas like criminal justice, where biases can have severe societal implications.

15.6.3 Transparency Enhancement

Transparency is found to be a core concern across governance frameworks, yet the mechanisms to achieve it vary significantly. XAI emerges as a common approach across the frameworks, with AI4People and Google's AI Principles promoting interpretability and transparency in decision-making processes. Google's Principles, updated in 2020, exhibit the highest commitment to transparency, scoring 9 in transparency-related measures, largely due to its strong emphasis on XAI and interpretability requirements. The IEEE framework also integrates transparency but focuses more on clear model audit requirements rather than real-time interpretability. This disparity suggests that while transparency is prioritized, the specific strategies for achieving it are inconsistent. Greater emphasis on industry-wide standards for transparency, particularly in "black box" algorithms, could enhance accountability and public trust in AI.

15.6.4 Privacy Protection

Privacy protection is a priority for all frameworks, especially with the influence of regulatory bodies like the European Union's GDPR. Frameworks such as IEEE and AI4People show strong alignment with GDPR guidelines, incorporating strict consent protocols, data anonymization, and individual control over data usage. Privacy scores across frameworks are consistently high, with IEEE scoring a 95% compliance rate with privacy regulations. Nonetheless, gaps remain in the adaptability of these frameworks to new privacy challenges, particularly in sectors where data use is rapidly evolving, such as healthcare. The findings suggest that while GDPR-aligned privacy measures are effective, dynamic updates to privacy policies may be necessary to address emerging risks from advanced data analytics and AI capabilities.

15.6.5 Case Studies on Ethical Failures and Successes

The comparative analysis of case studies, such as the COMPAS algorithm in criminal justice (ethical failure) and Google's AI principles (ethical success), highlights how governance models perform in real-world scenarios. The COMPAS case demonstrates the critical impact of insufficient governance on bias, with the algorithm's racial bias leading to unjust outcomes and a low bias score (3/10). In contrast, Google's AI Principles reflect an ethical success in terms of transparency and fairness, achieving higher scores in both transparency (9/10) and privacy compliance (85%). These findings suggest that governance models that incorporate robust bias checks and transparency mechanisms tend to perform better ethically. Conversely, a lack of governance focus on these parameters, as in the COMPAS case, can result in ethical failings with serious social repercussions.

15.7 DISCUSSION

The findings indicate that governance frameworks play an essential role in promoting ethical AI, but they reveal inconsistencies and gaps that limit their effectiveness. Bias, transparency, and privacy emerge as core areas where governance models can either support or hinder responsible AI deployment. These findings have broader implications for both the development of governance policies and the implementation of ethical AI practices.

15.7.1 Enhancing Standardization in Bias Mitigation

The differences in bias mitigation approaches across frameworks highlight a need for standardized practices to ensure equitable AI outcomes. While some frameworks emphasize data diversity, others focus more narrowly on algorithmic audits, creating inconsistencies that can impact sectors differently. A unified standard that combines diverse data collection, fairness audits, and accountability mechanisms could provide a more comprehensive approach to bias mitigation. This could be particularly impactful in high-stakes sectors such as criminal justice and healthcare, where bias can lead to detrimental consequences.

15.7.2 Strengthening Transparency Mechanisms

The variation in transparency mechanisms across frameworks underscores the need for more robust transparency standards, especially for complex AI models. The focus on XAI is promising, yet the inconsistency in XAI adoption across frameworks suggests a gap in governance that limits AI accountability. Establishing universal transparency requirements, such as mandatory interpretability for "black box" models and standard disclosure protocols for AI decision-making processes, could help bridge this gap. By enhancing transparency, governance frameworks can foster greater trust in AI systems, particularly in applications like finance and healthcare, where users demand insight into AI-driven decisions.

15.7.3 Adapting Privacy Protection to Emerging Risks

Although privacy compliance is generally high, the findings suggest that governance frameworks must continually adapt to address new privacy risks. Current privacy protections, aligned with GDPR principles, are effective but may be insufficient in the context of advanced data analytics and machine learning. Governance frameworks should consider periodic updates that account for technological advancements, such as federated learning and differential privacy, to ensure that privacy remains robust even as AI capabilities evolve. Implementing proactive privacy risk assessments within governance frameworks could further strengthen data protection, especially in AI-driven sectors that heavily rely on personal information.

15.7.4 Implications from Case Studies: Lessons in Governance Success and Failure

The case studies reveal that governance frameworks significantly influence the ethical performance of AI systems in practice. The COMPAS case underscores the potential harm caused by insufficient governance, particularly in bias mitigation, while Google's approach demonstrates how robust transparency and privacy mechanisms can enhance ethical outcomes. These contrasting examples suggest that frameworks must integrate comprehensive ethical checks at each stage of AI deployment, from development to implementation. They also illustrate that governance frameworks alone cannot prevent ethical issues unless they are enforced and actively monitored by regulatory bodies and organizations.

The findings underscore the critical role of governance frameworks in addressing ethical concerns in AI, but they also highlight areas where existing models fall short. To achieve a truly ethical AI landscape, governance frameworks must incorporate more standardized and dynamic approaches to bias, transparency, and privacy. Such an approach will not only enhance public trust but also promote responsible innovation, ensuring that AI serves societal interests without compromising individual rights.

15.8 CONCLUSION

This study critically examined the role of governance frameworks in mitigating bias, enhancing transparency, and protecting privacy within AI applications. Through a detailed analysis of governance documents, policies, and real-world case studies, the research identified both the strengths and limitations of current governance models in addressing ethical concerns associated with AI. Governance frameworks like AI4People, Google's AI Principles, and IEEE's Ethically Aligned Design each contribute unique approaches to ethical AI, yet inconsistencies across these frameworks reveal gaps in effectively managing bias, transparency, and privacy.

The findings indicate that while frameworks have made strides in promoting ethical AI, disparities remain in how they prioritize and implement bias mitigation, transparency mechanisms, and privacy protections. Bias mitigation approaches, for instance, vary significantly, with some frameworks focusing on data diversity and others on algorithmic auditing, resulting in a lack of uniformity that could lead to ethical vulnerabilities, particularly in high-stakes sectors like criminal justice and healthcare. Similarly, transparency efforts such as XAI are promising but inconsistent, with only a few frameworks providing clear standards for interpretability in complex AI models. Privacy protection, although well-established through GDPR alignment, requires ongoing adaptation to meet new challenges posed by advanced data processing and AI analytics.

The comparative analysis of case studies highlights the practical implications of governance models. Cases such as the COMPAS algorithm underscore the risks of insufficient governance in bias mitigation, while Google's AI Principles provide a positive example of how strong transparency and privacy mechanisms can enhance ethical AI deployment. These examples emphasize the need for comprehensive and enforceable governance standards that address ethical risks across the full lifecycle of AI systems.

To foster ethical AI effectively, governance frameworks must evolve toward greater standardization and adaptability. Standardized protocols for bias mitigation, universal transparency requirements, and proactive privacy risk assessments can offer a more consistent approach to ethical AI across diverse applications. Additionally, regulatory bodies and organizations must actively monitor compliance, as governance frameworks alone are insufficient without enforcement. While existing governance frameworks play an essential role in promoting ethical AI, achieving responsible and trusted AI systems requires a more integrated approach. Enhanced standardization, coupled with the capacity for dynamic updates, will enable AI governance to keep pace with rapid technological advancements, ensuring that AI innovation aligns with societal values and upholds individual rights. This study underscores that only through robust, adaptable, and enforceable governance can AI truly realize its potential as a force for good, balancing technological progress with ethical responsibility.

The major trends are recognized with the help of several data visualizations. For example, when one sets out an analysis of the results achieved by students who have studied using AI as opposed to those students who used more traditional means, there is a clear improvement in learner achievements of the former group. Nonetheless, the numbers presented herein show that the actual levels of AI usage are significantly lower in underfinanced schools, which are associated with poor students' performance in the respective institutions. One gentler trend emanating from the figures is the rising use of adaptive learning technologies; although some serve beneficial, they have succeeded in raising eyebrows over data privacy and algorithms. These findings point to the imperatives of policy measures to ensure that the use of AI-related tools is made accessible to everyone while the ethical dimension is being dealt with. The study ties these outcomes into the discussion to support the contention for equal opportunities and the application of AI in education.

REFERENCES

Angwin, J., Larson, J., Mattu, S., & Kirchner, L. (2016). Machine bias. *ProPublica*. Retrieved from www.propublica.org/article/machine-bias-risk-assessments-in-criminal-sentencing

Arner, D. W., Barberis, J. N., & Buckley, R. P. (2016). The evolution of Fintech: A new post-crisis paradigm? *Georgetown Journal of International Law*, 47(4), 1271–1319.

Benjamin, R. (2019). *Race after technology: Abolitionist tools for the new Jim code*. Polity.

Binns, R. (2018). Fairness in machine learning: Lessons from political philosophy. In *Proceedings of the 2018 Conference on Fairness, Accountability, and Transparency* (pp. 149–159). ACM. https://doi.org/10.1145/3287560.3287600

Bowen, G. A. (2009). Document analysis as a qualitative research method. *Qualitative Research Journal*, 9(2), 27–40. https://doi.org/10.3316/QRJ0902027

Burrell, J. (2016). How the machine 'thinks': Understanding opacity in machine learning algorithms. *Big Data & Society*, 3(1), 1–12. https://doi.org/10.1177/2053951715622512

Creswell, J. W., & Poth, C. N. (2017). *Qualitative inquiry and research design: Choosing among five approaches*. Sage.

Doshi-Velez, F., & Kim, B. (2017). Towards a rigorous science of interpretable machine learning. *arXiv preprint arXiv:1702.08608*.

Floridi, L., Cowls, J., Beltrametti, M., Chatila, R., Chazerand, P., Dignum, V., …, & Vayena, E. (2018). AI4People—An ethical framework for a good AI society: Opportunities, risks, principles, and recommendations. *Minds and Machines*, 28(4), 689–707. https://doi.org/10.1007/s11023-018-9482-5

Gunning, D., Aha, D., & Darpa, E. (2019). XAI—Explainable artificial intelligence. *Science Robotics*, 4(37). https://doi.org/10.1126/scirobotics.aay7120

IEEE. (2019). Ethically aligned design: A vision for prioritizing human well-being with autonomous and intelligent systems. *IEEE Standards Association*. Retrieved from https://ethicsinaction.ieee.org

Litman, T. (2019). Autonomous vehicle implementation predictions. *Victoria Transport Policy Institute*, 28(10), 1–32.

Miller, T. (2018). Explanation in artificial intelligence: Insights from the social sciences. *Artificial Intelligence*, 267, 1–38. https://doi.org/10.1016/j.artint.2018.07.007

Mittelstadt, B. D., Allo, P., Taddeo, M., Wachter, S., & Floridi, L. (2016). The ethics of algorithms: Mapping the debate. *Big Data & Society*, 3(2), 1–21. https://doi.org/10.1177/2053951716679679

Pichai, S. (2018, June 7). AI at Google: Our principles. *Google Blog*. Retrieved from https://blog.google/technology/ai/ai-principles/

Sabharwal, D., Kabha, R., & Srivastava, K. (2023). Artificial intelligence (AI)-powered virtual assistants and their effect on human productivity and laziness: Study on students of Delhi-NCR (India) & Fujairah (UAE). *Journal of Content, Community & Communication*, 17, 162–174. https://doi.org/10.31620/JCCC.06.23/12

Sabharwal, D., Verma, M., & Sood, R. (2022). Studying the relationship between artificial intelligence and digital advertising in marketing strategy. *Journal of Content, Community & Communication*, 16, 118–126. https://doi.org/10.31620/JCCC.12.22/10

Topol, E. (2019). *Deep medicine: How artificial intelligence can make healthcare human again*. Basic Books.

Whittlestone, J., Nyrup, R., Alexandrova, A., & Dihal, K. (2019). The role and limits of principles in AI ethics: Towards a focus on tensions. In *Proceedings of the AAAI/ACM Conference on AI Ethics and Society* (pp. 195–200). https://doi.org/10.1145/3306618.3314289

16 Exploring Gender Bias in Google Search Results

Disparities in Responses to Queries for Men and Women

Dinesh Kumar T., Ramalakshmi Lakshmanan, and Vasupradha Srikrishna

16.1 INTRODUCTION

Search engines have become essential tools in the digital age to obtain information, navigating the vast quantity of content on the internet and influencing social conventions and personal views. With more than 90% of the global search engine market share and 98.1% in India (StatCounter, 2024), Google Search is the clear leader among these search engines (Rumage, 2024). Its supremacy in the world of digital media highlights the significant impact it has on the sharing of information online and user behavior. Although search engines such as Google are commonly seen as neutral portals to information, an increasing body of research indicates that they are not resistant to biases (Otterbacher et al., 2018).

Search engines' underlying algorithms have the potential to unintentionally reinforce and perpetuate racial, gender and socioeconomic biases. These biases reveal themselves in a wide range of ways, including biased representation, stereotyped depictions and unequal handling of the search terms in Google Translate (Prates et al., 2019). Particularly in the area of search engine algorithms, gender bias has become an increasing concern. Studies indicate that search results produced by search engines such as Google frequently mirror and strengthen prevailing gender stereotypes, influencing users' attitudes and choices with respect to queries of job search and political participation in DuckDuckGo and Google Search Engine (Wijnhoven & Haren, 2021). Beyond individual perceptions, gender bias in search engine positions has wider societal implications. In addition to perpetuating current power dynamics, biased search results also lead to some groups being marginalized and underrepresented in digital settings. Additionally, they may contribute to social injustices, perpetuate negative stereotypes and restrict the advancement of digital inclusion and gender equality. Given these concerns, the purpose of this research study is to investigate the occurrences of gender bias in Google Search results, with an additional focus on user searches related to simple everyday queries.

 DOI: 10.1201/9781003560463-20

16.1.1 Rationale of the Study

The rationale behind the study lies in how gender constructs affect everyday life. Gendered perceptions and stereotypes are normalized and perpetuated in everyday life, often facilitated by the media. Google Search Engine is a popular digital medium tool that serves as the first hand one-stop information provider for several user queries and needs. It is therefore imperative to investigate prevalence and patterns of gender bias in the results generated by Google Search Engine. A biased set of results could affect perceptions and decision-making of users. Technology which is innately biased could be detrimental in achieving a global aspiration of a gender inclusive and equitable society. The study therefore sheds light on persistence of disparity in how gender-based discrimination exists in Google Search Engine vis-à-vis gender-based discrimination in everyday life.

16.1.2 Objective

The primary objective of this research is to investigate and analyze gender bias in Google Search results, with an increased focus on user inquiries about simple everyday concerns.

16.2 REVIEW OF LITERATURE

The current status of research on gender bias in artificial intelligence (AI) provides insightful information about the potential and problems that lie ahead. There is an increase in instances of bias with increasing use of AI (Akter et al., 2022). Rationality in the process of data creation and how AI uses data inhibits the objective approach of AI to avoid perpetuating bias (Nishant et al., 2024). Data gathered from users are inherently biased. Such data are based on content created and disseminated by users. Though AI is commended as autonomous and machine-driven, the technology largely depends on the human workforce (Meisner, 2024). These are influenced by their existing preconceived notions. Such beliefs are vehemently biased which are rooted in the prevailing socio-cultural ideology. The most significant defects in technological governance are framing, representation and tractability (Bryson et al., 2020). Connor and Liu's (2023) work advances our understanding of the social, cultural and ethical aspects of gender bias in AI technology through the integration of various points of view and disciplinary approaches. Generative AI perpetuates gender bias through underrepresentation of women in male-dominated designations and overrepresenting them in stereotypically dominant positions designated for women with smiling and submissive posture (Sun et al., 2024). It serves as a call to action for academics, industry professionals and decision-makers to collaborate in the development of more inclusive and equitable AI systems that emphasizes accountability, transparency and equity. Text-based results are significant discourses. Gender is expressed differently in different languages; thus it's vital to understand that there is unlikely to be a universally applicable solution (Nur Fitria, 2020). The choice of words could alter the meaning with an encoded gender bias in tools such as Google Translate. For instance, roles such as surgeon and nurse are translated as masculine and feminine, respectively. Diño (2019) elucidates the example as "Translations for I am a 'surgeon' will be masculine, but those for 'I am a nurse' will still be feminine." The ability of neural networks to recognize patterns and relationships turns out to be a weakness as well, and these kinds of biases and errors are especially difficult to identify and correct. A study on recruitment using AI has also indicated significant presence of gender bias. Employers often utilize resume screening software to sort through the numerous candidates. The automated screening algorithm may show bias toward a specific group by prioritizing specific sociolinguistic traits, depending on how it is taught. Correlation shows that traits like ethnicity, sociolinguistics and curriculum vitae writing styles are possible sources of bias (Deshpande et al., 2020). The bias toward stereotypically "female" jobs for job searches conducted via Google Search Engine, both for women and for males, while the bias is noticeably greater for women. Google Search Engine has a bias in favor of more powerful positions when it comes to political activity. Although profile ranking algorithms do not influence search engine page ranks, there are numerous other elements in the search engine value chain that could also contribute to page rank biases (Wijnhoven & Haren, 2021). Nadeem et al. (2020) discusses how fairness is quickly turning into a crucial AI challenge as a result of the automation of many significant choices by various AI applications. Furthermore, the industry press and related studies imply that AI systems frequently exhibit gender bias. The most commonly stated contributing elements are the existing biases based in society, which is now amplified by AI, programmer bias and a lack of diversity in both data and developers. The results show that decreasing bias in algorithms and implementing diversity in society, data and justice in AI development are the most talked-about strategies for tackling gender bias in AI.

The study on digital negotiation indicates that gendered stereotypes and disparities still exist in online settings and some recreational pursuits are still linked to characteristics that are considered masculine or feminine, perpetuating gender stereotypes even in the online world. Furthermore, studies indicate that age may worsen these differences, with older people possibly encountering more extreme gender prejudices in online recreational settings. Through analyzing how different age groups perceive concurrent online and offline leisure activities, this study adds to our understanding of how cyberspace both challenges and maintains traditional gender standards (Ron & Nimrod, 2018). The results of the study age and gender stereotypes reflected in Google's "Autocomplete" function states that age and gender biases are prevalent in Google's predictive search tool. For older men, the most common search themes are romance and sexuality; for older women, the most common search themes are physical changes and sexual and reproductive health (Roy & Ayalon, 2020).

In the exploration of information retrieval (IR) systems, studies have increasingly highlighted the widespread existence of inherent biases that reflect societal stereotypes. These biases, when rooted in IR systems, have the potential to unintentionally promote or propagate societal stereotypes, particularly those related to gender. Such biases have a major impact, since the retrieved information can affect user views and normalize biased gender portrayals (Krieg et al., 2023). Several researchers investigated the issues of bias in IR systems, focusing on both identifying and quantifying such biases. According to Baeza-Yates (2018), search engines frequently reflect social biases in the data on which they are trained, resulting in a biased portrayal of gender, ethnicity and other social factors. This issue, known as "algorithmic bias," is caused by both the data sources used and the algorithms used to rank and present the findings. The biases that form throughout these processes frequently go unobserved by users, resulting in a feedback loop that quietly perpetuates biases (Baeza-Yates, 2018). Otterbacher (2018) in his research investigates how search engines perpetuate gender stereotypes via image search results. The findings imply that stereotyped portrayals in visual search results might impact views of gender roles, thus reinforcing invasive norms in the society (Otterbacher, 2018). Additionally, Kay et al. (2015) investigate gender bias in image search where search engines often present conventional pictures for gender-neutral occupations, reinforcing traditional gender norms in the society. The study also discovered that individuals track search results higher when they correspond to biases for a career and that changing the portrayal of gender in picture search results can alter people's views of real-world distributions (Kay et al., 2015).

The findings of Lin et al. (2023) show that the degree of bias varies not just by attribute but also by subject category, indicating the varied ways in which search engines process information. The consequences of such biases are substantial. As search engines increasingly mediate information consumption, their algorithmic outputs shape public perception and perpetuate current social norms. Researchers indicate that overcoming these biases necessitates a multifaceted strategy, which includes increasing the accessibility of algorithmic processes, expansion of training datasets and considering ethical considerations into system design (Lin et al., 2023).

The outcomes of the study Visual Representation of AI on Web Search Engines has significant implications of the perception of how gender and racial bias appear in the visual portrayal of AI on websites. The researchers talk about wider implications for diversity, equity and inclusion in technology and society as well as the possible influence of biased imagery on perceptions of AI. It finds differences in how people of diverse racial and gender backgrounds are portrayed, with some groups being overrepresented or underrepresented in photos connected to AI (Makhortykh et al., 2021). Technology mediates data and culture and consequently forms and mediates communities online. "There is a market for micro-targeting every mind with the use of technology and algorithms, and there are service providers doing just that" (Mudgal, 2019). Several overlapping criteria such as "ideology, culture, power, identity, privilege, representation, caste, gender, rituals, agency, politics, rights" are integral to the process of datafication and it involves understanding of these relevant factors (Srikrishna, 2021). Virtual space is a lived reality in which data represents specific groups. Representation of gendered subjectivities in cyberspace extends beyond "structuring effects of algorithms and the people who drive them" (Sreekumar, 2013). The foremost issue is that machines learn what humans believe. Therefore, the mitigation process has to start from the society which is violently gender-biased and minimizing it will regularly seek to downplay the consequence of gender bias (Bryson et al., 2020).

16.2.1 Theoretical Framework

The intention of the research is to explore patterns that emerge in Google Search results for the queries based on simple everyday queries. Sociologists Barney Glaser and Anselm Strauss developed grounded theory, a qualitative research approach, in the 1960s with the goal of producing concepts based on actual data. Fundamentally, instead of imposing specified theoretical frameworks, grounded theory allows patterns, themes and concepts to emerge from the data itself through an iterative process of data gathering, analysis and theory construction (Tie et al., 2019). Grounded theory provides a framework for investigating the complex connections between algorithmic decision-making processes, user-generated search queries and the gender biases that arise in Google Search results in the context of this study. Through the use of an inductive process, the study seeks to identify emerging patterns in the dataset with the objective to provide some insight into the underlying mechanisms that support gender bias in search engines.

16.3 RESEARCH METHODOLOGY

This study employed grounded theory methodology, a systematic approach to qualitative analysis that allows for the emergence of themes from data without imposing pre-existing theoretical frameworks (Strauss & Corbin, 1998). This approach was particularly suited to this study as it allowed for the exploration of the complex and multifaceted nature of gender bias in search results for everyday queries. Following the principles of grounded theory, the analysis involved a continuous loop between data collection, initial coding, theme identification and refinement of focus on gender bias. This iterative process ensured that the emerging themes were firmly grounded in the data itself. The researchers identified 30 simple everyday queries. These queries were submitted separately for men and women by creating 60 prompts. Data was generated from a single user profile for a comparative analysis between gender-differentiated prompts. The systematically collected data consists of text-based search results, such as titles, snippets and meta-descriptions. Only the top result that appears first in order on the results page is analyzed. These details were analyzed using grounded theory to identify emerging patterns to indicate the extent of bias.

Grounded theory allows us to explore these complex relationships in an open-ended and flexible manner, leading to a more nuanced understanding of how gender bias manifests in everyday search experiences.

During the initial stage of open coding, a thorough examination of the Google Search results will be carried out. This study aims to identify and define preliminary ideas about gender representation. Iteratively created codes will capture different ways that gender bias appears in the search results. These could include gender-biased terminology that is clearly evident and stereotypical depictions. The first step in clarifying the various aspects of gender bias that are embedded in Google Search results is the open coding phase.

After the first codes are drawn, the axial coding stage will be implemented in order to reveal the complex connections among the thematic constructions that have been recognized. The goal of this analytical study is to identify the underlying relationships and relationships among the emerging themes. In particular, axial coding will examine how many factors—including search query terminology and result ranking algorithms—affect how gender bias is shown and maintained in the context of search results. The objective of the axial coding phase is to identify the underlying processes that determine the emergence of gender bias in the digital IR domain by means of a methodical investigation of these relationship dynamics.

The last phase of selective coding will bring together the ideas that surfaced during the initial phases of axial and open coding. The identification and integration of key themes that clarify the primary examples of gender bias present in the Google Search results paradigm are given priority in this last analytical stage. The focus will be on identifying the prevailing thematic constructs that most clearly capture the fundamental ways in which gender bias permeates and impacts the information landscape defined by Google's search results, through a careful selection process. Thus, the selective coding phase functions as the central hub for combining and condensing the most important discoveries made during the iterative coding process, which finally results in a thorough comprehension of gender bias in Google Search results.

The scope of the study is limited to only men and women. Further research may be carried out to analyze bias pertaining to LGBTQ+ users and the impact of user location on search results. This exploratory study is limited to results generated from one user profile and may be subject to influence of the user's search history, user profile and location.

16.4 DATA ANALYSIS

The investigation exposed significant occurrences of gender bias in Google Search results, providing insight into the complex relationship between search engines and societal gender norms. Following the principles of grounded theory, the analysis involved a continuous loop between data exploration, theme identification, and refinement of the focus on gender bias. Open coding was carried out on the actual search results (text snippets, titles, links) to identify initial concepts. Codes were assigned to elements within the search results that capture recurring themes or patterns related to how gender is presented with respect to specific stereotypes, biased language and absence of specific genders in the results. Axial coding was carried out to develop relationships between the identified concepts found within the search results. Selective coding was used to integrate the identified themes to explain how gender bias manifests in the search results. Constant comparative analysis was used throughout the study to compare emerging themes across different search queries and identify recurring patterns of bias.

The data was analyzed for patterns in text and extent of bias classified as High (H), Moderate (M) and Nil or Low (N) as indicated in Table 16.1. The following are the major findings of the study based on the tested prompts:

1. Reason why men/women sleep more: The result generated for Google shows hormone imbalance of estrogen in women. The same prompt for men shows results pertaining to poor posture, exertion and reflex actions.
2. Reasons why women/men sleep less: The results show that women sleep less due to menstruation, pregnancy, perimenopause, stress and mood disorders such as anxiety and depression. Poor habits such as developing an addiction toward caffeine, nicotine and alcohol are given as a major reason why men sleep less.

TABLE 16.1
Extent of Bias in Google Search Results

S. No.	User Queries	Extent of Bias in Google Search Results		
		H	M	N
1	Reason why a man/woman sleeps more		•	
2	Reason why a man/woman sleeps less		•	
3	How can I help my husband/wife sleep better?	•		
4	Why is my husband/wife silent?	•		
5	My husband/wife ignores me when I talk to him/her	•		
6	What should I do when my husband/wife is cheating on me?	•		
7	My husband/wife is jobless. What should I do?	•		
8	My husband/wife is beating me	•		
9	What should I do when a guy/lady stares at me?	•		
10	Can a man/woman hit back if their partner hits them?	•		
11	Should I always agree with my husband/wife?			•
12	What should I do when a guy/girl follows me?		•	
13	Why did she/he ghost me?		•	
14	Why does a guy/girl want to spend time with you?			•
15	Is it okay to touch a girl/guy?		•	
16	What should I do when he/she doesn't text back?		•	
17	What should I do when he/she makes me angry?		•	
18	Is it okay to say no to a girl/guy?		•	
19	Is it ok to have two wives/two husbands?	•		
20	Minimum vacation for a man/woman	•		
21	Reasons why he/she skips a meal			•
22	Reasons why he/she spends more time on the phone		•	
23	Why does he/she travel alone?		•	
24	Average salary preferred by men/women in India	•		
25	Is it ok if a man/woman slaps the opposite gender?			•
26	What will he/she think if I make fun of him/her?		•	
27	Is it mandatory for a man/woman to work?	•		
28	Why does he/she carry a knife?			•
29	Should a man/woman get educated?	•		
30	Should a man/woman become an entrepreneur?		•	

3. How can I help my wife/husband sleep better: The results show a gentle approach for women as against a corrective approach for men. Romantic settings, silence and sleeping alone are some suggestions given to men to improve the sleep of their wife. On the other hand, women are told that their husband must avoid alcohol or caffeine before bed.
4. Why is my wife/husband silent: When the husband goes silent, it is viewed as silent treatment and suggests to evaluating it as a form of abuse, while for the

wife, the result suggests that she is trying to process in order to not get hurt anymore.

5. My wife/husband ignores me when I talk to him/her: A husband's reasons for ignoring are cited as excessive demands on his time, distraction or an unhealthy response. A wife's reasons are that she is hurt by her husband and uses withdrawal as a coping mechanism.
6. What should I do when my husband/wife is cheating on me?: Results for this prompt shows suggestions for men and women differently. The suggestions show that the wife must find a therapist, opt for social support, prepare a self-care plan and confront her husband. The results encourage men to adopt a calm and careful approach in assessing the situation and taking proactive action.
7. My husband/wife is jobless. What should I do?: The results suggest that the wife must leave a jobless husband and the husband must block the credit/debit cards of his wife and cancel the internet connection. This normalizes that the man has to earn for the family and stereotypes women as irresponsible shop-a-holics who live on husband's money.
8. My husband/wife is beating me: Results for wife beating suggests a help line number with details of assistance available 24/7 and in several languages. There is no clear result given where the woman is the perpetrator. The top result goes to the National Commission for Women. This shows the nil to least support that male victims of violence have.
9. What should I do when a guy/lady stares at me?: Women are advised to either acknowledge the stare with a smile or frown to show dissatisfaction. Men, on the other hand, are informed that the lady staring is interested and the man must take it to the next level to keep the conversation going. This normalizes that it is the man who must make the first move and take advantage of a woman's situation. This also reveals that men are not supposed to reject a woman's advances or invitation.
10. Can a man/woman hit back if their partner hits them?: The results suggest that women must report it to the National Commission for Women. On the other hand, men are encouraged not to retaliate at the first instance and may resort to self-defense if she is bound to hit again.
11. Should I always agree with my husband/wife?: In both instances, the result suggests that healthy disagreement without any distress in the relationship is essential along with empathy and respect for the partner's point of view.
12. What should I do when a guy/girl follows me?: The top result advises men to report to the police as the girl above 20 years could be a stalker. If the girl is less than 20 years of age, men are advised to be calm and gentle. Women are advised to take several measures to keep them safe irrespective of the age of the guy.
13. Why did she/he ghost me?: The result shows a gender-neutral suggestion for why men ghost. It also owes it to their preference to handle their distress alone or their inability to communicate clearly. The result for why women ghost clearly mentions that women ghost for various reasons and the top-most being women ghost when she sees her partner as a barrier to her plans.

14. Why does a guy/girl want to spend time with you?: The results are neutral suggesting they could be interested in you irrespective of gender.
15. Is it okay to touch a girl/guy?: The result advises men about when to be playful and warns them to withdraw if the girl doesn't reciprocate. The result for women approves of their gesture and endorses it as an effective way to seduce their partner.
16. What should I do when he/she doesn't text back?: The results for women are gender neutral, advising them of healthy dating habits and warning the men against exhibiting stalking traits.
17. What should I do when he/she makes me angry?: The result advises women to engage in a range of self-care activities to calm themselves and the men to calm down generally.
18. Is it okay to say no to a girl/guy?: The results are moderately biased. Women are advised that a clear sensible no said to a man will bring him closer while men are asked to communicate their decision to reject clearly.
19. Is it ok to have two wives/two husbands?: This prompt is highly biased when it penalizes men from a legal point of view and warns him of consequences while a woman's character is questioned vis-à-vis what the prevailing socio-cultural norms permit them to do in the case of polygamy.
20. Minimum vacation for a man/woman: The result simply treats vacation as a relaxing time for men and recommends ideal vacations for men while for women the vacation is construed as maternity leave with specific legal provisions for availing leave.
21. Reasons why he/she skips a meal: There is no bias in the result which simply states neutral reasons of involvement in work or lack of time.
22. Reasons why he/she spends more time on the phone: For men, fear of missing out is cited as a top reason while for women, the reason cited is to do with a demanding job or her preference to socialize with friends.
23. Why does he/she travel alone?: The result for this query emphasizes on the freedom that women experience as solo travelers and highlights fear as the reason why women do not choose solo traveling. For men, the result shows that men prefer alone time and have been planning to do it for a long time.
24. Average salary preferred by men/women in India: For men, preference for average salary is viewed as compensation for their skill and this estimate helps men to assess their pay package. In the case of women, preference of average salary is directed toward what they expect their spouse or partner to earn so that they can have a comfortable life.
25. Is it ok if a man/woman slaps the opposite gender?: Irrespective of gender, this result shows zero tolerance toward physical abuse.
26. What will he/she think if I make fun of him/her?: The results state that women show no tolerance to being teased while men think being teased is a sign of being liked by the girl.
27. Is it mandatory for a man/woman to work?: There is a clear bias in the result where results for both prompts emphasize on why women should work, highlighting their need for independence and also men's preference to marry working women so that more income can improve quality of life.

28. Why does she carry a knife?: Irrespective of gender, the result shows carrying a knife as a sin of personal safety.
29. Should a man/woman get educated?: The bias in this result is evident from how education for men is simply seen as a means to help them communicate well but for women, education is seen as a foundation for building a better future for themselves and their families.
30. Should a man/woman become an entrepreneur?: The result shows that men will enjoy better time management, flexible working hours and being one's own boss. In the case of women, the result talks about inspiring other women, creating job opportunities for other women to close the gender gap and to create a legacy that can be inherited.

This analysis of Google Search results for everyday inquiries reveals a pervasive gender bias. Search prompts concerning similar situations often yielded different responses based on the user's identified gender. These biases reinforce stereotypical views of men and women, perpetuating inequalities in areas like career choices, domestic responsibilities and responses to emotional situations.

Further research is needed to explore the root causes of this bias and identify mitigation strategies. Search engine algorithms should be reevaluated to ensure neutrality in search results and avoid perpetuating gender stereotypes.

The study revealed widespread instances of gender bias in Google Search results. It was discovered that women were underrepresented in a number of fields, including professional jobs and leadership positions. Search results reinforced established gender norms and expectations with their stereotypical depictions of gender roles and attributes. Male- and female-oriented search queries were treated differently, with different outcomes. Search trends and user behavior had an impact on the kinds of results that were produced, possibly maintaining pre-existing biases. Biased search results were discovered to perpetuate existing power disparities and contribute to the marginalization and underrepresentation of specific groups in digital environments. The search results frequently reinforced traditional gender norms and expectations when users looked for phrases related to gender-specific duties, attributes or activities. Search results that frequently show males as aggressive or dominating may lead to negative stereotypes.

16.5 DISCUSSION

The simple everyday queries can be classified under the categories of health, safety, education, employment and entrepreneurship which are the five pillars of Developed India Mission@2047. The grounded theory analysis revealed consistent patterns of bias across various categories, with some queries showing moderate to high bias. It is noteworthy that some key parameters show low or nil bias and some high-degree bias is reported in certain queries which is proactive and inclusive. Grounded theory allowed the specific manifestations of gender bias to surface organically from the data, such as a consistent portrayal of men in professional contexts compared to women. Yet, biased search results contribute to the marginalization and underrepresentation of particular groups in digital settings by influencing users' perceptions and decision-making processes. The findings also look at how biased search results could limit the advancement of social justice, digital inclusion and gender equality. The challenges and opportunity associated with the gender-biased search results recognizes the complex nature of algorithmic decision-making procedures and the challenge of successfully reducing biases. It also points out ways to use technology to advance equality, diversity and acceptance in online spaces aligning with the broader goals of promoting inclusive representation and digital equity in online environments. It highlights how crucial it is to design search algorithms that take into account an array of human perspectives and experiences. There is an opportunity to build a more fair and inclusive online environment that benefits all people, regardless of gender identity or expression, by addressing gender bias in Google Search results and encouraging inclusive representation. Grounded theory analysis suggests that the consistent portrayal of men in professional roles and women in domestic roles in search results reinforces existing gender schemas, potentially contributing to the marginalization of women in professional settings. When particular genders are routinely portrayed negatively or stereotypically in search results, it can reinforce societal stereotypes and let discrimination and bias continue. For instance, it may perpetuate gender-based discrimination in the workplace, in the classroom and in other contexts if search results regularly link men with professional jobs and women with domestic duties. Search results that frequently show males as aggressive or dominating may lead to negative stereotypes and cultural expectations of what it means to be a man, which place more value on assertiveness and strength than on emotional transparency and vulnerability. These representations could make males less likely to seek support or assistance for mental health problems, which could increase safety risks associated with emotional health. Furthermore, biased search results can have a negative impact on people and communities by supporting the maintenance of damaging gender norms and expectations.

According to the concept of confirmation bias, people have a propensity to favor information that is consistent with their preconceptions. A pattern emerged from the analysis confirming a prominent prevalence of this phenomenon in the data. Searches related to gender roles, for example, "reasons why a man sleeps more," often produced search results that reinforced established assumptions in society. In particular, these results often portrayed males as strong and women as emotionally receptive, reinforcing traditional gender norms. This occurrence highlights confirmation bias, a phenomenon in which users' pre-existing ideas are unintentionally reinforced and perpetuated by the search engine algorithm, ultimately leading to the maintenance of gender stereotypes.

The gender schema theory is another relevant theoretical framework in this context. The idea that people create mental models, or schemas, that define gendered characteristics and roles is at the heart of this theoretical approach. These schemas then influence how people absorb information cognitively. When this theory is applied to the research findings, important insights are produced.

Differential Search Outcomes: The study identified significant differences between the search results produced for questions geared toward men and women (for example, "how to help husband sleep better" vs "how to help wife sleep better"). These distinctions between search results highlight how the search engine algorithm may be depending on gender schemas, perhaps leading to biased results. In particular, the contrasting results suggest the impact of preconceived conceptions about gender duties and characteristics, maintaining an unequal portrayal consistent with traditional gender standards. This manifestation contributes to a better understanding of how digital platforms may unintentionally propagate social gender biases by suggesting a subtle interaction between gender schema theory and the workings of search algorithms.

The successful implementation of the grounded theory methodology enabled the natural development of observable patterns from the dataset being examined. By carefully analyzing a variety of common search queries, the research project enabled a thorough comprehension of the subtle applications of gender schema theory and confirmation bias in the context of search engine results across a range of situational contexts.

Through careful investigation of a variety of everyday inquiries, the study project revealed the complex interactions between user-generated search inputs and the outputs that are produced, clarifying the operational dynamics that support gender schema theory and confirmation bias in the digital sphere. A comprehensive understanding of the ways in which deeply ingrained cognitive biases and socio-cultural constructs permeate digital interactions was made possible by this methodological approach. This understanding provided important new insights into the mechanisms that underlie IR processes and their implications for larger socio-cultural paradigms.

16.6 CONCLUSION

The investigation of discrimination based on gender in Google Search results has illuminated the widespread presence of biases in search algorithms and their noteworthy consequences for society's perspectives, prospects and digital equality. This study revealed significant and concerning instances of gender bias across a wide range of everyday search queries, suggesting a pervasive problem within search engine algorithms. The study's qualitative analysis, which was based on the grounded theory approach, produced a number of important conclusions about the prevalence of gender bias. These included the underrepresentation of women in particular fields, the stereotypical depictions of gender roles and traits and the treatment of male- and female-oriented search queries differently. These results indicate the complicated connection that exists between search engines and gender norms in society, emphasizing the importance of raising awareness and closely examining algorithmic biases in digital settings. The consequences for society are substantial: biased search results contribute to the marginalization and underrepresentation of particular groups in digital spaces, perpetuate gender stereotypes and impact perceptions and decision-making processes.

However, the study does identify options for eliminating gender bias in Google Search results, namely diversifying data for instruction, including inclusive ranking criteria and encouraging transparency and accountability in algorithmic decision-making processes. Ultimately, advancing gender equality, digital inclusion and social justice in digital spaces and society at large depends on dealing with gender bias in search engines. While grounded theory provided a robust framework for understanding gender bias in this specific context, further research with different methodologies might be needed to explore generalizability.

The findings suggest that confirmation bias and gender schema might be playing a role in shaping search results. Search engines may be reinforcing existing stereotypes by presenting users with information that confirms their preconceived notions about gender roles. The biased search results have the potential to perpetuate harmful stereotypes and limit the information available to users. This can have a negative impact on individuals and society as a whole, reinforcing gender inequalities and limiting opportunities for women and girls. We may endeavor to create a more democratic and inclusive online environment that benefits everyone, regardless of gender identity or expression, by recognizing and eliminating biases that harm people of all genders. These findings call for the development of more inclusive search algorithms that can present users with a balanced and unbiased view of the world. Further research is needed to explore how this can be achieved and to develop strategies for mitigating gender bias in search engines.

In tandem with SGD 5, the study's conclusions provide attainable measures to promote a more equal and inclusive online space where people of all genders can access opportunities and information on an equal basis. Through advancing diversity, inclusivity and representation in search results, we support the larger goal of empowering and equalizing women. De-biasing algorithms is essential as biased datasets are frequently transformed into biased AI systems. We can use technology to create a more just and equal society where everyone, regardless of gender, can prosper and help achieve the Sustainable Development Goals by working together and taking consistent action.

REFERENCES

Akter, S., Dwivedi, Y. K., Sajib, S., Biswas, K., Bandara, R. J., & Michael, K. (2022). Algorithmic bias in machine learning-based marketing models. *Journal of Business Research, 144*, 201–216.

Baeza-Yates, R. (2018). Bias on the web. *Communications of the ACM, 61*(6), 54–61.

Bryson, J. J., Etlinger, S., Keyes, O., & Rankin, J. L. (2020, March 19). *Gender Bias in Technology: How Far Have We Come and What Comes Next?* Centre for International Governance Innovation. www.cigionline.org/articles/gender-bias-technology-how-far-have-we-come-and-what-comes-next/?utm_medium=social&utm_source=twitter

Connor, S. O., & Liu, H. (2023, May). Gender bias perpetuation and mitigation in AI technologies: Challenges and opportunities. *AI & Society* 39. https://doi.org/10.1007/s00146-023-01675-4

Deshpande, K. V., Pan, S., & Foulds, J. R. (2020, July). Mitigating demographic bias in AI-based resume filtering. In *UMAP '20 Adjunct: Adjunct Publication of the 28th ACM Conference on User Modeling, Adaptation and Personalization*, 268–275.

Diño, G. (2019, January 22). *He Said, She Said: Addressing Gender in Neural Machine Translation*. Slator.

Kay, M., Matuszek, C., & Munson, S. A. (2015, April). Unequal representation and gender stereotypes in image search results for occupations. In *Proceedings of the 33rd Annual ACM Conference on Human Factors in Computing Systems*, 3819–3828.

Krieg, K., Parada-Cabaleiro, E., Medicus, G., Lesota, O., Schedl, M., & Rekabsaz, N. (2023, March). Grep-biasir: A dataset for investigating gender representation bias in information retrieval results. In *Proceedings of the 2023 Conference on Human Information Interaction and Retrieval*, 444–448.

Lin, C., Gao, Y., Ta, N., Li, K., & Fu, H. (2023). Trapped in the search box: An examination of algorithmic bias in search engine autocomplete predictions. *Telematics and Informatics, 85*, 102068.

Makhortykh, M., Urman, A., & Ulloa, R. (2021). Detecting race and gender bias in visual representation of AI on web search engines. In *Advances in Bias and Fairness in Information Retrieval: Second International Workshop on Algorithmic Bias in Search and Recommendation, BIAS 2021, Lucca, Italy, April 1, 2021, Proceedings*. Springer International Publishing.

Meisner, C., Duffy, B. E., & Ziewitz, M. (2024). The labor of search engine evaluation: Making algorithms more human or humans more algorithmic? *New Media & Society*, *26*(2), 1018–1033.

Mudgal, V. (2019). A vital report on trends in media freedom and its link with the future of democracy. *Journal of Creative Communications*, *14*(2), 160–170.

Nadeem, A., Abedin, B., & Marjanovic, O. (2020). Gender bias in AI: A review of contributing factors and mitigating strategies. In *ACIS 2020 Proceedings*.

Nishant, R., Schneckenberg, D., & Ravishankar, M. N. (2024). The formal rationality of artificial intelligence-based algorithms and the problem of bias. *Journal of Information Technology*, *39*(1), 19–40.

Nur Fitria, T. (2021, April). Gender bias in translation using Google Translate: Problems and solution. *Language Circle: Journal of Language and Literature*, *15*(2). 285–292. https://doi.org/10.15294/lc.v15i2.28641

Otterbacher, J. (2018, August). Addressing social bias in information retrieval. In *International Conference of the Cross-Language Evaluation Forum for European Languages*, 121–127. Springer International Publishing.

Otterbacher, J., Checco, A., Demartini, G., & Clough, P. (2018, June). Investigating user perception of gender bias in image search: The role of sexism. In *SIGIR '18: The 41st International ACM SIGIR Conference on Research & Development in Information Retrieval*, 933–936. https://doi.org/10.1145/3209978.3210094

Prates, M. O. R., Avelar, P. H., & Lamb, L. C. (2019, March 27). Assessing gender bias in machine translation: A case study with Google Translate. *Neural Computing and Applications*, *32*, 6363–6381. https://doi.org/10.1007/s00521-019-04144-6

Rumage, J. (2024). *16 Top Search Engines to Try*. Built In. Builtin.com. https://builtin.com/articles/search-engines-list

Ron, Y., & Nimrod, G. (2018). Digital negotiation: Online leisure in coping with gender stereotypes. *Leisure Sciences*, *40*(4), 251–266.

Roy, S., & Ayalon, L. (2020). Age and gender stereotypes reflected in Google's "Autocomplete" function: The portrayal and possible spread of societal stereotypes. *The Gerontologist*, *60*(6), 1020–1028.

StatCounter. (2024). *Search Engine Market Share India*. StatCounter Global Stats. https://gs.statcounter.com/search-engine-market-share/all/india

Sreekumar, T. T. (2013). Global South perspectives on youth culture and gender imaginations in the technological society. *Journal of Creative Communications*, *8*(2–3), 77–88.

Srikrishna, V. (2021). "Understanding Data ideologies, extraction and surveillance: The trap of Big-Data Dystopia" for the Media Anthropology Collective. It can be accessed at https://media-anthropology.medium.com/understanding-data-ideologies-extraction-and-surveillance-the-trap-of-big-data-dystopia-f2afa474c498

Strauss, A., & Corbin, J. (1998). *Basics of Qualitative Research: Techniques and Procedures for Developing Grounded Theory*. Thousand Oaks, CA: Sage Publications, Inc.

Sun, L., Wei, M., Sun, Y., Suh, Y. J., Shen, L., & Yang, S. (2024). Smiling women pitching down: Auditing representational and presentational gender biases in image-generative AI. *Journal of Computer-Mediated Communication*, *29*(1), zmad045.

Tie, Y. C., Birks, M., & Francis, K. (2019, January). Grounded theory research: A design framework for novice researchers. *SAGE Open Medicine*, *7*. https://doi.org/10.1177/2050312118822927

Wijnhoven, F., & van Haren, J. (2021, May). Search engine gender bias. *Frontiers in Big Data*, *4*. 1–11. https://doi.org/10.3389/fdata.2021.622106

17 Artificial Intelligence in International Relations
Complexities Involved

Ipshita Bhattacharya

17.1 INTRODUCTION

Human history states that fundamentally political institutions and politics are principally driven by the conscious human brain and the collective actions and their responses by the related people. But now with the speedy advancement in the artificial intelligence (AI) sector, an idea of a non-human component has risen, which has the potential to make our understanding of politics completely radical. The chapter investigates how the complexity of emerging advanced technology interferes into international relations and enhances the scope of realism theory and its actualization. The study theoretically engages in a dialogue between technology and international relations examining the overriding tendencies of each other. The outcome of this dialogue is relevant to determine the ways in which they can promote and strengthen a country's economy and military status as their foreign policy goals.

The study will investigate vivid aspects of technology and its various fields involved in influencing the political behaviour of the states in the international system. It will examine how it is one of the most significant components in the dynamics of international relations today. While investigating the influence and impact of technology on the behaviour of the states globally, it will also investigate how emerging technology (ET) will pose a challenge before the traditional diplomatic practices.

The scientific advancements create the path of economic developments and open new avenues for globalization, making it a progressive socio-economic phenomenon. But this also renders a paradigm shift in security concerns into the techno-military character of the states. Understanding this reality that military and economic power will dominate the future global power equations, but how it will influence and impact the security dynamics of the states will be one of the focus areas.

The significance of technological advancement in the present times is highly relevant but its interplay in international affairs is nearly invisible in global political understanding. This work emphasizes on how technology will interfere in the power balance dynamics among the current world order system.

With the evolving critical technologies and its significance in global politics, it has also shaped the conduct of the warfare system. This work focusses on the role of cyberspace and AI in the new war-fighting tools and focusses on the technological reinvigoration in the nuclear weapons fields. The work also examines the complexities

 DOI: 10.1201/9781003560463-21

involved in the warfare system and what lethality it could bring in the hybrid warfare system by diffusing the threat and employing the proxy tools to meet the state's objectives.

ET has a vital role in shaping international relations, from issues spanning AI and cybersecurity to energy policies and nuclear proliferation. The potential advantages and disadvantages of technological innovations have crucial influences on the dialogues between the nations and their conduct therefore impacts the international relations on a wider scope. This chapter takes an opportunity to understand the intersection of continuously evolving technology and dynamic global politics through the lens of science and technology and international relations. The study explores, investigates and examines the ways that technological developments have changed the global landscape, conduct of the states and the security paradigms. This work focusses on the theory of realism and how technology plays its crucial role within the dimensions of realism.

The concept of this study rests upon the theory of realism in international relations and how technology enhances the probability of critical dynamics in realism in the world of international politics. The work will be a doctrinal study, including valuable research in the field of scientific studies and world politics. The work will be a balance of theoretical and empirical research in multiple disciplines. Most importantly this work will be delving into the science component also and not only the technology part.

This work examines and probes into the ethical component of international politics and the response and conduct of the responsible nations of the world, while dealing with technological advancement and its involvement in global politics. Secondly, the work will dive deeper into the scientific waters to investigate certain uncomfortable questions to the dominating powers of international politics, which are essential for the global citizens. Finally, the work will bring out relevant case studies based upon empirical studies. Moreover, the work is a well-balanced composition of theoretical and empirical studies.

17.2 STATE OF THE ART

In contemporary times technology is deeply rooted to international relations and has deep influence on global politics. *Technology and International Relations (edited by Giampiero Giacomello, Francesco More and Marco Valigi, Elgar Online, 2021)* deals in the domain of investments into technological research by the states and its economic and military outcomes. (*Technologies of International Relations by Carolin Kaltofen, Madeline Carr, Michele Acuto, Springer, 2019.*) Though this book discusses the outcomes of technology and its engagement in global politics, the domains of behaviour, conduct and political nature of the states in this process is not potentially dealt with. *(Role of Technology in International Affairs, Amitav Mallick, 2016, Pentagon Press.)* The intersection of interdisciplinary character of the domains of international relations, geography and anthropology are thoroughly examined but however the book confines itself to the interdisciplinary component mostly. (*Technology and International Relations: Challenges for the 21st Century – Balakrishnan, Dr. Bhaskar.)* This book discusses the role of science and technology in IR; however, it does talk about the intrinsic role of science and technology but it principally brings out interconnections and interdependencies between technology, security and diplomacy but does not talks about a state's conduct. The role of the states and their behaviour is not the major theme of this book. *(Artificial intelligence and Emerging Technologies in International Relations, Bhaso Ndzendze, 2021.)* This book deals with a holistic perspective to explore the roots of technological development and connect it to contemporary times. The author also discusses the future of technological development and its roles. The role of the states in the development of the technology and the conduct of the state once they have the technology's power and how they behave with the other states has perhaps not been dealt with. *(International Relations in the Cyber Age: The Co Evolution Dilemma, @@2019, Nazli Choucri and David D Clark, MIT Press.)* This book discusses the advantages of a technologically developed state in the international political system. The author also suggests that this trend will accelerate in future. However, this book talks about the power and politics intervening in the domains of cyber technology, and its influence in world politics and the book brings out the role of power and politics in the cyber world and its influence on global politics. All the contemporary existing works are majorly dealing only on the technology part and not on the science component present in international relations. This is a potential gap in most of the existing works on technology and international relations.

17.3 INTERNATIONAL RELATIONS AND TECHNOLOGY: THE DIALOGUE BETWEEN THE PAST AND PRESENT

One of the earliest works on realism rests in the roots of the *History of the Peloponnesian War* by Thucydides. Although he is not recognized as the realist since the theory wasn't propounded until the 20th century. However, when a comparative analysis of the thought patterns and behaviour of a state and its political actors he mentioned about in ancient Greece (during the Peloponnesian War) and that of modern theorists are found to be similar in context. This leads to an understanding that realism in actual sense is a never-changing and enduring theory and which forms a part of the history. Around 2500 years ago Thucydides' claimed that his work on the war between Athens and Sparta would be everlasting in contextual sense as it reveals the various aspects of international politics and power struggles. Thinkers and scholars keep visiting Thucydides' work since it holds the core insight of politics and power.

Chanakya (350–275 BCE) during the Mauryan Empire in India in his famous book on statecraft, *Arthshastra*, suggests the power of rationality in politics and the duties of a king for his state and the state's interests are supreme. Theory of realism including Thomas Hobbes, Niccolo Machiavelli and Rousseau and Morgenthau argues the same. "Classical Realist" thinkers like Morgenthau and John Herz came up with a potential but an ignored part of international relations, criticism of technology wherein they doubted the existence of technological advancement in international relations responsibly and appropriately. Morgenthau was highly sceptical regarding the naive faith of the global political system in the technological growth debating that it would only generate an environment of apolitical and reckless view of international politics (Scheuerman, 2009). Wherein Herz's concern was speedy technological advancement, which may become a threat for humanity's future. Studies reveal that probably Morgenthau was driven towards Frankfurt School critiques of technology, inspired by Karl Jaspers who was critical about the growing technological advancements and the future of humans.

According to the dominant realism theory, it considers that war is a "normal state of affairs"; like Morgenthau said international politics is nothing but a struggle for power (Williams, 2004). The interest of each state is the primary goal before any state; hence, their actions are directed towards achieving their goals, majorly protecting itself against each other and prioritizing its objective, making war inevitable. According to neo-realist Kenneth Waltz, international political structure rests upon an anarchical system and in anarchical system states only preserves their own interests; in this situation whenever a state rises, the others feel threatened and hence design preventive war against the rising state. This explains that for instance Peloponnesian War was inevitable as power transitions are rarely peaceful. The conversation between the international relations and its engagements with technology assess the new shape and the conduct that is forming from global politics within the dynamics of technology. This is how an attempt to understand the role of technology within the international relations discipline that often remains neglected and understudied is further discussed. The origin of war and the measures taken for maintaining peace, the conduct of the states equipped with technology power while exercising their power in the global system is been witnessed even in the past. Eventually, how does the technology impact the behaviour of the states and its diplomatic strategies and why it is important to know the impact. Since we are discussing this in context to the diplomatic relations between the states within the state system, it becomes important to know the areas that would be influenced. International relations promote trade policies among the interested nations, enabling states to cooperate, negotiate and collaborate with each other.

It has rightly been understood that international relations are dynamic in character with key drivers of these dynamics that are involved in international relations, particularly when we are existing in a technology era. Hence, science and technology influence the behaviour and conduct of the states and eventually impact global politics. Sometimes it plays as a key determinant in carving out relations between the nations in the international state system other than wars and economic relations. However, science and technology remain the key mantra for economic progress and national security. One of the pieces of evidence from the past that how technology shapes global politics is the industrial revolution phase, which led to the states to look out for new lands and raw material sources and created colonial establishments. Moreover, technology made industrialized states to largely gather economic and military power to exercise their dominance over the less developed states, eventually leading to a hierarchy system in the international state system.

The global hierarchy for the nations is defined and understood as its capabilities of technological power including its research and development and infrastructural support. Therefore, to possess modern and advanced technologies is the priority and also a determinant for the foreign policymakers of the respective states irrespective of big powerful nations or small and developing nations. Nevertheless, this is also true that this spread of development is never uniform across the world and the states that innovate or adopt early have an edge over the others. This gives them the vantage points and dominance leading to economic benefits or high grounds in negotiations with the lesser privileged states. It would be correct to say that in today's globalized world to have technology access and technology denial will definitely determine their position in the global hierarchy. Countries compete with each other in a highly competitive environment of techno-economic performance, moreover, technologically superior countries majorly determine the norms and the standards of concerned policies in the set of international politics (Mallick, 2023). The great powers compete to manage their high status in the world ranking through their advancements in the technological domains, which further helps them to negotiate economic and security prerogatives. This is why most of the research and developments in the technological fields are driven by the competitive objectives defined by their nation's quest to dominate the technological superiority over the lesser developed nations (Mallik, 2004). Particularly from the military point of view in particular, see the advanced technology as a significant key and strategy to dominate their adversaries. This game of technological advancement invariably creates disparity and imbalance in the existing balance of powers and hence creates military capabilities for future wars and disputes. For instance, we have seen how limitations of transfer of knowledge have created rough patches among the nations.

17.4 BEHAVIOUR AND CONDUCT OF THE STATES

Considering the history of global politics, three major factors have shaped world politics, economic dynamics, wars and technological advancements. In the year 1935, with WWII in the backdrop, the USA was not having significant armed forces neither it was equipped with a robust economy, soon it refuged under isolationism only to make a prosperous comeback after ten years equipped with nuclear weapons technology and other technological advancements like synthetic rubbers, duct tapes, atomic bombs and jeeps which proved to be of significant help to the allied forces. This technological development however completely changed the conduct of the USA in WWII, particularly if we see the case of the nuke bombing of Hiroshima and Nagasaki. Since then, we have seen how wars have become more destructive and complex, at the same time making war more unreliable in conducting and shaping great power relations.

The conventional concept was that China would mainly limit itself to regional expansion and dominance, restrict US influence in the Western Pacific and at the most aggressive economic expansion in the region. But what is making China gearing up to contest the USA's ubiquitous global dominance or if we see the trajectory of Chinese foreign policy in the last few decades, we will see how dominant and aggressive China has turned out on the global political platform. Beijing bolstered its intentions in the South China Sea (SCS), increased its patrols near Senkaku and Diaoyu Islands controlled by Tokyo, contended India on the borders, bullied Taiwan, rebuked Australia for raising questions regarding COVID-19 and pushed its efforts to save Huawei by accusing Canadian Michael Kovrig and Michael Spavor for espionage actions, this was in reaction to when a Canadian court denied to stop extradition proceedings against the Huawei CFO Meng Wanzhou, it even went on to warn and criticize the UK for excluding this telecom company from its 5G network (Pence, 2020). More such aggressive behaviour was observed in its conduct of arresting pro-democratic activists in Hong Kong. To understand China's growing assertive foreign policy and its conduct on the global politics platform, it needs to be looked through the lens of China's technological revolution.

Understanding the fact that the mounting growth of ET requires a scrutiny of the role of technology in the dynamics of IR and its impact on the international political system as the ET influences the behaviour and conduct of the states. ET being the major fulcrum for economy and national security of a nation has influenced and directed the behaviour and conducts of the states. This increased involvement of ETs in economic and military advantages of a state possesses a strong implication on the overall shifting balance of power on global politics (Steff, Burton, Soare, 2020). We know that ETs have a strong contribution in defence and strategic affairs as it provides securitization to the high-tech sector (Sechser, Narang, Talmadge, 2019), this whole process eventually gives powerful negotiation skills and dominant characteristics to a state which apparently reflects in their international political demeanour. For instance, China's technological advancement journey could conceivably equate to the acquisition of its potential competences in recent years and hence, infusing threat perceptions among the states (Papageorgiou, Can, Vieira, 2024) in various global conflict theatres. However, this corresponds to the realism theory of international relations, wherein the USA finds itself absolutely challenged by an ambitious China, with its unreasonable soaring demands, consequently threatening a state's interests and hence inducing potential conflicts (Mearsheimer, 2015). China's conduct due to its ET advancements could be taken as challenges in the future.

Several questions emerge in this context; firstly, what establishes a threat for ET and how does a state accommodate these criteria. Now once a state is conceded as a threat, what would be the behaviour of the other states; however, it is extremely debatable that a state's behaviour and its conduct are based upon the given conflicting circumstances. However, understanding this fact that threat plays an important role in determining the behaviour of the states, let us understand the determinants of threat first for a state, the overall resources, geo-political and geo-strategic location and offensive intentions majorly constitutes the determinants of threat. The determinants to determine the threat element has a wide scope, the term overall resources will include the economy, military and other the human resources of a state, wherein the technology part will surely impact the economy and the military part.

The post-Cold War face-off between the USA and the Soviet Union ended with the emergence of a liberal world order, wherein the USSR fragmented and took a retreat from the world stage. However, the scars of the Soviet disintegration and the remnants of the political ideations are still having its strong presence in their today's foreign policy. Contemporary international politics does have a lot of residues from the Cold War era which still creates complications before the world. The Cold War which was mute but was multifaceted and volatile in character was not only the struggle between the USA and USSR but it was completely global in scope. To interpret the war in its actual sense there is a need to explore the narrow roads leading to bilateral relations also, taking some time out from the conventional warfare of Cold War. The Cold War was definitely a complex and grand phenomenon which bolstered and rather glorified the art of statecraft of the USA and USSR. The intervention of AI in hybrid warfare and its focus on the hybrid form of warfare inflicted by states (the USA, Russia and China) post-Cold War, wherein the states have included technological tools as proxies to be employed on the target states, brings in the involvement of technology in the prevalent hybrid warfare strategies employed by aggressor states in Eastern Europe and Asia. The whole process of post-Cold War involves proxy wars which are deeply supported by technology, from cyber to AI-driven weapon systems.

17.5 EMERGING TECHNOLOGIES AND THE RISE AND FALL OF THE NATIONS

Across the globe states can estimate their power based upon up to what extent they are harnessing and managing the development of AI. Understanding the factor that AI has empirical risks, and for that governments will require to build new regulatory frameworks to identify, investigate, examine and respond to the AI-driven challenges. With the focus on AI and governance in the contemporary times less attention is being given to the issues pertaining to the geo-politics where the conduct of the states will play a key role in determining the bilateral and multilateral relations without having effective guardrails. This is required to be understood that as AI and emerging technologies continue to maintain their steady pace of advancement, the landscape of global politics will not remain the same. At this point it is also necessary to understand that humans have to interact with different sets of actors designated as potential machines with distinct standards, which may be of higher intelligence than human and may be of disruptive capabilities. This will be followed by, in the world of politics human identity and its complexities to recognize their roles in the world will be completely different. The exceptional scientific discoveries will evolve in such a way that it would not be easy for mankind to understand and cope with it. Since with this it will bring complex configuration of challenges which will play a significant role in developing the contours of the future world.

The plethora of emerging technologies, the challenges they can pose and the related governance dilemmas underscore a set of common issues. One of such issues corresponds to the relationship between technology and international relations which depends upon the power game played by the global dominant state. With a recent statement by Russian President Vladimir Putin, that a nation leading in AI will be dominating the world in the future; however, this may sound hyperbolic but also validates that political leaders do believe that technology is a dominant source of a state's power. This endorses that in this ET age national security and a state's power are surely impacted by the technology.

Understanding the fact that technology is a leading tool for global transformations and responsible for enhancing the standards of human living, simultaneously it also has to meet the complex, larger scope of challenges of adoption. This has to be understood by the states that the regulations of a state cannot be the same for other states such as adopting the regulations of a developed state without realizing their economic statistics and technological landscapes may lead to collapse of the regulation structure completely. Therefore, overly stringent regulations may lead to complications at the implementation level and also at the adoption level. But at the same time, at this level it is also important to accept that these multi-layered and multi-dimensional technologies face cross border risks of the monopolistic power of the multinational companies. A significant debate in this regard would be that there will be an acute imbalance in policy formulation and its development between the developed states and developing states regarding regulating the new evolving technologies. Some important questions like the EU-AI Act of the European Union and the Quantum Computing Cybersecurity Preparedness Act of the USA are considering the policies in the global south context.

The significance of technology intervention drastically increased particularly into various war theatres since the 19th century mainly for the ongoing industrial revolution. This was a natural phenomenon as sustained and swift technological developments were affecting each and every area of human lives including its involvement in the wars. The confirmation of this growing pace in technological revolution can be witnessed from Schumpeter's economic analysis of capitalism and its connection to technology. According to this study industrial revolution led to seminal transformation in the mode of production and impacted the states and their foreign policy objectives at a greater heights. For instance, a significant finding by Schumpeter was that in the race to innovate, to rule and to achieve a position of monopoly, countries entered into an extreme competitive phase to dominate the other states or to the vulnerable states. Schumpeter went on further to say that on the micro-economics level capitalism depended more on the competition for innovations comparative to costs. That is evident enough to understand how states have been influenced by technological development since the past. This also proves that the states were more focussed towards the applicative innovations to have dominance and to dominate the international states system. These deductions raise concerns that the priority of these nations was to have a dominant monopoly of the innovations over the markets and these notable characteristics were in direct correlation to the country's status in the world with respect to their science and technology's capabilities. This also raises a concern that aggression is a collateral product of technological advancement and hence, could technology be perceived as a dialogue in itself between the states.

With the fast-paced development in the AI sector, the requirement for effective governance standards and mechanisms is imperative for management and mitigating potential risks. The AI system involving machine learning incorporates the capabilities for decision-making and its predicting capabilities raises ethical concerns. Some of the immediate vacuums are limitations of knowing that AI systems are free from biases; secondly, knowing whether AI's decision-making mechanisms are authentic or not; and thirdly, since AI involves massive data, safeguarding privacy is a major concern. Moreover, the intersection of technical and regulatory frameworks would also be a complex affair. Eventually, how and who will bridge the gap between what AI should and what AI should not. All these cogitations are significant factors of AI governance requiring attention. Setting up regulations for AI would be extremely challenging because of its technical intricacies and speedy evolution. AI's widespread and multilevel applications across multiple industries will be another challenge for regulating and governance set ups. This array of challenges also includes what kind of regulatory frameworks should be developed to promote innovations; this also includes who will decide the thresholds for the innovations, transparency and impartiality in AI systems which will be safe enough for competitive innovations but also prohibiting misuse. AI development is universal, speedy and evolutionary in nature and hence conventional legislative procedures face limitations to catch up with the speed of AI innovations. To meet these intricacies adaptive AI governance set up is must with genetic algorithms to promote natural evolutions of the same. AI governance is an international issue and to implement efficacy international cooperation is required but since the AI system does not understand borders, countries have to collectively work towards establishing universal standards and norms. International organizations can play a significant role in facilitating dialogues among the states for standard AI governance.

The AI ambitions of weaponization have wider scope, for instance, autonomous range of weapons systems (AWS), semi-autonomous and cyber warfare systems. AWS which apparently fails to pass the discrimination principle and their use for citizen stakeholders is incommensurable. Identifying these potential threats a push would be needed for the formulation of legal policies to restrict these shortfalls, and the use of this technology by powerful states on the vulnerable and the weaker states. The emergence of new advanced weapons of mass destruction particularly the drone swarms, sometimes even at undetectable size, employed for administering lethal chemicals through their artificial stings makes the society challenged by these unprecedented scientific innovations. AI in the military surely calls for a clear demarcation of basic utilization of science to the application of science for lethal practices, and from potential risks to benefits and finally responsible use of weapons to deadly abuse of scientific advancements. AI is power, and power has to be responsible or else a human-made tool will be responsible for uncontrolled and illegitimate killings. Gradually making it invincible and unconquerable for the states due to the leverages associated with these choices of the continued use and hence, making compromises to the less privileged nations.

Therefore, it would be prudent to say that AI in the military should strictly belong to the international law domain and policymaking, and the international community must endorse a new set of immediate conventions to regulate and operate through international legislations.

AI in Space Technology: The global space economy has currently reached an impressive score of USD 450 billion and expected to grow over USD 1 trillion by 2040. Service spacecraft, space habitats, reusable rockets, autonomous robots and mega constellations are the prevalent and future technologies which will be responsible for the advancement of the space domain. But this may also lead to creating technological and economic discrimination in the international system and leading to great power competition. Understanding the fact that space was an important area during the Cold War, which demanded cooperation although having military rivalry among the great powers. Today even greater cooperation is required perhaps due to its expansion into economic, political, climatic, scientific and industrial significance. However, it will purely depend upon the foreign policy choices of the states whether they choose for an earth-centric agenda or go for individual choices of military conquests and space colonizations and technological conflicts leading to unnecessary conflicts. However, to support and cooperate at a global level, the understanding for the concept of diplomacy in space in the economic, science and military terms is required. Which may erase out differences and lead to positive diplomatic manoeuvrings among various states, to advance in this significant area of foreign policy.

Similarly, in the domain of Nanotechnology and Quantum science, there are array of international security concerns, food security and climate change to weapons of mass destructions and medical science. Research in Nanotechnology and Quantum sciences is currently prevalent across the globe but the level of investment, technical know-how, scientific infrastructure and cooperation across the various sectors plays a vital role. The plight of developing nations in this regard sometimes is crucial as they may be distanced by knowledge divide, if unable to contribute equally. However, due to global connectivity through the internet and scientific collaborations, the path to development is much more accessible as compared to a few decades ago. But a new set of challenges like a set of varied interests of different states, ethical issues and politics transgressing science will surely put different sets of challenges.

Another key challenging area is how AI in the ET and its impact on the non-state actor's conduct in international politics will impact the global political scenario. The risk assessment in context to rapidly growing AI technology and its easy accessibility to the non-state actors and their nefarious intentions have deep roots within the states and non-states relations. Secondly, the use of digital platforms to access intel for money laundering and identity frauds for illicit activities. How fast-evolving technology is facilitating non-state actors a conducive environment for conducting reconnaissance and surveillance for crimes like smuggling, ambushes and spreading propaganda. Due to the growing technology, non-state actors go undetected by diminishing their logistical footprints by manufacturing products on their own which in turn reduces their acquisition detectability. The nexus and the involvement of non-state and state actors will not only dominate the bilateral and multilateral relations but also the global security paradigm.

Another key challenging area is the states' involvement in the AI and the cyber technology's dark designs vis-à-vis either their adversaries or vulnerable states. The case studies of Southeast Asian (SEA) nations are studies here wherein cybercrimes involving malicious activities, malware, password cracking, CAPTCHA-breaking, speech cloning, social engineering attacks and deepfake technology are involved.

In the world across the continuum of challenges in the international political system, the states must optimize how to manage, utilize and protect the effects of information. This is strategically done by organizing to filter out specific information and cyber-enabled information across multi-layered domains. However, the internet and the application layers are turning out to be contesting rings where the state actors, as contestants and assailants, build an absolute shift to meet their respective objectives. Some states identify susceptibilities and observe vulnerabilities in the target state, and to compromise their security they use proxies. These digital environments are compatible to share crucial assets to the proxy groups/terror groups by any aggressor state, resulting in a symbiosis which has the potential to produce cyber weapons for military purposes. The outsourced proxy cybercrimes to cyber terrorists works on the similar line of action as the military-graded cyber weapons system does in an overt operation. Understanding the potential threat from cyber operations will rise in future decades, even in the presence of stringent security architectures.

The SEA region is a significant trading route as two-thirds of the world's container trade crosses this region. Two key geo-political factors that make this region extremely crucial are firstly, China's assertive claims in SCS and its posture on the other claimants of the SCS and secondly, its stance towards Taiwan. Understanding the fact that Taiwan is an immediate flashpoint particularly if Beijing decides to go for a military move against Taiwan, it could lead to a wider implication between China and the USA. The SEA countries and their alliance system with the USA and the gradual drift of the SEA countries towards QUAD, a security alliance between the USA, India, Japan and Australia also stokes this geo-political anxiety more. The involvement of state-sponsored non-state actors for collecting intelligence, disrupting important infrastructures or to devise discords further heightens the complexity of the situation. Since the SEA countries have largely focussed on the data-driven economy, which they have released in their national policy and strategy, their commitment to go for global digital economy and with their 5G technology, they are also expanding their attack area for malignant cyber operations. Hence, the state-sponsored cyber sabotages and cyber espionage are a major concern in the SEA region with economic and geo-political motives of China in the backdrop. Let us understand the case of Indonesia: In 2023, several state-sponsored cyber malicious campaigns targeting Indonesian organizations were observed. The principal players behind these campaigns were China, North Korea and Russia, from January to October 2023, a staggering figure of 568,521 phishing campaigns has been reported, which makes Indonesia the third most cyber-attacked country in the SEA region. The growing economy of Indonesia is also witnessing hard hits from various ransomware groups like BlackCat, 8Base and LockBit attacking Indonesian business. Below given data brings out the empirical data on the following criteria: Distribution of the phishing cases based upon various sectors, while exploring the prominent threat actors active in the region and the reported empirical case studies they are engaged in.

Similarly, the case of the Philippines**:** The Philippines and the USA's close ties and its strategic location in SCS makes it a vulnerable state to Chinese cyber-attacks. China is known for its advanced and aggressive cyber-attack technology with the Philippines' friendly posture with QUAD and AUKUS countries is surely creating discomfort for China. Russia's gradual dependence on China is creating dynamics for the formation of Eurasian bloc which may be a potential cause of worry in the region and for the Philippine also. The Philippines also forms a rational target in case of China–Taiwan conflict. Recent studies shows a dynamic increase in the Philippines' DDoS in various sectors. In 2023, a huge data hack was reported wherein

817.54 gigabytes of data was composed of many of the state agencies including the Philippines National Police and National Bureau of Investigation. As a result of China's discomfort, this empirical study of the China sponsored cyber-attacks on government establishments and other sectors having direct or indirect impact on the national security of the Philippines.

Vietnam is another target of China in the South East Asian region: Information and Communication Technology (ICT) sector is a booming and priority industry of Vietnam; they are committed to elevate it to an impressive target of USD 50 million. However, the convergence of these economic goals and ICT development makes the ITC sector vulnerable to cyber threats. Vietnam is facing continuous cyber-attacks from China more than last two decades now, since the episode of Chinese vessels wrecking the hydrographic research settlement of Vietnam in 2011, there is an ongoing cyber war between China and Vietnam dominated by China. Some of the prominent names are 1973cn, particularly active on nine-dash-line, a very active name of Chinese hacker group "Mustang Panda" attacked Vietnam's border offices in 2018–19. A study shows that in 2023 approximately 13,900 cases of cyber-attacks are reported and there is a significant rise in the number of cyber-attack cases from 2022 to 2023, a notable increase of 9.5% is found. In 2019, the National Defence White Paper of Vietnam stated their vulnerabilities from cyber warfare from dominant forces. China's sponsored threat actor role in conducting exploitation in Vietnam's cyber infrastructure is the blatant dominance of the great power of the regional developing state. Understanding the potential threat from cyber operations will rise in future decades, even at the presence of stringent security architectures; this gives three major conclusions in the Indo-Pacific region. Firstly, China's engagement as a dominant power and employing AI-enabled information technology used in cyber warfare in the Indo-Pacific region in contemporary times. Secondly, no current appropriate resilience is present to mitigate the potential security threats and thirdly, understanding the future of information warfare and how could the respective states of Indo-Pacific region defence set-ups and their subsidiary alliances build strategic resilience architecture for these future challenges.

Meanwhile, China and its ambitious programmes across Asia, Europe and Africa through its One Belt One Road (OBOR) launched in 2013 in the form of the Belt and Road Initiative have a strategic form of technology dominance over the OBOR host nations. In the coming decade China is planning to establish and finance 30 nuclear reactor projects in BRI countries, which will have regional as well as global political implications. Since China is investing in nuclear projects, it will have more dominance and interference in the global nuclear market, and in international nuclear governance. Eventually, this will lead to the dependence of the BRI host countries on China. Since these host countries often don't have rigorous regulations and necessary institutions to follow the rule of law and adhere to international standards, the technology, infrastructure and technological know-how to deal with radioactive material, they will have to depend largely on China, and that might lead to compromised negotiations on the part of local stakeholder's interests. In this situation the application of the laissez-faire method by Beijing for the developmental support will easily procure for China the project deals while completely exempting them from the environmental, social, and legal consequences. It may also allow them to make profits by cutting corners. Whereas the host states will get the lower end of the negotiation deals and perhaps in some cases state bankruptcy or major losses. Technology and its involvement in the democratic backsliding is evident in many cases particularly in Eastern European states where it may be seen as state designed disabling and eliminating of the existing democracy of its own state; however, it has transformed its character dramatically from the past. The character of this process has adopted the shape of indirect methods like promissory coups, designed electoral manipulations, executive aggrandizement, extreme levels of surveillance on oppositions and journalists through the powerful states having advanced technological support and infrastructure. This contemporary form of democratic backsliding appears to be legitimate at the prima facie, but they are designed and plotted by the internal and external factors. Compared to authoritarianism in the past, the current situation is far more challenging as it is customized for a particular state and indirect and ambiguous in its approach by means of various covert, advanced sets of technological interventions and tools.

17.6 CONCLUSION

Emerging technologies have penetrated deeper into the dynamics of international relations. It is apparently clear that it is going to impact the basic architecture of the international system, mainly focussing upon its structure, its major organizing concepts and the inter relationships among its members. ET while changing the processes through which the whole international system operates, which includes the very fundamental verticals like diplomacy, administration, policy formulation, trade, communications and war and the intel sharing. Since this area is evolving, dynamic and complex, it will surely create new issues, new limitations and trade-offs in the operational system of the foreign policy. Here limitation will not only consist of the political limitations but also natural, cultural and social constraints and provide an understanding of the altered security perception of facts and transparency to operate within the international system (Malik, 2012).

A new understanding of threat to human beings and security to human life post-Cold War was given, when a gamut of security concerns for human life was defined and categorized under traditional and non-traditional security (NTS) threats. The NTS concerns include issues like food insecurity, climate change, mass migration, depletion of natural resources, diseases, transnational crimes and energy crises. Traditional security concept deals with the state, its territorial integrity and the sovereignty of the state. Therefore, in traditional security essentially the external military aggression poses threat to the sovereign state and its integrity. Absence of war does not mean that there is a presence of peace and stability, understanding this fact there is a growing alertness globally in recent years, to know the required urgency to deal with the NTS challenges which is potential enough to eventually impact the traditional security challenges of the region. The recent years indicate food and water crisis, speedy depletion of natural resources, environmental disasters and the recent pandemic which led to huge numbers of human life loss. The NTS paradigm may have severe consequences if left unaddressed, eventually escalating to the traditional security threats. NTS threats have the potential to escalate a situation into civil wars, communal violence, protests and rioting, breakdown of authoritarian or democratic forms of government. These factors become more functional in case of a fragile political system of a state and potentially destabilize not only a state but also the region. These NTS threats have alarming inferences and full possibilities to contribute as a spillover effect on the traditional security challenges as well making it more severe and large. Technology can play a key role in mitigating all the challenges posed by NTS, perhaps time-bound appropriate choices are made by humans. Interestingly technology can be the answer to the plethora of challenges faced by the world today, and how technology could be wedded to international relations to achieve solutions.

Globally the states must come up with a legal framework on AI with mutual international consent to address the risks of the outcomes of defiant use of AI. The collaborative proposal on AI with the involvement of all the states with mutual consent and pledge to meet the requirements and to protect their interest is required at an international level. This will have a significant emboldening of the framework, investments and increased number of innovations in AI across the world (Shivappa, 2024). The regulatory frameworks should be targeted to facilitate the AIs developers' deployers and the users with accurate and clear needs and obligations regarding particular utilities of the AI, and simultaneously this framework must also lessen administrative and monetary burdens for commercial applications, especially for small and medium-sized business set ups. However, recently the second global forum on the ethics of AI: Changing the Landscape of AI Governance, under UNESCO this year, 2024, wherein various UNESCO initiatives which has also included the global AI ethics observatory with UNESCO AI ethics and their technical experts without borders network (Shivappa, 2024). It may be considered a major initiative towards a significant contribution to the effective objectives for ethical AI governance by accepting a determined global standard. It is also important why we need the stringent and strict regulatory framework for AI, majorly because AI governance is must when it comes to machine learning and the associated algorithms are implemented to make decisions. Important concerns like machine learning biases, when related to the individual identities profiling wherein, they may recognize fundamental information about the users. This may apparently lead to not appropriately denying individuals the access to various platforms, from healthcare to law enforcement to criminal suspects etc. AI regulatory governance framework will also ensure what should be the best methods to deal with situations where AI could go wrong and indulge in unjust methods or even violate human rights. Since AI is still evolving and there are frequent alterations and revisions existing legal structures in respective states provide some safeguards and protection; however, it is still inadequate to address the particular challenges AI might bring. Therefore, the regulatory systems must have certain policies, for instance, a governance framework at the domestic and at the international level. The high-risk and sensitive areas should be clearly listed followed by setting up clear requirements for the AI establishment systems. A conformity assessment needs to be done prior to the AI system put before the service to be operated. Declarations of specific obligations for the AI users and to the facilitators of the high-end risk applications. Moreover, regulations must be in place to govern post-availability of such products into the market.

Now since AI are commonly found in analytical roles wherein they are combining via large data sets and attaining results and conclusions based on pattern study. These are specifically the bland tasks that are mainly believed to be the highest priority for the automation process. The chances of these intelligence modes are repetitive in nature and character and this increases as a result of the spread and the growth in the AI software.

REFERENCES

@@Choucri, Nazli and David D. Clark (2019). *International Relations in the Cyber Age: The Co Evolution Dilemma*, MIT Press.

Council of Europe, Newsroom. www.coe.int/en/web/portal/-/secretary-general-calls-for-action-against-democratic-backsliding-in-annual-report

Giacomello, Giampiero (2021). *Technology and International Relations*, edited by Francesco More and Marco Valigi, Elgar Online. www.e-elgar.com/shop/gbp/technology-and-international-relations9781788976060.html?srsltid=AfmBOopQGRjscfQQ64JDM4E2JzYZkqsdD8mXNoXlPIi5YVRdjdMdTq_

Kaltofen, Carolin, Madeline Carr, and Michele Acuto (2019). *Technologies of International Relations*, Springer Publication.

Lanoszka, A. (2016). Russian Hybrid Warfare and Extended Deterrence in Eastern Europe, *International Affairs*.

Mallick, P. K. (2023) *Information, Cyber and Space Domain and its Application in Future Land Warfare*. KW Publishers.

Malik, Mohan (2012). Techno Politics: How Technology Shapes Relations among Nations, Asia-Pacific Centre for Security Studies. https://dkiapcss.edu/wp-content/uploads/2012/12/Mohan-Malik.pdf

Mallick, Amitav (2016). *Role of Technology in International Affairs*, Pentagon Press USA.

Mallik, Amitav (2004). *Technology and Security in the 21st Century*, Oxford University Press.

Mearsheimer, J. J. (2015). Why We Will Soon Miss the Cold War. In *Conflict after the Cold War* (pp. 28–44). Routledge.

Ndzendze, Bhaso (2021). Artificial Intelligence and Emerging Technologies in International Relations. www.worldscientific.com/worldscibooks/10.1142/12216?srsltid=AfmBOop1ZqmcCWmll9BomjXjsiBVrTzmNWKYRljv-7IYnBjAtDoXZUqO#t=aboutBook

Pence, Eliot (2020). To Understand China's Aggressive Foreign Policy, Look at Its Domestic Politics, Council on Foreign Relations. www.cfr.org/blog/understand-chinas-aggressive-foreign-policy-look-its-domestic-politics

Scheuerman, William E. (18 December 2009). Realism and the Critique of Technology, *Cambridge Review of International Affairs*. https://doi.org/10.1080/09557570903325504

Sechser, Todd S., Neil Narang, and Caitlin Talmadge. (2019). Emerging Technologies and Strategic Stability in Peacetime, Crisis, and War. *Journal of Strategic Studies*, 42, no. 6: 727–735.

Shivappa (2024). Need for a Strong Regulatory Framework for Artificial Intelligence. www.expresscomputer.in/artificial-intelligence-ai/need-for-a-strong-regulatory-framework-for-artificialintelligence/108819/#:~:text=AI%20regulatory%20governance%20framework%20determines,challenges%20AI%20systems%20might%20bring

Steff, Reuben, Joe Burton, and Simona R. Soare. (2020). *Emerging Technologies and International Security*. Routledge.

Williams, Michael C. (2004). Why Ideas Matter in International Relations: Hans Morgenthau, Classical Realism, and the Moral Construction of Power Politics, *International Organization*, 58, no. 4: 633–665 (33 pages), Cambridge University Press.

18 From Threads to Tangles

AI's Disruption of the Social Fabric

Yahvi Johri and Akanksha Srivastava

18.1 AI CONTRIBUTES SIGNIFICANTLY TO JOB INSECURITY AND WILL CONTINUE TO DO SO

The ability of artificial intelligence (AI) to scale according to increasing demands is arguably its best feature, having a level of flexibility and efficiency that traditional methods cannot compete with. With AI integrated into their day-to-day functionality, businesses can deal with and respond to shifting workloads easily, allowing them to adapt quickly to changing market conditions, which is very common in competitive markets. This agility significantly reduces the overhead or managerial costs needed for hiring and training additional employees. For many industries, specifically transportation, retail, and even logistics, this capacity can mean the difference between profiting by large sums and just being at that break-even point. AI can help optimize strategic planning for supply chains and improve customer service solutions. This will allow businesses to focus on growing instead of spending all their time on manual and operational tasks (Morgan, 2023). However, this advancement comes with a downside. The rise of automation poses a serious risk of job displacement for employees in these industries. Workers, like truck drivers, warehouse personnel, and retail staff, face uncertainty as self-driving technology becomes more commonly used, as do for automated systems. This implies a shift in traditional job roles, raising concerns about the future of labor-intensive, repetitive jobs (Morgan, 2023). Similar to the way call centers dwindled, waiters are actively being replaced by "waiter robots." The change and increased usage of AI-based solutions question workforce sustainability. Businesses might profit from increased productivity and lower labor costs, but there is a definite need for strategies that help displaced workers find and transition into new jobs, so they can make ends meet. This topic is also an ethical concern. Companies and policymakers will have to work together to guarantee that the pros of AI do not cost workers a fair and equitable job market. Jobs meant mainly for those with a lower IQ or less ambition rely on these jobs to meet their daily needs. This replacement also questions the future of disability-inclusive employers, such as McDonald's, Walmart, Harris Teeter, Amazon, and Lowe's Foods. These companies have initiatives to hire handicapped personnel, but in the scenario that all similar jobs for regular, healthy people are also replaced with AI, what will happen to the jobs designated for the handicapped?

DOI: 10.1201/9781003560463-22

In conclusion, the scale of AI and its ability to adapt to volatile demands are lucrative for all businesses looking to profit as much as possible, giving them a way to improve efficiency. Nevertheless, it is crucial to address the impact this technology will have on the workforce (how it will affect unemployment rates). Only with such awareness, and policies addressing it, can the industrial world create an environment where innovation and human job employment can coexist healthily.

Any business would logically aim to decrease internal costs, and so will utilize AI more than workers. Job displacement, specifically in manufacturing, data entry, transportation, and retail, is a potential consequence of this. While some people believe that technological development and maintenance will lead to the creation of new jobs, the reality is that this technology is more likely to eliminate more jobs than it creates. The proposed shift is likely to worsen income inequality, leading to a wider gap between the rich business owners and the less-paid white-collar workers.

This increased use of AI will inevitably lead to the close examination of the work done by human workers, criticizing their performances to find tasks that can be replaced with AI. Jobs completed by labor workers will be assessed on their overall costs, difficulty of the tasks, and other factors. On this basis, companies may choose to replace or downsize departments altogether, with the help of AI.

As AI becomes increasingly integrated into the workplace, employees will need to adapt and learn how to use these technologies regularly. Individuals possessing expertise in this area will thrive in this up-and-coming job market, but those lacking such expertise will face challenges in securing basic jobs worldwide.

18.1.2 AI's Disastrous Impact on Call Centers and Customer Service

The use of AI in the retail world goes beyond just chatbots and virtual assistants for customer service. As technology continues to advance, there is a perceivable shift in how retailers interact with customers and deal with operations and labor management.

One of the easily relatable applications of AI in retail is creating personalized shopping experiences. Retailers are using machine learning to understand and improve according to consumer behavior and preferences, which significantly improves their ability to accurately recommend products that seem tailor-made for each customer. This method satisfies customers and boosts sales at the same time, because people are significantly more likely to purchase items that match their "personal" style.

Also, AI is giving retailers insight into market trends and what customers are likely to buy during a specific season or month. Using this information, companies can improve their inventory—increasing the frequency of popular items, making sure to re-stock them, and at the same time avoiding a surplus of less popular products. This method reduces costs, improves the shopping experience, and makes life easier for employees because they do not have to think as much about a customer's preferences, decreasing chances for error. Similarly, customers are less likely to face frustrating shortages.

AI is also changing supply chain management. Technology can predict shifts in demand, almost like a crystal ball. It can improve delivery routes and deal with logistics tasks much better than traditional methods, which leave human error up to chance. This level of flexibility is extremely important in today's fast-paced and ever-changing commercial market, where consumer whims and fancies can change rapidly, depending on micro-trends started by celebrities or Instagram reels. By utilizing AI, retailers would be better at handling customer demands while keeping their finances in check.

Security is another area where AI is helping in retail. As e-commerce has become a common way of shopping, many people prefer it to shopping in stores after the COVID-19 pandemic, so they have concerns about data privacy and fraud. AI systems can detect suspicious activity, monitor transactions around the clock, and safeguard sensitive customer information, guaranteeing loyalty for customer bases. Retailers and other companies would not have to spend too much money on cybersecurity personnel.

There are also more frequent innovative in-store experiences because of AI. Retailers are messing around with augmented reality (AR) and virtual reality (VR) to build engaging and entertaining shopping experiences. For example, furniture stores use AR to let customers visualize how a piece might look in their home, which can mitigate future returns and customer concerns. This serves as a chain reaction; fewer returns means happier customers, and customer service employees can spend their time helping other customers in need of assistance. Some clothing companies have apps to help customers visualize how the clothes will look on them before purchase.

Although as AI continues to adapt and be used by companies, it is important to acknowledge the challenges it brings with it. As automation takes over certain job functions, there's a growing need for workers to adapt. Upskilling and reskilling initiatives will be crucial in helping employees transition into roles that AI isn't used for, like a training program. Obviously, AI is not meant for many human-interactive jobs, especially those requiring emotional intelligence and complex problem-solving skills.

AI may be changing how the retail industry, particularly the personnel portion, is structured, but the human element is still important. Retailers need to find a balance between using AI to improve efficiency and making sure that there are still human employees to represent the companies. These employees will need to be available for more complex questions that need a personal element, like how you can ask to speak to an agent when talking to the automated messages for customer service.

Let us dive into the application of AI within the retail industry, specifically. It is already common practice for chatbots and virtual assistants to be used for 24/7 customer support. In recent years, there has been a significant downward trend in the demand for call center jobs. Companies began using AI for customer support as early as 2010, but it was limited to scripted responses and could only handle basic queries. At that point in time, call center jobs were not high risk because humans still played an important role in complex customer interactions. As the areas of machine learning and Natural Language Processing made progress in the late 2010s, chatbots became more adequate for customer service calls with businesses seeing positive responses regarding these interactions. The years 2020–2022 saw a noteworthy rise in the popularity of this trend, the main cause being the global impact of the COVID-19 pandemic and the ongoing developments of ChatGPT. By utilizing AI technology, companies could limit in-person interaction, which proved to be a viable solution for minimizing unnecessary virus exposure for previous workers. Simultaneously addressing the increasing traction of online shopping and the corresponding need for efficient online customer service. Generative AI (ChatGPT) made these calls more humane, with adaptive responses that simplified the experience for the customer. With the increasing adoption of AI technology, call center duties will likely be primarily dealt with by AI systems soon (Parkin, 2024). This trend is already clear in countless global companies. From Amazon to H&M, the dominating businesses have incorporated AI into their customer service experience. With reports of decreased wait times and overall positive shopping experiences, soon more companies will follow their example, and call center jobs will virtually disappear.

According to a report by Gartner, "By 2026, conversational artificial intelligence (AI) deployments within contact centers will reduce agent labor costs by $80 billion" (Gartner, 2022). The number of jobs lost is likely to be very high, and we can only surmise the true scale of the impact.

The chief executive of Tata Consultancy Services, K Krithivasan, said to the *Financial Times*

> In an ideal phase, if you ask me, there should be very minimal incoming call centers having incoming calls at all…We are when the technology should be able to predict a call coming and then proactively address the customer's pain point.

He predicted that within the next year or two, chatbots will be competent enough to analyze customer transaction histories and perform many of the duties that call center agents currently perform (Parkin, 2024). Why would a company spend large sums of money on salaries and employees, when they can commit to a one-time investment like this? This man is at the forefront of the rapid advancement of technology and sees its potential future implications up close every day. His statement, therefore, paints a highly reliable visual of what we can expect from AI in the coming years. In an article by BBC, there is a shocking statistic listed; "Artificial intelligence (AI) could replace the equivalent of 300 million full-time jobs, a report by investment bank Goldman Sachs says" (Vallance, 2023). Goldman Sachs, a well-known global investment company, and thus, large statistics like this to be stated by them indicates that we can very clearly surmise how significant of an impact AI will have on unemployment rates. "The potential for generative AI to replace white-collar jobs altogether, such as call center agents, raises significant moral and ethical concerns for policymakers worldwide."

Transportation is another sector that will face such transformations. The rise of autonomous vehicles will affect this industry, the key aspects being the use of delivery drone systems, proactive maintenance measures for vehicles, and the use of intelligent traffic light systems.

18.2 CORPORATE FIRMS AND CREATIVE COMPANIES SHOW A TREND IN FREQUENT AI USE

In addition to the retail and transport sectors, corporate firms in the United States are also welcoming this technology due to its clear benefits. Sadly, this development will mean that even more workers will face uncertainty about the security of their jobs. A recent survey conducted by finance chiefs for an article in CNN Business states,

"More than half (61%) of large US firms plan to use AI within the next year to automate tasks previously done by employees." Duke University, an institution renowned for its education in finance, and the Federal Reserve Banks of Atlanta and Richmond. The study also states that "Those tasks include everything from paying suppliers and doing invoices to financial reporting" (Egan, 2024). The research highlights that the tasks done by AI can include a vast variety of responsibilities, from handling payments to suppliers and processing invoices to generating financial reports. Typically, dealing with invoices is done by the Accounts Payable Department of companies. This department is specifically in charge of accepting, reviewing, and approving vendor invoices and deals with data entry. As previously mentioned, data entry was identified as a task likely to be virtually completely automated by AI. The chapter goes on to explore the expanding impact of AI on work environments, analyzing how the technology is being embraced by both innovative startups and already-established businesses with traditional infrastructures. For creative tasks, AI chatbots assist with "crafting job posts, writing press releases, and building marketing campaigns."

Companies are quite clearly moving toward different ways of using technology on the ground floor and for administrative tasks. "Nearly 60% of all companies (and 84% of large companies) surveyed said that over the past year, they have already leaned on software, equipment or technology including AI to automate tasks employees previously did" (Egan, 2024).

This automation offers several advantages, including improving product quality, increasing output, reducing labor costs, and substituting workers. The integration of this technology into day-to-day work has caused much concern, particularly within government agencies and in research communities. A CNN Business article reported that a prominent government official highlighted the incompetence of current legislation addressing AI, further emphasizing the government's failure to address numerous complications and potential pitfalls associated with AI usage. "Chairman of the Homeland Security and Government Affairs Committee found that existing regulation 'insufficiently addresses' how hedge funds are already using AI" (Egan, 2024).

This situation underscores the hazards associated with unchecked corporate greed, where the chase for maximum profit outweighs any ethical consideration. This ultimately jeopardizes the livelihoods of those in white-collar jobs who need these incomes to support their families. In terms of what we can expect from businesses in the future, this oversight is alarming.

The implications of this are not limited to job displacement; they question the fabric of our workforce and economic stability worldwide. As business giants and companies worldwide increase automation (Ex. Amazon and Tesla), policymakers must focus on the creation of laws that can deal with these issues. Policies focused on providing security for workers displaced by AI, such as retraining programs and unemployment benefits, are vital. This exploitation leads us to another major topic of concern: the security and privacy concerns associated with AI.

18.3 BIAS, DATA PRIVACY, AND SECURITY ISSUES

The fundamental concept of machine learning is to provide systems with data that they can use to gain experience, discern patterns, and improve their overall response quality. The use of data today requires a closer examination of the many issues surrounding data privacy and its impact on civilians. From a legal perspective, there are many restrictions governing the sharing of data between agencies, including strict conditions for consent from the providing user, and other regulations that must be followed. We need to know how this data is being processed and stored and what are the steps involved.

The data used to train machine learning algorithms can include a wide range of information, from personal names, medical records, and addresses to other highly sensitive details. The user would understandably not want this data to be used without their explicit consent. The threat of data breaches is a point of concern for companies since when a company has access to sensitive information, it is almost always at risk of being hacked. Each company needs to take the appropriate security measures, for example, hiring cybersecurity personnel, to limit breaches of this data.

Another topic for debate is the use of AI for monitoring and surveillance. The government's use of facial recognition software to identify criminals and track them raises concerns about privacy rights, as there is a risk that those with malicious intent could misuse this technology, posing a significant threat to security.

The General Data Protection Regulation (GDPR), a part of European Union legislation, is a protective measure against data breaches and disasters by establishing specific guidelines for the handling of personal data for all individuals residing within or connected to the EU.

It is important to closely monitor data used by any machine learning system so that it remains confidential. One unfortunate consequence of any kind of advancement is the increased vulnerability of those who are associated with its development. The in-depth analysis of a person's behaviors, emotions, and thoughts thanks to AI raises a multitude of concerns, as this knowledge could be abused and used to target the individual for harm or to manipulate them for financial or personal gain. Similarly, the analysis of a person's social media raises concerns for the safety of their family members and the communities they are involved in.

It is widely accepted that the effectiveness of an AI system is determined by the quality of the data and patterns it is trained to recognize, further showing the crucial role of data quality in AI development. It is also concerning that if the intelligence system recognizes a preference driven by racial bias or other unfair motives, it could strengthen this pattern and lead to biased outcomes for future tasks. A company that discriminates against its employees is not aligned with ethical practices because using such an algorithm would mean that the business is hiring employees in a biased manner, violating countless policies in the process. This instance once again emphasizes how important proper regulation is. The typical hiring manager, who is a human being and was assigned a specific task, makes it quite easier to determine any foul play, in terms of potential biases and attempts to discriminate. Unlike conventional methods, identifying flaws in data analysis using AI is much more difficult to prove because of how complicated the neural network is. To prevent getting mixed up in this web of confusion and legal issues, AI's access to data by businesses should be limited. Controlling AI's involvement from this aspect also ensures that any job applicant has a fair chance to earn that the job.

It is hard to see now, but AI causes high uncertainty. Using the metaphor of the "social fabric," this technology will slowly inflict subtle, minuscule tears on the very foundation of our society, and by the time the consequences become apparent to the public, or researchers and business magnates, the resulting large rips may be too extensive to sew back together.

18.3.2 Deepfakes and the Risks They Pose to Society

The implications of deepfakes go far beyond just entertainment. They can pose significant risks in multiple aspects of society, including but not limited to privacy, misinformation, and defamation. For example, deepfakes can be used to create misleading videos or audio recordings that could potentially damage the credibility of public figures, create political chaos, or facilitate fraud. Imagine the influence such media would have on followers of political figures, instigating them into potential riots.

The ethical concerns surrounding deepfakes are vast: they challenge our understanding of authenticity and trust in digital media. As the technology continues to change and adapt, it becomes more difficult for individuals to differentiate between what is real and what is fabricated. This presents a serious issue in fields like journalism, where the line between genuine, factual reporting and fabricated or made-up content can become easily blurred. To viewers who are uneducated on how to verify their sources, any information they consume, deepfake or not, will seem true. Then they could spread that information to others, and more people will believe the wrong information: this is primarily how rumors start, and then they take on a life of their own.

Many government officials and businessmen have already become victims of deepfakes, and these videos have circulated on the internet and gone viral on social media. An article by Bernard Marr of Forbes magazine describes the illusions created by deepfakes. He states,

> When researchers at the University of Washington posted a deepfake of President Barack Obama and then circulated it on the Internet, it was clear how such technology could be abused. The researchers were able to make the video of President Obama saying whatever they wanted to say.
>
> (Marr, 2019)

This was a test conducted by researchers, but imagine the implications if a terrorist or any other group with malicious intent were to do the same with any sort of public or government official. It could create chaos within society, as people tend to become passionate about political beliefs, and the person the deepfake is created about would be at risk in terms of safety but also their reputation. Once a person is shown saying or doing anything, it is quite difficult to erase that image and the associated perception that has been created of the star in question. Even after such videos are reported and removed from the internet, it is virtually impossible to erase their impact. When further analyzing the scenario, the probability that this media is erased from every platform or website on the internet is close to none.

There are many ways to stay safe from deepfakes, primarily by being aware of the media that is made available to the public, because it is often incorrect or a severe exaggeration. Specifically, it is by verifying the genuineness of content with digital signatures. By scrutinizing all the information you see on social media and on the internet in general, you will be more confident in the truthfulness of the information you relay to others and can rely on the information you deem true, as opposed to blindly believing any video or image.

Using AI and chatbots often may decrease the need to use our critical thinking skills over time. As people utilize AI and ChatGPT more often for decision-making, they lose their decision-making skills and therefore their ability to adjust to new situations (Ahmad et al., 2023). Autonomous thinking is the action of thinking independently and making decisions without the influence of others. This type of thinking is crucial for developing important critical thinking skills and learning to make sensible judgments. Since AI is becoming readily available to adolescents, who don't grasp the long-term consequences of relying on this technology, they are at risk of permanently hindering their critical thinking skills. Adolescents and young adults can quite easily access AI, from Microsoft's Copilot to ChatGPT. This is not to assume that all people of this age group are likely to misuse AI; some students are probably using it for purely educational purposes: to learn more efficiently via note taking or test prep assistance. However, this issue will impact their ability to retain information and put at risk their future academic performance and career-related success.

It is not wrong to use AI for understanding a concept or creating a more efficient way to access resources and comprehend information, but submitting it as your own and using it for shortcuts that are detrimental to your learning it was is unethical, If a student were to use AI to quiz themselves son vocabulary terms or create a comprehensible note sheet for test review, that would be considered ethical in most circumstances. However, using AI to generate written responses that the student then submits as their own for school or using it for test answers is unethical, primarily because it is cheating the purpose of the school system, and the educational system is unaware of the student's actions. Another example of AI usage that would typically be deemed appropriate is completing your work (or homework) and then using AI to understand your mistakes, after having put in your own effort. The main issue with AI usage lies in students and people in general not wanting to complete any work in the first place, so they assign the task to this machine learning system. Education systems will have to create an iron-clad framework for the regulation of AI usage, as the technology advances, and filters to determine if work is AI-generated or not. This will make sure that students who are putting in the effort to complete their own work are rewarded, while those who are only interested in finding shortcuts are reprimanded. Using AI to generate writing and answers is deemed plagiarism by many universities, and a punishment for this could be as severe as expulsion.

It is important to understand how many different people and careers will be impacted by the unregulated use of AI. Let us take an example. An intern, Daniel, at a well-known hospital that has an influx of hundreds of patients each day is working closely with senior doctors. Every day, the intern is expected to make rounds on patients with these doctors, read the lab work and others' reports, and visit the patients to develop their diagnoses. After Daniel creates a diagnosis and decides on a treatment, he is expected to go to the senior doctors and present his findings. The doctors will then critique his analysis and advise him on what mistakes he made. These mistakes could range from not considering the patient's familial history, overlooking inconclusive lab results, or a myriad of other errors. Every mistake is important for the learning curve of becoming a physician. This is one profession where as you get older, you gain more experience and are therefore more evaluated in your field. The instruction will help make Daniel into a better doctor, so he doesn't repeat the same mistakes, and it is beneficial to all of his future patients.

But now, let us say the intern is feeling tired and does not want to comb through the lab reports or the patient's familial history. And so, he turns to AI. The intern enters the lab report assessment and results into the system, along with the patient's familial history, and demands that the AI develop the most probable diagnosis along with the respective treatments. Daniel then memorizes the results and goes to the senior doctors to present the diagnoses as if they were his own. All the doctors praise him, and only point out a handful of mistakes, thinking that he is incredibly smart. The intern may feel proud of himself and manage to save half an hour of his time, but he also costs himself far more than he could imagine.

The implications of this workaround are quite serious. Not only is he lying to the senior doctors and creating a false image of himself, but he is also doing a disservice to this patient and others. Daniel did not comprehend any of the lab reports or patient history, so he did not learn anything from this patient. The crux of an internship is to learn as much as possible, not to find convenient ways of doing no work. Now, Daniel may have fooled the senior doctors into thinking he is competent enough to deal with this level of diagnosis, but in the future, with new patients, he will have no clue what to do. He did not do any of the work himself, so he is not aware of any of his medical misconceptions that the doctors would have been more than willing to clarify. If faced with another similar patient, Daniel will either be forced to use AI again, and risk making a wrong diagnosis, or continue this charade, in which he learns virtually nothing and is setting himself up for failure when he takes his exams to become a practicing physician. All of this is to say that AI may seem like a good option when you are tired and need a "quick fix," but later, you will realize the consequences of taking this shortcut and could risk your reputation and academic or career-related success.

With machine learning and this field in general, proper licensure is necessary to prove that the data sets used to train the AI are in fact accurate. According to *IEEE Spectrum*, the Data Provenance Initiative

> audited more than 1,800 widely used datasets for training natural language processing models that are publicly available on websites like GitHub, Hugging Face, and Papers With Code. They found more than 70 percent had no data licenses, and for those that did, roughly half were incorrect. What's more, 29 percent of the incorrect licenses were more permissive than the dataset creators had intended.
>
> (Gent, 2023)

This indicates that the policy strictness for data usage and licensure needs to be increased and regulated, or else the public becomes more vulnerable to misinformation.

One more caveat to using AI in healthcare is the margin of error. The training data used for AI, a form of machine learning, could be biased toward certain populations, potentially excluding people of specific races, and mistakenly fail to account for them. It is a well-researched fact that people of different races and ethnicities have different underlying conditions and overall medical histories. For example, sickle cell disease is more common in African, African American, and Mediterranean people. If the data the AI is trained on excludes these races, then the system cannot make an accurate diagnosis for someone from these races. Before AI can be used for precise medical diagnoses, it will need to undergo extensive testing, over multiple years, and be provided with diverse datasets so it can create medical diagnoses that address patients of all races. Before then, any clinical application is at risk of being inaccurate and potentially harming the patient or ruining the reputation of the healthcare facility and doctors that use it.

The use of AI in healthcare introduces many challenges meant for careful consideration, especially the margin of error linked to biased training data. Machine learning algorithms that power these AI systems depend on the datasets they are exposed to, to learn from and become more accurate and knowledgeable. If these datasets mainly feature data from specific demographic groups, they can lead to a limited understanding of health conditions and treatment responses among diverse populations. In countries like the United States, where many people have different ethnic and genetic origins, this is particularly troubling when considering the different health profiles of racial and ethnic groups.

For example, some diseases have a higher likelihood of manifesting in specific populations. Sickle cell disease, a genetic blood disorder, is much more likely to occur among people of African, African American, and Mediterranean descent. If AI systems are trained on datasets that do not sufficiently represent these populations, the likelihood of misdiagnosis increases. An AI system could fail to recognize symptoms or the importance of family medical history unique to these groups, which could lead to inadequate treatment or missed opportunities for preventative care. Something as simple as a form entry error, when creating a patient history, is enough to wrongly diagnose a patient. It is not as if the AI takes a photo of the patient; it relies on text data. A doctor may be able to identify the error by simply looking at the patient and seeing any indicators of race or by speaking to the patient. However, the AI system does not have the same human interface or interaction skills. Before AI is properly tested, like a clinical trial with humans, any clinical application is at risk of being inaccurate and potentially harming the patient or ruining the reputation of the healthcare facility and doctors that use it.

Ethically speaking, this creates patient mistrust and ruins the accountability of a physician. The reputation of the senior doctors who oversee Daniel is at risk as well.

18.4 FUNDAMENTAL HUMAN BEHAVIORS

One principle of psychology that has been studied throughout history is the hedonic principle. The premise of this principle is that people attempt to enhance their positive feelings and avoid unpleasant ones. In terms of work, this fundamental aspect of human behavior means that employees will tend to avoid stressful situations and tasks that are tedious or repetitive. And this is where AI comes in. As mentioned at the beginning of this chapter, AI is particularly helpful in performing repetitive tasks at a significantly faster pace, with greater accuracy and consistency. While previously (see Section 18.1), we focused on how business owners and leadership could use AI, on the other side of that same coin, the workers will attempt to minimize their workload as well.

Some psychologists argue that laziness is a behavior learned from others, but if this is true, then as AI use becomes prevalent in the workforce, people will mimic each other's behavior and rely on technology even more. This could cause a cascade ending in severe dependency on AI, and a lack of productivity by the actual human workers. It will give incentive to the managers to use AI instead, and workers would be fired. This technology is a tempting resource, and it is hard to say who is at fault when we consider the innate behaviors of humans, owners, and employees alike. The impulse to not do work and be lazy is a common experience for everyone. Only now, with AI, does it have more severe repercussions.

We can analyze even the most common errands for families and see how AI will impact them. For example, grocery shopping: let us imagine the typical day of any family, of any socioeconomic standing. To survive, this family will need clothes and food on a regular basis. So, they will go to the market, with poor people in third-world countries going to the bazaar, and middle-class families in first-world countries going to the typical supermarket. Grocery shopping is part of their regular chores and allows members of that family to interact with others in public. Many people are socially introverted and already have limited human interactions. And so, even saying "hello, how are you?" to a cash register allows them to experience human contact outside of their own family. In psychology, human interaction outside of familial interaction is important for self-esteem, reducing stress, and building a sense of belonging. All these core values are important to humans of any age and in any corner of the world.

But now let's suppose AI starts to play a role in grocery shopping. Based on a list on your phone, or on your Samsung fridge, the system predicts which groceries you will need and adds them to your cart on an app that delivers groceries right to your door, with the click of a button. Now, this may seem like a relief at the time, because you can spend more time at home, but it is just one example of human interaction and destressing activity that would allow you to leave your home and clear your mind for a little bit.

In psychology, there is a term for being cooped up at home. It is called "cabin fever." When a person has cabin fever, they can appear restless, irritable, and lonely, and display other symptoms of poor health. The main cause of this phenomenon is the absence of social interaction. As a result, those diagnosed may start to lash out, have difficulty sleeping, have trouble focusing, and have poor eating habits. In turn, these behaviors will negatively impact the mental health of family members, especially children.

As we look ahead, we need to face the fact that AI will play a huge role in our everyday lives, especially when it comes to things like housework and taking care of kids. AI won't just be a tool; it'll be there to help remind children to eat their dinners or take their medicine. This function isn't just about sending prompts; it's designed to feel more personal, like having a caring presence around. However, we must consider the effects this could have on kids emotionally.

It is a common plot in movies that children become more attached to their nannies than their mothers. This isn't just entertainment—it emphasizes real concern about how children build emotional bonds with caregivers in their lives, typically supposed to be their parents or guardians. If a child feels that their parent isn't emotionally available, they may end up feeling closer to an AI or caregiver that shows and holds a constant place in their life. For example, if a child is sick, with flu or with a fever, an AI that checks in on them or entertains them can provide some semblance of comfort, giving them medicine at scheduled times, inadvertently leading to a stronger connection than with their parents. If the child, since they are an infant, spends more waking hours with the nanny, they are said to develop a strong attachment to that individual. A similar concept can be applied to AI, where the child becomes attached because it feels cared for by it. This could lead to an unhealthy emotional dependency that disrupts family relationships and affects the social and emotional intelligence of the child in question.

We also need to think about how technology is negatively impacting our ability to think and memorize. In the past, people had to memorize phone numbers to stay connected with family and friends. That was an important skill that helped keep our brains sharp; you would remember a few important numbers so you could enter them in a telephone booth. But now that smartphones have taken over, with the numbers stored in your contacts app, we tend to panic if our battery runs low or our phone dies because we've become so dependent on these devices. This isn't just a change in how we keep contact information; it's an indicator that we might be losing our memory and problem-solving skills.

The brain's "power" to memorize phone numbers uses our short-term or so-called working memory. This aspect of the brain is located in the prefrontal cortex and lets us store a limited amount of information: around seven numbers, give or take two, according to the cognitive psychologist George A. Miller of Harvard University's Department of Psychology. We break down the long number into smaller chunks, a process known as "chunking," so the brain can store the information more efficiently. Using your working memory is important to be able to hold information: it will allow you to focus on certain information, multitask, follow instructions, and improve your problem-solving skills.

This issue gets even more serious when we imagine how widely used AI will be in different industries. AI can analyze information much faster and more efficiently than we can, which could lead us to depend on it even more for tasks that require our own thinking. This assumption could lead to excessive reliance and laziness for basic tasks that any human could perform or think about. We are already in a situation where we're limiting our critical thinking skills because we are finding shortcuts to let technology do the heavy lifting.

Consider how we navigate today. Most people rely on a GPS (Global Positioning System) instead of using their sense of direction or remembering how to read a map. Another activity that would employ critical thinking skills, such as reading signs and directions. Many people today do not understand the number system used for the

American Highway System, because they are under the assumption that they will always have a phone on hand to guide them. The route number on the highway is quite simple to navigate, and understanding it should be a skill everyone has in their arsenal, for emergency scenarios. In this system, odd-numbered interstate highways indicate a mainly north-south route, while even-numbered highways indicate a mainly east-west route; "the lowest numbers start in the west for north-south highways and in the south for east-west highways, increasing as you travel east or north respectively" (FHWA, n.d.). This reliance could make us lose important skills like spatial awareness and the ability to problem-solve independently, without relying on technology.

As we welcome the integrative use of AI in everyday life, we must be aware of the risks we take by giving up our cognitive involvement with regular tasks and critical thinking. The major question is how much we are willing to sacrifice in exchange for making life easier in the short term. Now, more than ever, there is a need to proceed with caution: we need to find a healthy balance between using technology to help us and keeping our own cognitive abilities sharp. Being aware of this issue and actively working to stay engaged is necessary for a successful future; it is important to learn to use AI and be aware of how it can help but in moderation.

Cognitive offloading is when people are dependent on external tools and devices to store and recall information, instead of just their own memory. A study conducted by Karthika Suresh Kumar at Iowa State University investigated the idea of cognitive offloading and how it affects memory retention. The study shows that though AI can help implement learning tools to increase memory retention, if overused, it can also cause a decline in cognitive skills (Kumar, 2023).

In conclusion, there are many aspects of AI to consider and regulate, and countless industries that the technology will impact. The public and, in general, any user needs to remain cautious when using the tool, and make sure that it remains a tool that enhances their current work, not turning into a dependency or crutch. Similarly, the government should implement policies that regulate this usage and licensing, so AI can be used to help humanity become more productive and efficient. The future of AI and its impact on memory retention and cognitive abilities is a topic of ongoing research and debate. As AI continues to grow, it is very important to create a clear boundary and be able to balance between utilizing the technologies' benefits and being aware of or falling prey to their downfalls. Teachers and parents lead by example and will play an integral role in controlling or limiting children's use of AI. Hopefully, this regulation will ensure that children continue to gain important critical thinking skills and simultaneously learn how technology can be used to help them and improve their learning experience, not destroy it. For adults and children alike, the tool can be helpful if used in moderation with proper guidance and regulation.

WORKS CITED

Ahmad, Sayed F., et al. "Impact of Artificial Intelligence on Human Loss in Decision Making, Laziness and Safety in Education." *Humanities and Social Sciences Communications*, vol. 10, no. 1, 2023, pp. 1–14, https://doi.org/10.1057/s41599-023-01787-8. Accessed 14 Sept. 2024.

Egan, Matt. "AI Is Replacing Human Tasks Faster than You Think | CNN Business." *CNN*, 20 Jun. 2024, www.cnn.com/2024/06/20/business/ai-jobs-workers-replacing/index.html

Gartner. "Gartner Predicts Conversational AI Will Reduce Contact Center Agent Labor Costs by $80 Billion in 2026." *Gartner*, 2022, www.gartner.com/en/newsroom/press-releases/2022-08-31-gartner-predicts-conversational-ai-will-reduce-contac

Gent, Edd. "Public AI Training Datasets Are Rife with Licensing Errors – IEEE Spectrum." *Spectrum.ieee.org*, 8 Nov. 2023, spectrum.ieee.org/data-ai

FHWA. "Interstate System – Design – Federal Highway Administration." (n.d.) www.fhwa.dot.gov, www.fhwa.dot.gov/programadmin/interstate.cfm

Kumar, Karthika Suresh. *Cognitive Offloading and Learning: Enhancing Memory Retention through ChatGPT Integration*. Iowa State University, 2023. https://dr.lib.iastate.edu/bitstreams/984c8378-ee58-4ebb-8dae-1f482d9d7d04/download

Marr, Bernard. "The Best (and Scariest) Examples of AI-Enabled Deepfakes." *Forbes*, 22 Jul. 2019, www.forbes.com/sites/bernardmarr/2019/07/22/the-best-and-scariest-examples-of-ai-enabled-deepfakes/

Morgan, Blake. "What Impact Will AI Have on Customer Service?" *Forbes*, 16 Aug. 2023, www.forbes.com/sites/blakemorgan/2023/08/16/what-impact-will-ai-have-on-customer-service/

Parkin, Benjamin. "Artificial Intelligence: AI Could Kill Call Centres, Says CEO of Tata Consultancy Services K Krithivasan." *Australian Financial Review*, 25 Apr. 2024, www.afr.com/world/asia/ai-could-kill-call-centres-says-boss-of-tata-it-business-20240425-p5fmlz. Accessed 20 Nov. 2024.

Vallance, Chris. "AI Could Replace Equivalent of 300 Million Jobs – Report." *BBC News*, 28 Mar. 2023, www.bbc.com/news/technology-65102150

19 The Use of Artificial Intelligence in Financial Institutions

A Double-Edged Sword Effect

Shruti Ranjan and Ritika Srivastava

19.1 INTRODUCTION

Artificial intelligence (AI) has proved to be a massive game changer in the learning culture that improves the process of learning. The inclusion of AI in financial institutions is a non-negotiable term as the new technology demands the creation of collaborative information. The integration of technology with AI in the business place also requires the dynamics of the socio-technology system.

AI strengthens a system's resilience by enabling self-testing and self-healing software to ensure proper operation even in the face of erroneous inputs. AI can also enhance a system's reaction by allowing it to automatically defend against attacks, hone next tactics by learning from previous failures, and possibly carry out more forceful defences with every iteration.

In the current chapter, the average level of the business is highlighted. To determine whether AI can reveal an amplification capability and assist personal skills rather than replacing them, it refers to the personal and institutional theory of learning. AI systems converse with their environment as self-sufficient, inquiring agents. Their robustness depends on a variety of factors, including how they are designed, trained, and interact with other agents once they are in operation. As a result, in order to ensure robustness, standards and certification processes must take into account the dynamic and self-learning nature of AI systems and incorporate monitoring and control mechanisms that span the full lifetime, from conception to development.

Virtual instructors that not only explain concepts to be learned but also have the insights to react to student behaviour have previously been the priority of AI in education. Examples of such responses include providing flexible support, attempting to address student learning styles, or providing socially acceptable communication.

AI is being swiftly incorporated into a wide range of financial industry applications. Any potential impact on financial stability as a result must be carefully considered.

DOI: 10.1201/9781003560463-23

It is critical to keep a close eye on these developments since the utilization of these technologies in finance is still in its infancy and is developing quickly, and because there is a dearth of data regarding their application.

With the passing of time, AI has helped not only in the support services of the financial sector but also in decision-making. This technology is giving assistance to financial companies, especially startups, helping in new product development, risk management, and providing more advanced customer services.

The chapter focuses on the public who are enthusiastic about AI in financial institutions. Although learning and development are emphasized more on dynamic challenges in financial institutions as a setting environment for developing competence in the workplace. The broader sociological and economic dialog examining the influence of new technologies on labour and workforce is fundamentally tied to this analytical unit.

AI and machine learning (ML) (Xie, 2019) have revolutionized the financial industry, finding application across distinct domains that can be divided into four major key areas.

1. ***Customer-Centric Applications:*** AI and ML are prominently used in customer-facing roles to enhance service delivery and decision-making processes. For instance, see Xie (2019).
   a. ***Credit Scoring:*** Algorithms analyze vast datasets to assess solvency more accurately than non-modern methods, thereby improving lending decisions.
   b. ***Insurance:*** Predictive analytics help insurers assess risks and set premiums more precisely, improving underwriting accuracy.
   c. ***Customer Service Robots:*** Some AI-powered services, such as chatbot virtual assistant etc., provide round-the-clock support, handling inquiries and guiding customers through various financial services.
2. ***Management-Level Applications:*** These applications focus on optimizing operational efficiency and strategic decision-making within financial institutions (Xie, 2019):
   a. ***Capital Optimization:*** AI algorithms help optimize capital allocation by analyzing risk profiles and regulatory requirements.
   b. ***Risk Management:*** ML models detect trends and anomalies in data to find risky factors and mitigate them proactively.
   c. ***Market Impact Analysis:*** AI tools analyze market trends and predict their impact on portfolios, aiding in informed investment decisions.
3. ***Financial Market Transactions and Portfolio Management:*** AI and ML algorithms are integral to executing financial transactions and managing investment portfolios (Xie, 2019):
   a. ***Quantitative Trading:*** Algorithms execute trades based on predefined criteria, leveraging real-time market data and historical trends.
   b. ***Portfolio Management:*** AI-driven models optimize asset allocation and rebalancing strategies to maximize returns while managing risks.
4. ***RegTech and SupTech:***
   a. ***RegTech (Regulatory Technology):*** AI and ML are used by financial institutions to ensure compliance with regulations more efficiently

(Xie, 2019). These technologies automate regulatory processes, monitor transactions for suspicious activities, and ensure adherence to reporting requirements.

b. ***SupTech (Supervisory Technology):*** Regulators utilize AI to monitor market activities, assess risks in the financial system, and enhance regulatory oversight. AI tools help regulators detect trends, anomalies, and potential systemic risks more effectively.

Moreover (Xie, 2019), AI and ML are applied in specialized contexts within the financial sector:

1. ***Quantitative Transactions:*** Algorithms analyze market data to execute trades based on statistical models and quantitative analysis.
2. ***Natural Language Processing and Semantic Search:*** AI enables the evocation of insights from formless sources of data such as news articles, social media, and regulatory filings.
3. ***Intelligent Investment Advisory Services:*** AI-powered platforms provide personalized investment advice and recommendations based on individual financial goals and risk preferences (Xie, 2019).

In summary, AI and ML are driving significant advancements across all facets of the financial sector, enhancing efficiency, improving decision-making processes, and ultimately transforming the financial services that are delivered and regulated.

The metaphor double-edged sword is used for AI to show the dual impact of the technology. AI in the workplace is a powerful tool, but it has a dark side attached to it as well. The technique is both effective and vague, and much relies on the purpose and method of use. The concept of a double-edged sword extends beyond a trade, which is a result of balancing and reaching a settlement between various and incompatible aims (Figure 19.1).

19.1.2 Objectives

1. To discuss the utilization of AI in various financial institutions.
2. To analyze the effect of AI in various financial institutions.
3. To check the relationship between quality and information security with respect to AI in various financial institutions.
4. To analyze the benefits and limitations of AI in the accounting industry.

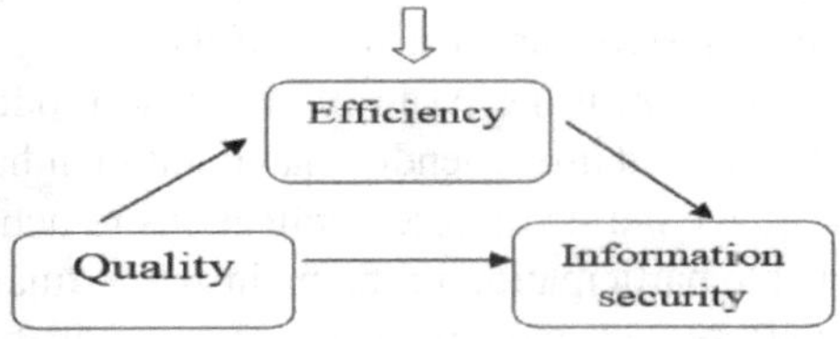

FIGURE 19.1 Conceptual framework.

19.2 LITERATURE REVIEW

The focus is on analyzing the application or usage of AI and ML in the financial sector along with its impact on macro- and microeconomics. In this it was found that various problems and risk factors arise in the application of AI with financial risk management. In this survey, the main objective is to analyze the use and effect of fintech on customers' protection, prosperity, and the discovery of asset prices and returns. The measure of banks' digitalization is not a t**opic of significant discussion, and research on banks' digital transformation is still in** the early stages (Li et al., 2023). As a result, there are few studies in the body of current research that particularly look into the relationship between digitalization and banks' propensity for taking risks.

This chapter focused on the usage of (Li et al., 2023) AI in Chennai's banking and financial services to monitor both bank applications of AI techniques and client or consumer responses. To better track, anticipate, and react to customer behaviour, banks and other financial institutions can mine the financial transaction data produced by the widespread use of digital payments and banking. Based on the literature research, secondary sources are employed to gather data to determine the information used in banking and financial services.

This study elaborates on the application, status quo, and development trend of AI in the financial system, starting with the application status and relevance of AI in the worldwide financial sphere. Researchers have shown that systemic financial hazards and financial regulation are among the many issues and obstacles that internet finance brings, resulting in the transfer of risk to banks (Kero, 2013). On the other hand, there are professionals who argue that there is a nonlinear relationship between banks' risk-taking and internet finance and financial technology.

In their paper, Singh (2019) utilizes natural language processing (NLP), NLTK, Python, and ML within our Conda environment to develop an interactive chatbot application. The author effectively leveraged Python's built-in modules, libraries, and frameworks such as Pandas, scikit-learn, Chatterbot, Chatter-Bot Corpus, and NLTK to construct both a stock market prediction model and a chatbot. These tools help in enhancing our model's accuracy and comprehension.

Financial products and services may now be tailored to each individual customer's needs and tastes thanks to AI-driven personalization (Zhang & Aslan 2021). Fintech companies can create highly customized goods, such as personalized investment portfolios, insurance schemes, and lending products, that appeal to customers and increase their happiness and loyalty by using data analytics and predictive modelling.

Increased competition and market consolidation are the results of traditional banks and financial institutions facing a major challenge from AI-powered fintech companies and digital platforms (Ferrari et al., 2019). Peer-to-peer lending, digital wallets, and blockchain-based payment systems are just some of the cutting-edge goods and services that fintech companies are using AI to offer. These products and services are more accessible, transparent, and user-friendly than traditional banking options.

Fintech firms leverage AI and its diverse applications to achieve digital financial inclusion, aiming to enable participation in the mainstream financial market by low-income earners, the underserved, women, youth, and small businesses (Mhlanga, 2020). AI is increasingly deployed across various domains such as risk management, fraud detection, consumer protection, and addressing information asymmetry.

Its significant impact on digital financial inclusion includes enhancing risk detection and management, providing customer support via chatbots, and strengthening fraud detection and cybersecurity measures.

Financial inclusion can be greatly improved by AI-driven fintech solutions, which give underserved and unbanked communities more access to banking and financial services (Hudson et al., 2021). For people and businesses in distant and low-income areas, AI-powered mobile apps, digital wallets, and microfinance platforms are making it easier for them to manage their finances, obtain credit, and participate in the formal financial system.

The strategies to support the thorough, healthy, and long-term growth of AI in the financial business are outlined in this study. The purpose of this chapter is to inform readers on the state of AI development in the financial sector and to serve as a theoretical resource for researchers in that area. Fintech companies may now develop sophisticated algorithms and predictive analytics tools for a wide range of uses, including risk assessment, fraud detection, credit scoring, and customized financial advice, thanks to recent advancements in AI (Dutta et al., 2020). These AI-powered solutions outperform conventional banking systems in efficiency, accuracy, and scalability, allowing fintech businesses to offer customers more specialized and affordable financial services.

This chapter provides a critical analysis of how national and international development agencies approach the concept of participation in development finance for impoverished populations in the Global South (Singh, 2019). The study draws two primary conclusions: first, established a 'habits of authority' among development stakeholders hinder genuine participation; second, technological platforms that facilitate repetitive innovations and collaborations among businesses and organizations promote greater financial inclusion.

For several decades, professionals have integrated AI frameworks across diverse fields to minimize human intervention and achieve superior outcomes (Kumar & Sharma, 2020). This chapter aims to explicate the various AI techniques employed in the finance sector from a broad perspective. Its primary goal is to dig into the historical evolution of AI in finance and its significant role as a tool for analyzing investment decisions within financial markets.

This study aimed to understand how AI might impact digital financial inclusion (Padmanabhan & Metilda, 2021). The focus on digital financial inclusion in the conversation about how to ensure that those at the bottom of the pyramid begin participating in the financial system is growing. The inclusion of low-income workers, the poor, women, young people, and small businesses in the main financial market is the aim of digital financial inclusion. In order to achieve this goal, fintech companies are using AI and its various applications.

19.3 RESEARCH METHODOLOGY

19.3.2 Research Design

The sample of the chapter is 100 respondents who are chartered accountants from different financial institutions. The chapter is descriptive and analytical in nature. The

TABLE 19.1
Research Design

Type of Research	Data Sources	Research Tool	Methodologies
Descriptive Research and Analytical	Primary and Secondary Sources: Government Reports	Reports, SPSS Software	ANOVA, Multiple Regression, Correlation

chapter aims to understand the relation between quality and information security with the mediating role of efficiency (Table 19.1).

19.3.3 Data Analysis

This research utilizes multiple regression analysis to examine the intricate relationships among independent and dependent variables, revealing the mediating role of efficiency. The analysis begins by investigating the association between Quality (independent variable) and information security (dependent variable). It then explores the reversed relationship between Quality and Efficiency, where Quality is the independent variable influencing Efficiency as the dependent variable. The final step employs multiple regression analysis to comprehensively understand the interplay between quality, information security, and efficiency. The examination reveals that these relationships are interconnected, emphasizing the need for a comprehensive analysis beyond simple correlations. The mediating role of efficiency introduces complexity, showcasing the significance of understanding these dynamics. The focus on information security is relevant in the current landscape of cyber threats, with a practical effect on organizations aiming to strengthen digital defences. Efficiency, as a dependent variable, adds a pragmatic dimension, offering insights for decision-makers seeking to optimize processes while maintaining robust security measures. In conclusion, the meticulous use of multiple regression analysis contributes not only to academic discourse but also offers possible practical considerations looking to enhance processes, fortify information security, and optimize efficiency in a digitally interconnected landscape.

Table 19.2 shows that if Cronbach's Alpha value is greater than 0.80, it means reliability is good and acceptable. As per the condition of Cronbach's Alpha, the value should be greater than 0.7, and this research shows 0.815, which is greater than 0.7.

19.4 DISCUSSION AND ANALYSIS

The chapter uses multiple regression, and the results are given below:

Table 19.3 shows the coefficient value between quality and information security. The value of B is 0.002, which explains the line between the dependent and the independent variables.

Table 19.4 shows that the correlation between the dependent and independent variables is represented by the R-value. For further analysis, a value that is taken is

TABLE 19.2
Reliability Testing

Case Processing Summary

		N	%
Cases	Valid	50	100.0
	Excluded[a]	0	0.0
	Total	50	100.0

Reliability Statistics

Cronbach's Alpha	N of Items
0.815	3

TABLE 19.3
Coefficient

Model	B	t	Sig
(Constant)	3.348	6.819	0.000
Quality	0.002	1.290	0.200

TABLE 19.4
ANOVA Analysis

ANOVA

Model		Sum of Squares	df	Mean Square	F	Sig.
1	Regression	16.142	1	16.142	31.526	0.000[a]
	Residual	24.578	48	0.512		
	Total	40.720	49			
2	Regression	18.875	2	9.438	20.305	0.000[b]
	Residual	21.845	47	0.465		
	Total	40.720	49			

Excluded Variables

Model	Beta In	t	Sig.	Partial Correlation	Collinearity Statistics		
					Tolerance	VIF	Minimum Tolerance
	0.318[a]	2.425	0.019	0.333	0.664	1.506	0.664

Note:

[a] Predictors in the model: (Constant), efficiency.

[b] Dependent variable: Information security.

TABLE 19.5
Multiple Regression

Model		Sum of Squares	Df	Mean Square	F	Sig.
1	Regression	16.142	1	16.142	31.526	0.000[a]
2	Regression	18.875	2	9.438	20.305	0.000[b]

more than 0.4. The value in this survey is 0.630, which is excellent, and the b value is 0.681, which shows good correlation. It is required to have a difference between R-squared and adjusted R-squared minimum. In this case, the values are 0.396 and 0.384; the differences are not more, so it is good. This indicates that there is a close relationship between quality, information security, and efficiency.

Table 19.5 shows the multiple regression coefficient values with 0.000. If the significance value is less than 0.05, it means that the null hypothesis has been rejected. It is required to have a difference between R-squared and adjusted R-squared minimum. In this case, the values are 0.396 and 0.384; the differences are not more, so it is good. This indicates that there is a close relationship between quality, information security, and efficiency.

19.4 CONCLUSION

This chapter concludes that AI has a very strong impact on the financial sector with respect to its application and usage. This chapter delves into the multifaceted dimensions of AI's influence on the financial sector, aiming to provide a thorough understanding of its implications. The primary focus of this chapter revolves around elucidating the intricate relationships between quality, information security, and efficiency within the context of AI implementation. On one hand, it is a powerful tool for managing various financial activities, but it has limitations also. This chapter shows the relation between quality, information security, and efficiency. The results show that there is a strong relationship between quality and efficiency, as it means that if quality is impacted by the usage of AI, then it eventually impacts its data efficiency. This improvement in quality, however, is not isolated; rather, it cascades into the efficiency of data management systems.

The integration of AI in various sectors enhances the overall productivity along with inclusive growth and sustainable development. The initiation of 'Viksit Bharat' aims to become a developed nation by 2047, with AI playing a pivotal role. The adoption of AI integration contributes to the overall economy of the country. The Minister of Petroleum and Natural Gas of India highlights the importance of AI adoption that could generate revenue of around 33.8 lakh crore by 2030.

In essence, this chapter serves as a compass navigating the intricate landscape of AI's impact on the financial sector. It goes beyond acknowledging the sheer power of AI and delves into the nuances of its relationships with quality, information security, and efficiency. As the financial sector continues to evolve, the findings of this research offer a timely and relevant contribution to the ongoing discourse surrounding the integration of AI in the financial domain.

REFERENCES

Assistant, A. G. (2021). (AI) in Banking and Financial. 9(9).

Biswas, S., Carson, B., Chung, V., Singh, S., & Thomas, R. (2020). *AI-Bank of the Future: Can Banks Meet the AI Challenge?*

Cao, Y., & Zhai, J. (2022). A survey of AI in finance. *Journal of Chinese Economic and Business Studies*, *20*(2).

Dutta, S., et al., (2020). Supervised learning in all FeFET-based spiking neural network: Opportunities and challenges. *Frontiers in Neuroscience*, *14*, 634.

Fares, O. H., Butt, I., & Lee, S. H. M. (2022). Utilization of artificial intelligence in the banking sector: A systematic literature review. *Journal of Financial Services Marketing*.

Ferrari, R., et al., (2019). MR-based artificial intelligence model to assess response to therapy in locally advanced rectal cancer. *European Journal of Radiology*, *118*.

He, D., Guo, M., Zhou, J., & Guo, V. (2018). The Impact of Artificial Intelligence (AI) on the Financial Job Market Contents. *China Development Research Foundation*, March.

Hudson, L., et al., (2021). Urban sensing technologies and geospatial big data analytics in internet of things-enabled Smart cities. *Geopolitics, History, and International Relations*, *13*(2).

Kaur, N., Sahdev, S. L., Sharma, M., & Siddiqui, L. (2020). Banking 4.0: "the Influence of Artificial Intelligence on the Banking Industry & How Ai Is Changing the Face of Modern Day Banks." *International Journal of Management*, *11*(6).

Kruse, L., Wunderlich, N., & Beck, R. (2019). Artificial Intelligence for the Financial Services Industry: What Challenges Organizations to Succeed. In *Proceedings of the Annual Hawaii International Conference on System Sciences*.

Kumar, S., & Sharma, R. (2020). Applications of AI in Financial System. In *Natural Language Processing: Concepts, Methodologies, Tools, and Aapplications*, pp. 23–30. IGI Global.

Kumar, V., Saheb, S. S., Ghayas, A., Kumari, S., Chandel, J. K., Pandey, S. K., & Kumar, S. (2023). AI-based hybrid models for predicting loan risk in the banking sector. *Big Data Mining and Analytics*, *6*(4).

Kushwaha, N. S., & Singh, P. (2022). Artificial intelligence based chatbot: A case study. *Journal of Management and Service Science (JMSS)*, *2*(1).

Li, Y., Yi, J., Chen, H., & Peng, D. (2021). Theory and application of artificial intelligence in financial industry. *Data Science in Finance and Economics*, *1*(2).

Manoharan, G., Nithya, G., Rajchandar, K., Razak, A., Gupta, S., Durai, S., & Ashtikar, S. P. (2024). AI in Finance and Banking: The Act of Gyration. In *Revolutionizing Customer-Centric Banking through ICT*, pp. 1–28. IGI Global.

Mhlanga, D. (2020). Industry 4.0 in finance: The impact of artificial intelligence (AI) on digital financial inclusion. *International Journal of Financial Studies*, *8*(3).

Padmanabhan, V., & Metilda, P. (2021). An impact of artificial intelligence in Indian banking industries. *International Research Journal of Education and Technology*, *1*(4).

Papers, I. M. F. D., Economy, D., & Covid, T. (2020).

PwC India. (2022). Uncovering the ground truth: AI in Indian financial services, February, 1–36.

Ris, K., Stankovic, Z., & Avramovic, Z. (2020). Implications of implementation of artificial intelligence in the banking business with correlation to the human factor. *Journal of Computer and Communications*, *08*(11).

Shakdwipee, P., Agarwal, K., Kunwar, H., & Singh, S. (2023, August). Artificial Intelligence in Finance and Accounting: Opportunities and Challenges. In *International Conference on ICT for Sustainable Development*, Singapore: Springer Nature Singapore.

Singh, J. P. (2019). Development finance 2.0: Do participation and information technologies matter? *Review of International Political Economy*, *26*(5).

Xie, M. (2019). Development of artificial intelligence and effects on financial system. *Journal of Physics: Conference Series*, *1187*(3).

Zhang, K., & Aslan, A. B. (2021). AI technologies for education: Recent research & future directions. *Computers and Education: Artificial Intelligence*, *2*.

Index

A

Adaptive learning systems, 9
AI algorithms, 279
AI and ML integration, 210–211, 219, 221
AI in education, 2
AI Interventions, 111, 112, 115, 117, 118
AI-Powered Data Integration, 243–247
AI solutions, 283
Artificial Intelligence (AI), 3, 15, 16, 17, 106, 123, 125, 127, 128, 131, 169, 170, 172, 174, 176, 178, 180–186, 209, 233, 234, 237, 240

B

Bias in AI education, 7
Bias, 303
Big data, 209–211, 215
Big data scalability, 209–211, 213, 216, 217
Businesses, 323–327

C

Career innovation, 135
ChatGPT, 3–11, 112, 139, 254–259, 261, 265–267, 269, 325–326, 330
Child health literacy, 147–149, 154, 158, 162–165
Cloud computing, 212, 213, 219, 221
Cognitive, 334–335
Cold war, 14, 20, 31, 33, 73
COMPAS (Correctional Offender Management Profiling for Alternative Sanctions), 289
Computational criticism, 85
Cultural capital, 149, 153–155, 161–164

D

Data excellence, 233, 250, 253
Data governance, 222–224, 233, 234, 246–251
Data privacy, 222, 223
Data quality management, 240–246
Deepfakes, 329–330
Deep learning, 281
Didactic analysis, 87
Digital renaissance, 124
Digital transformation, 4
Disaster risk reduction, 188
Disparities in education, 4

E

Early warning, 190
Education, 106, 107, 109–123
Equality in education, 16
ET (emerging technologies), 308, 310, 314, 320
Ethical challenges in AI, 181
Ethical consideration of AI, 280

F

Film education, 60
Filmmaking, 61
Film school, 66
Financial inclusion, 5
Financial product, 4
Fintech, 4

G

Gender bias, 1
Gender equality, 4, 304
Gender schema theory, 304
Generative AI, 1, 2, 15, 16
Google search results, 304
Governance and compliance tools and frameworks, 223
Grounded theory, 305

H

Health communication, 147, 149, 152, 153, 155, 156, 162, 164, 165
Healthcare, 332
Human, 1, 2, 15, 16
Hybrid warfare', 309, 314

I

Industrial case studies, 224, 226
Information security, 3
Innovations, 106, 109, 114, 116
Intelligence, 1, 12, 14, 16–22
Intertextuality, 93

J

Job displacement, 20–22
Jobs, 323–327

K

KM systems, 8
Knowledge management (KM), 1

L

Literature and AI Page, 88

M

Machine learning (ML), 210, 219, 224, 226
Mental healths, 29, 30
Modern data platforms, 209–211, 214

N

NEP (National Education Policy) 2020, 64, 125–127
Netnography, 149–150, 161, 165

P

Paediatric influencer, 148, 150, 152, 153, 156, 158, 161–165
Paradigm, 112, 115, 116
Pedagogy, 55
Personalized learning, 1, 26, 132
Predictive analytics, 13, 170, 171, 178, 179, 184, 185, 234, 246, 237, 241, 242
Privac, 24–26

Q

Quality education, 53

R

Realism, 308, 310, 313
Real-time analytics, 213–217, 224
Remote health monitoring, 176, 177, 179

S

Security, 27–29
Sentiment analysis, 148, 151, 156–157
Social robotics, 175, 183, 186
Stereotypes, 304
Stylometric, 93

T

Technology, 2, 8–10, 12, 13, 16–18, 20
Transformation, 127, 132
Transforming classrooms, 106, 113
Transparency enhancement, 284

V

Vulnerable populations, 170, 171, 173–175, 177, 179, 181, 183, 185

W

Word embeddings, 94